Algebra & Geometry

An Introduction to University Mathematics

Algebra & Geometry

An Introduction to University Mathematics

Mark V. Lawson

Heriot-Watt University
Edinburgh, UK

CRC Press
Taylor & Francis Group
Boca Raton   London   New York

CRC Press is an imprint of the
Taylor & Francis Group, an **informa** business

A CHAPMAN & HALL BOOK

CRC Press
Taylor & Francis Group
6000 Broken Sound Parkway NW, Suite 300
Boca Raton, FL 33487-2742

© 2016 by Taylor & Francis Group, LLC
CRC Press is an imprint of Taylor & Francis Group, an Informa business

No claim to original U.S. Government works

Printed on acid-free paper
Version Date: 20160315

International Standard Book Number-13: 978-1-4822-4647-6 (Paperback)

Visit the Taylor & Francis Web site at
http://www.taylorandfrancis.com

and the CRC Press Web site at
http://www.crcpress.com

Printed and bound by CPI Group (UK) Ltd, Croydon, CR0 4YY

In Memoriam
WILLIAM HENRY LAWSON
24th April 1915, Ypres
CHARLES PEIRCE ECCLESHALL
6th June 1915, Gallipoli

Contents

Preface

The aim of this book is to provide a bridge between school and university mathematics centred on algebra and geometry. Apart from pro forma proofs by induction at school, mathematics students usually meet the concept of proof for the first time at university. Thus, an important part of this book is an introduction to proof. My own experience is that aside from a few basic ideas, proofs are best learnt by doing and this is the approach I have adopted here. In addition, I have also tried to counter the view of mathematics as nothing more than a collection of methods by emphasizing ideas and their historical origins throughout. Context is important and leads to greater understanding. Mathematics does not divide into watertight compartments. A book on algebra and geometry must therefore also make connections with applications and other parts of mathematics. I have used the examples to introduce applications of algebra to topics such as cryptography and error-correcting codes and to illustrate connections with calculus. In addition, scattered throughout the book, you will find boxes in smaller type which can be read or omitted according to taste. Some of the boxes describe more complex proofs or results, but many are asides on more advanced material. You do not need to read any of the boxes to understand the book.

The book is organized around three topics: linear equations, polynomial equations and quadratic forms. This choice was informed by consulting a range of older textbooks, in particular [4, 15, 26, 27, 56, 71, 84, 96, 122, 126], as well as some more modern ones [7, 25, 33, 40, 44, 47, 53, 94, 110, 116], and augmented by a survey of the first-year mathematics modules on offer in a number of British and Irish universities. The older textbooks have been a revelation. For example, Chrystal's books [26, 27], now Edwardian antiques, are full of good sense and good mathematics. They can be read with profit today. The two volumes [27] are freely available online.

Exercises. One of my undergraduate lecturers used to divide exercises into *five-finger exercises* and *lollipops*. I have done the same in this book. The exercises, of which there are about 250, are listed at the end of the section of the chapter to which they refer. If they are not marked with a star (*), they are five-finger exercises and can be solved simply by reading the section. Those marked with a star are not necessarily hard, but are also not merely routine applications of what you have read. They are there to make you think and to be enjoyable. For further practice in solving problems, the Schaum's Outline Series of books are an excellent resource and cheap secondhand copies are easy to find.

Prerequisites. If the following topics are familiar then you probably have the background needed to read this book: basic Euclidean and analytic geometry in two and three dimensions; the trigonometric, exponential and logarithm functions; the arithmetic of polynomials and the roots of the quadratic; experience in algebraic manipulation.

Organization. The book is divided into two sections. *Section I* consists of Chapters 1 to 4.

- Chapters 1 and 2 set the tone for the whole book and in particular attempt to explain what proofs are and why they are important.

- Chapter 3 is a reference chapter of which only Section 3.1, Section 3.8 and the first few pages of Section 3.4 need be read first. Everything else can be read when needed or when the fancy takes you.

- Chapter 4 is an essential prerequisite for reading Section II. It is partly revision but mainly an introduction to properties that are met with time and again in studying algebra and are likely to be unfamiliar.

Section II consists of Chapters 5 to 10. This is the mathematical core of the book and the chapters have been written to be read in order. Chapters 5, 6 and 7 are linked thematically by the remainder theorem and Euclid's algorithm, whereas Chapters 8, 9 and 10 form an introduction to linear algebra. I have organized each chapter so that the more advanced material occurs towards the end. The three themes I had constantly in mind whilst writing these chapters were:

1. The solution of different kinds of algebraic equation.

2. The nature of the solutions.

3. The interplay between geometry and algebra.

Wise words from antiquity. Mathematics is, and always has been, difficult. The commentator Proclus in the fifth century records a story about the mathematician Euclid. He was asked by Ptolomy, the ruler of Egypt, if there was not some easier way of learning mathematics than by reading Euclid's big book on geometry, known as the *Elements*. Euclid's reply was correct in every respect but did not contribute to the popularity of mathematicians. There was, he said, no royal road to geometry. In other words: no shortcuts, not even for god-kings. Despite that, I hope my book will make the road a little easier.

Acknowledgements. I would like to thank my former colleagues in Wales, Tim Porter and Ronnie Brown, whose *Mathematics in context* module has influenced my thinking on presenting mathematics. The bibliography contains a list of every book or paper I read in connection with the writing of this one. Of these, I referred to Archbold [4] the most and regard it as an unsung classic. My own copy originally belonged to Ruth Coyte and was passed onto me by her family. This is my chance to thank them for all their kindnesses over the years. The book originated in a course I taught at Heriot-Watt University inherited from my colleagues Richard Szabo and Nick Gilbert. Although the text has been rethought and rewritten, some of the exercises go back to them, and I have had numerous discussions over the years with both of them about what and how we should be teaching. Thanks are particularly due to Lyonell Boulton, Robin Knops and Phil Scott for reading selected chapters, and to Bernard Bainson, John Fountain, Jamie Gabbay, Victoria Gould, Des Johnston and Bruce MacDougall for individual comments. Sonya Gale advised on Greek. At CRC Press, thank you to Sunil Nair and Alexander Edwards for encouraging me to write the book, amongst other things, Amber Donley, Robin Lloyd-Starkes, Katy E. Smith, and an anonymous copy-editor for producing the book, and Shashi Kumar for technical support. I have benefited at Heriot-Watt University from the technical support of Iain McCrone and Steve Mowbray over many years with some fine-tuning by Dugald Duncan. The *TeX-LaTeX Stack Exchange* has been an invaluable source of good advice. The pictures were created using Till Tantau's Ti*k*Z and Alain Matthes' *tkz-euclide* which are enthusiastically recommended. Thanks to Hannah Carse for showing me how to draw circuits and to Emma Blakely, David Bolea, Daniel Hjartland, Scott Hunter, Jian Liao, John Manderson, Yambiso Marawa, Charis Peters, Laura Purves and Ben Thompson (with a 'p') for spotting typos.

Errata, etc. I shall post these at the following page also accessible via my homepage

http://www.macs.hw.ac.uk/~markl/Algebra-geometry.html

Mark V. Lawson
Edinburgh, Summer Solstice and Winter Solstice, 2015

Prolegomena

"L'algèbre n'est qu'une géométrie écrite; la géométrie n'est qu'une algèbre figurée." – Sophie Germain

ALGEBRA began as the study of equations. The simplest kinds of equations are those like $3x - 1 = 0$ where there is exactly one unknown x and it only occurs to the first power. It is easy to solve this equation. Add 1 to both sides to get $3x = 1$ and then divide both sides by 3 to get $x = \frac{1}{3}$. We can check that this really is the solution to the original equation by calculating $3 \cdot \frac{1}{3} - 1$ and observing that this is 0. Simple though this example is, it illustrates an important point: to carry out these calculations, it was necessary to know what *rules* the numbers and symbols obeyed. You probably applied these rules unconsciously, but in this book you will need to know explicitly what they are. The method used to solve the specific example above can be applied to any equation of the form $ax + b = 0$ as long as $a \neq 0$. You might think this example is finished. It is not. We have yet to deal with the case where $a = 0$. Here there are two possibilities. If $b \neq 0$ there are no solutions and if $b = 0$ there are infinitely many. This is not mere pedantry since the generalizations of this case are of practical importance. We have now dealt with how to solve a *linear equation in one unknown*.

The next simplest kinds of equations are those in which the unknown x occurs to the power two but no more

$$ax^2 + bx + c = 0,$$

where $a \neq 0$. They are called *quadratic equations in one unknown*. Whereas solving linear equations is easy, ingenuity is needed to solve quadratic equations. Using a method called completing the square, it is possible to write down a formula to solve any such equation in terms of the numbers a, b and c, called the *coefficients*. This formula yields two, one or no solutions depending on the values of these coefficients.

Quadratic equations are not the end; they are only a beginning. Equations in which x occurs to the power three but no more

$$ax^3 + bx^2 + cx + d = 0,$$

where $a \neq 0$, are called *cubic equations in one unknown*. Solving such equations is much harder than solving quadratics, but using considerable algebraic sophistication there is also an algebraic formula for the solutions. There are never more than three solutions, but sometimes fewer. Similarly, equations in which x occurs to the power four are called *quartics*, and once again there is a formula for finding the solutions and there are never more than four. More generally, a finite sum of powers of x, each multiplied by a real number, set to zero is called a *polynomial equation*. The highest power of x that occurs in such an equation is called its *degree*. Our discussion suggests that this can be viewed as a measure of the complexity of the equation. Until the nineteenth century, algebra was largely synonymous with the study of polynomial equations and culminated in three great discoveries.

1. Equations often have no solutions at all which is vexing. When methods for solving cubic equations were first developed, however, it was observed that real solutions could be constructed using chimeras involving real numbers and square roots of negative numbers. Over time these evolved into the complex numbers. Such numbers still have an aura about them: they are numbers with charisma.

2. With the discovery of complex numbers, it was possible to prove the fundamental theorem of algebra, which states that every non-constant polynomial equation has at least one solution. If this is combined with a suitable way of counting solutions, it can be proved that a polynomial equation of degree n always has exactly n solutions, which is as tidy a result as anyone could want.

3. The third great discovery deals with the nature of the solutions to polynomial equations. The results on linear, quadratic, cubic and quartic equations raise expectations that there are always algebraic formulae for finding the roots of any polynomial equation whatever its degree. There are not. For equations of degree five, the *quintics*, and those of higher degree, there are no such formulae. This does not mean that no formulae have yet been discovered. It means that someone has proved that such formulae are impossible, that someone being Evariste Galois (1811–1832)[1]. Galois' work was revolutionary because it put an end to the view that algebra was only about finding formulae to solve equations, and instead initiated a new structural approach. This is one of the reasons why the transition from school to university algebra is difficult.

The equations we have discussed so far contain only one unknown, but we can equally well study equations in which there are any finite number of unknowns and those unknowns occur to any powers. The best place to start is where any number of unknowns is allowed but where each unknown can occur only to the first power and in addition no products of unknowns are allowed. This means we are studying *linear equations* like

$$x + 2y + 3z = 4$$

where in this case there are three unknowns. The problem is to find all the values of x, y and z that satisfy this equation and so a solution is actually an ordered triple (x, y, z). For example, both $(0, 2, 0)$ and $(2, 1, 0)$ are solutions whereas $(1, 1, 1)$ is not. It is unusual to have just one linear equation to solve; more commonly there are two or more forming a *system of linear equations* such as

$$\begin{aligned} x + 2y + 3z &= 4 \\ x + y + z &= 0. \end{aligned}$$

---

[1] The James Dean of mathematics.

The problem now is to find all the triples (x, y, z) that satisfy both equations simultaneously. In fact those triples of the form

$$(\lambda - 4, 4 - 2\lambda, \lambda),$$

where λ is any number, satisfy both equations and every solution is of this form. Solving a single polynomial equation of degree one is easy and furthermore does not require the invention of new numbers. Similarly solving systems of linear equations in any number of unknowns never becomes difficult or surprising. This turns out to be the hallmark of linear equations of any sort in mathematics whereas non-linear equations are difficult to solve and often very surprising.

This leaves us with those equations where there are at least two unknowns and where there are no constraints on the powers of the unknowns and the extent to which they may be multiplied together. These are deep waters. If only squares of unknowns or products of at most two unknowns are allowed, then in two variables we obtain the *conics*

$$ax^2 + bxy + cy^2 + dx + ey + f = 0$$

and in three variables the *quadrics*. These are comparatively easy to solve however many unknowns there are. But as soon as cubes or the products of more than two unknowns are allowed the situation changes dramatically: hic sunt dracones. For example, equations of the form

$$y^2 = x^3 + ax + b$$

are called *elliptic curves*. They look innocuous but are not: their theory was a key ingredient in Andrew Wiles' proof of Fermat's last theorem; one of the Millennium Problems deals with a question about such equations; and they form the basis of an important part of modern cryptography. Just as the transition from quadratics to cubics required more sophisticated ideas, so too does the transition from conics and quadrics to *algebraic geometry*, the subject that studies algebraic equations in any number of unknowns. For this reason, conics and quadrics form the outer limits of what the methods of this book can easily handle.

The focus so far has been on the form taken by an equation: how many unknowns there are and to what extent and how they can be combined. Once an equation has been posed, we are required to solve it but, as we have seen, we cannot take for granted the nature of the solutions. The common or garden idea of a number is essentially that of a real number. Informally, these are the numbers that can be expressed as positive or negative decimals, with possibly an infinite number of digits after the decimal place, such as

$$\pi = 3.14159265358\ldots$$

where the three dots indicate that this can be continued forever. Whilst such numbers are sufficient to solve linear equations in one unknown, they are not enough in general to solve polynomial equations of degree two or more. These require the complex numbers. Such numbers do not arise in everyday life and so there is a temptation to view them as somehow artificial abstractions or of purely theoretical interest. This temptation should be resisted. The square root of two and the square root of minus one are both equally abstract, the only difference between them being the purely psychological one that the former is more familiar than the latter. As for being of only theoretical interest, quantum mechanics, the theory that explains the behaviour of atoms and so ultimately of how all the stuff around us is made, uses complex numbers in an essential way.[2]

It is not only a question of extending our conception of number. There are also occasions when we want to restrict it. For example, we might want to solve an equation using only whole numbers. It turns out that the usual high-school method for solving equations does not work in this case. Consider the equation

$$2x + 4y = 3.$$

To find the real or complex solutions to this equations, let $x = \lambda$ be any real or complex value and then solve the equation for y in terms of λ. Suppose instead that we are only interested in whole number solutions. In fact, there are none. You can see why by observing that the left-hand side of the equation is exactly divisible by 2, whereas the right-hand side is not. When we are interested in solving equations, of whatever type, by means of whole numbers we say that we are studying *Diophantine equations*, named after Diophantus of Alexandria who studied such equations in his book *Arithmetica*. It is ironic that solving Diophantine equations is often harder than solving equations using real or complex numbers.

We have been talking about the algebra of numbers but there is more to algebra than this. You will also be introduced to the algebra of matrices, and the algebra of vectors, and the algebra of sets, amongst others. In fact, the first surprise on encountering university mathematics is that algebra is not singular but plural. Different algebras are governed by different sets of rules. For this reason, it is essential in university mathematics to make those rules explicit.

Algebra is about symbols, whereas GEOMETRY is about pictures. The ancient Greeks were geometrical wizards and some of their achievements are recorded in Euclid's book or *The Elements*. This described the whole of what became known as *Euclidean geometry* in terms of a handful of rules called *axioms*. Unlike algebra, geometry appears at first sight to be resolutely singular since it is inconceivable there could be other geometries. But there is more to geometry than meets the eye. In the nineteenth century, geometry became plural when geometries were discovered such as hyperbolic geometry where the angles in a triangle never add up to two right angles. In the twentieth century, even the space we inhabit lost its Euclidean trappings with the advent of the curved space-time of general relativity and the hidden multi-dimensional geometries of modern particle physics, the playground of mathematics. Although in this book we only discuss three-dimensional Euclidean geometry, this is the gateway to all the others.

One of the themes of this book is the relationship between ALGEBRA and GEOMETRY. In fact, any book about algebra must also be about geometry. The two

---

[2]I should add here that complex numbers are not the most general class of numbers we shall consider. In Chapter 9, we shall introduce the quaternions.

subjects are indivisible although it took a long time for this to be fully appreciated. It was only in the seventeenth century that Descartes and Fermat discovered the first connection between algebra and geometry, familiar to anyone who has studied mathematics at school. Thus $x^2 + y^2 = 1$ is an *algebraic* equation that also describes something *geometric*: a circle of unit radius centred on the origin. Manipulating symbols is often helped by drawing pictures, and sometimes the pictures are too complex so it is helpful to replace them with symbols. It is not a one-way street. The symbiosis between algebra and geometry runs through this book. Here is a concrete example.

Consider the following thoroughly algebraic-looking problem: find all whole numbers a, b, c that satisfy the equation $a^2 + b^2 = c^2$. Write solutions that satisfy this equation as triples (a, b, c). Such numbers are called *Pythagorean triples*. Thus $(0,0,0)$ and $(3,4,5)$ are Pythogorean triples as is $(-3,-4,-5)$. In addition, if (a,b,c) is a Pythagorean triple so too is $(\lambda a, \lambda b, \lambda c)$ where λ is any whole number. Perhaps surprisingly, this problem is in fact equivalent to one in geometry. Suppose that $a^2 + b^2 = c^2$. Exclude the case where $c = 0$ since then $a = 0$ and $b = 0$. We can therefore divide both sides by c^2 to get

$$\left(\tfrac{a}{c}\right)^2 + \left(\tfrac{b}{c}\right)^2 = 1.$$

Recall that a *rational number* is a real number that can be written in the form $\frac{u}{v}$ where u and v are whole numbers and $v \neq 0$. It follows that

$$(x,y) = \left(\tfrac{a}{c}, \tfrac{b}{c}\right)$$

is a point with rational coordinates that lies on the unit circle, what we call a *rational point*. Thus Pythagorean triples give rise to rational points on the unit circle. We now go in the opposite direction. Suppose that

$$(x,y) = \left(\tfrac{m}{n}, \tfrac{p}{q}\right)$$

is a rational point on the unit circle. Then

$$(mq, pn, nq)$$

is a Pythagorean triple. This suggests that we can interpret our algebraic question as a geometric one: to find all Pythagorean triples, find all the rational points on the unit circle with centre at the origin. This geometric approach leads to a solution of the original algebraic problem, though not without some work.[3]

---

[3] Rational points on the unit circle are described in Question 9 of Exercises 2.3. The application of this result to determining all Pythagorean triples is described in Question 9 of Exercises 5.3.

I

IDEAS

The nature of mathematics

"Pure mathematics is, in its way, the poetry of logical ideas." – Albert
Einstein in an appreciation of Emmy Noether

First chapters, like this, tend to receive short shrift from readers since they are
viewed as little more than literary throat-clearing. But before we can do mathematics
in the right spirit, we must think about what mathematics is. To this end, here are
three assertions about the nature of mathematics that will be discussed here and in
the next chapter:

1. Mathematics is a living subject and not holy writ whose arcane lore is simply
   passed unchanged down the generations.

2. Mathematics is not confined to the ivory towers of academe but permeates the
   modern world.

3. Mathematics is based foursquare on proofs.

1.1   MATHEMATICS IN HISTORY

We can only guess at the mathematical knowledge of people in prehistory. Mon-
uments, such as Stonehenge, demonstrate clear evidence for practical skills in engineering and observational astronomy but it is a lot harder to infer what practical
mathematics they used. It is only with the advent of writing in Mesopotamia, in the
form of clay tablets, that we can actually read about mathematics for the first time.
What is striking is that this is already non-trivial and does not only deal with mundane problems. Two examples illustrate what we mean.

The first is probably the most famous mathematical clay tablet of them all.
Known as Plimpton 322, it is kept in the George A. Plimpton Collection at Columbia
University and dates to about 1800 BCE. Impressed on the tablet is a table of numbers written in base 60 arranged in fifteen rows. Its meaning and purpose is much dis-
puted, but one interpretation is that it consists of Pythagorean triples, that is triples of

whole numbers (a, b, c) such that $a^2 + b^2 = c^2$. The table is reproduced below though for convenience it has been turned on its side and scribal errors have been corrected in the usual way.

	1	2	3	4	5	6	7	8	9	10	11	12	13	14	15
b	119	3367	4601	**12709**	65	319	2291	799	481	4961	45	1679	161	1771	56
c	169	4825	6649	**18541**	97	481	3541	1249	769	8161	75	2929	289	3229	106

If you calculate $c^2 - b^2$ you will get a perfect square a^2. Thus (a, b, c) is a Pythagorean triple. For example,[1]

$$18541^2 - 12709^2 = 182250000 = 13500^2.$$

How such large Pythagorean triples were computed is a mystery as is the reason for computing them.

The second clay tablet, of roughly the same date, shows a square and its diagonals. One of the sides of the square is labelled 30 and one of the diagonals is labelled with two numbers that in base 10 are, respectively, $1 \cdot 41421\ldots$, which is the square root of 2 accurate to five decimal places, and $42 \cdot 426388\ldots$, which is the calculated value of $\sqrt{2}$ multiplied by 30. This tells us that a special case of Pythagoras' theorem was known nearly four thousand years ago along with ideas about proportion. Today, we might need to calculate $\sqrt{2}$ accurately to meet engineering tolerances, but at that time, buildings were constructed out of mud bricks and so it seems highly unlikely that there were any practical motives for calculating $\sqrt{2}$ so accurately. Perhaps they did it because they could, and it gave them intellectual pleasure to do so. Perhaps they found poetry in numbers.

Mesopotamian mathematics is the overture to Greek mathematics, the acme of mathematical achievement in the ancient world. The very word 'mathematics' is Greek as are many mathematical terms such as lemma, theorem, hypotenuse, orthogonal, polygon; the Greek alphabet itself is used as a standard part of mathematical notation; the concept of mathematical proof, the foundation of all mathematics, is a Greek idea.

The Greek Alphabet

α	alpha	A	ν	nu	N
β	beta	B	ξ	xi	Ξ
γ	gamma	Γ	o	omicron	O
δ	delta	Δ	π	pi	Π
ε	epsilon	E	ρ	rho	P
ζ	zeta	Z	σ	sigma	Σ
η	eta	H	τ	tau	T
θ	theta	Θ	υ	upsilon	Υ
ι	iota	I	ϕ	phi	Φ
κ	kappa	K	χ	chi	X
λ	lambda	Λ	ψ	psi	Ψ
μ	mu	M	ω	omega	Ω

---

[1]Four thousand years later, I carried out this calculation using a twenty-first century calculator.

This reflects the fact that the mathematics of ancient Greece is the single most important historical influence on the development and content of mathematics. It flourished in the wider Greek world that included most of the Mediterranean in the thousand or more years between roughly 600 BCE and 600 CE. It begins with the Presocratics, Thales of Miletus (ca. 585 BCE) and Pythagoras of Samos (580–500 BCE), only known from fragments, and is developed in the work of Euclid (ca. 300 BCE), Archimedes (287–212 BCE), Apollonius (262–190 BCE), Diophantus (ca. 250 CE) and Pappus (ca. 320 CE).[2] Any modern book on algebra and geometry has its roots in the books written by Euclid, Apollonius and Diophantus, these being, respectively, the *Elements*, the *Conics* and *Arithmetica*. Although Greek mathematics drew on the legacy of Mesopotamian mathematics, it transformed it. For example, we described above how the Mesopotamians knew good approximations to $\sqrt{2}$. In principle, they could have used their methods to approximate $\sqrt{2}$ to any degree of accuracy. The Greeks, however, took this much further in that they proved that no fraction could be the square root of two: in other words, that the approximations would always remain approximations and never exact values. It requires a different level of understanding even to think of asking such a question let alone finding a way of proving it.

The history of mathematics neither begins nor ends with the Greeks, but it is only in sixteenth century Italy that it is matched and surpassed for the first time when a group of Italian mathematicians discovered how to solve cubics and quartics. The principal players were Scipione del Ferro (1465–1526), Niccolo Tartaglia (ca. 1500–1557), Geronimo Cardano (1501–1576) and Ludovico Ferrari (1522–1565). The significance of their achievement becomes apparent when you reflect that it took nearly forty centuries for mathematicians to discover how to solve equations of degree greater than two. These technical advances were just the beginning. The first glimmerings of complex numbers appear in the work of Rafael Bombelli (ca. 1526–1573) who realized that the solution of cubics demanded this step into the unknown. Although their work left unsolved the question of how to solve quintics, equations of degree five, this very question became the driving force in the subsequent development of algebra. It was only finally resolved by Galois, marking a watershed in the development of algebra.

In seventeenth century France, the first connections between algebra and geometry were developed in the work of René Descartes (1596–1650), from which we get the term *cartesian* coordinate system, and Pierre de Fermat (1601–1665). It is this that enables us to view an algebraic equation such as $x^2 + y^2 = 1$ as also describing something geometric: a circle of radius one centred on the origin. It is worth noting that with the development of calculus by Sir Isaac Newton (1643–1727) and Gottfried Wilhelm von Leibniz (1646–1716), most of what constitutes school mathematics was in place.

---

[2]Of these, Archimedes stands out as the greatest mathematician of antiquity who developed methods close to those of integral calculus and used them to calculate areas and volumes of complicated curved shapes. Greek civilization was absorbed by the Romans, but for all their aqueducts, roads, baths and maintenance of public order, it has been said of them that their only contribution to mathematics was when Cicero rediscovered the grave of Archimedes and had it restored [112, page 38].

1.2 MATHEMATICS TODAY

In the nineteenth century there was a revolution in mathematics which has led to contemporary mathematics being almost a different subject from what is taught in school. Our conceptions of what we mean by number and space, which are treated as self-evident in school and in pre-nineteenth century mathematics, have been expanded in often counterintuitive ways. In the twentieth and twenty-first centuries, mathematics has continued to develop to such an extent that it is impossible for one person to encompass it all. To give some flavour of it, we briefly describe the work of two significant twentieth century mathematicians.

In the early decades of the twentieth century, mathematicians became interested in the foundations of their subject. What might have been of merely parochial interest turned out to be revolutionary for the whole of society. This revolution is indissolubly linked with the name of Alan Turing (1912–1954), a mathematician who is celebrated within mathematics and popular culture. In fact, he is unique amongst mathematicians in being the subject of both a West End play[3] and a movie, albeit a highly inaccurate one.[4] Turing is best known as one of the leading members of Bletchley Park during the Second World War, for his rôle in the British development of computers during and after the War and for the ultimately tragic nature of his early death. As a graduate student in 1936, he wrote a paper entitled *On computable numbers with an application to the Entscheidungsproblem*. The long German word means *decision problem* and refers to a specific question in mathematical logic that grew out of the foundational issues mentioned above. It was as a result of solving this problem that Turing was led to formulate a precise mathematical blueprint for computers now called *Turing machines* in his honour. Turing is regarded as the father of computer science.

Mathematicians operate on a different timescale from everyone else. We have already mentioned Pythagorean triples, those whole numbers (x, y, z) that satisfy the equation $x^2 + y^2 = z^2$. For no other reason than that of playful exploration, we can ask what happens when we consider sums of cubes rather than sums of squares. Thus, we try to find whole number solutions to $x^3 + y^3 = z^3$. There is no reason to stop at cubes; we could likewise try to find the whole number solutions to $x^4 + y^4 = z^4$ or more generally $x^n + y^n = z^n$ where $n \geq 3$. Excluding the trivial case where $xyz = 0$, the question is to find all the whole number solutions to equations of the form $x^n + y^n = z^n$ for all $n \geq 3$. Back in the seventeenth century, Fermat, whom we met earlier, wrote in the margin of his copy of the *Arithmetica*, that he had found a proof that there were no solutions to these equations but that there was no room for him to write it out. This became known as *Fermat's Last Theorem*. In fact, since Fermat's supposed proof was never found, it was really a conjecture. More to the point, it is highly unlikely that he ever had a proof since in the subsequent centuries many attempts were made to prove this result, all in vain, although substantial progress was made. The problem became a mathematical Mount Everest. Finally, on Monday, 19th September 1994, sitting at his desk, Andrew Wiles (b. 1953), building on over three centuries of work,

---

[3] *Breaking the code*, by Hugh Whitemore, 1986.
[4] *The imitation game*, directed by Morten Tyldum, 2014.

and haunted by his premature announcement of his success the previous year, had a moment of inspiration as the following quote from the *Daily Telegraph* dated 3rd May 1997 reveals.

> "Suddenly, totally unexpectedly, I had this incredible revelation. It was so indescribably beautiful, it was so simple and so elegant."

As a result Fermat's Conjecture really is a theorem, although the proof required travelling through what can only be described as mathematical hyperspace using the theory of elliptic curves. Wiles' reaction to his discovery provides a glimpse of the intellectual excitement of doing mathematics that not only engages the intellect but also the emotions.[5]

1.3   THE SCOPE OF MATHEMATICS

Box 1.1: Mathematics Subject Classification 2010 (adapted)

00. General   01. History and biography   03. Mathematical logic and foundations   05. Combinatorics   06. Order theory   08. General algebraic systems   11. Number theory   12. Field theory   13. Commutative rings   14. Algebraic geometry   15. Linear and multilinear algebra   16. Associative rings   17. Non-associative rings   18. Category theory   19. K-theory   20. Group theory and generalizations   22. Topological groups   26. Real functions   28. Measure and integration   30. Complex functions   31. Potential theory   32. Several complex variables   33. Special functions   34. Ordinary differential equations   35. Partial differential equations   37. Dynamical systems   39. Difference equations   40. Sequences, series, summability   41. Approximations and expansions   42. Harmonic analysis   43. Abstract harmonic analysis   44. Integral transforms   45. Integral equations   46. Functional analysis   47. Operator theory   49. Calculus of variations   51. Geometry   52. Convex geometry and discrete geometry   53. Differential geometry   54. General topology   55. Algebraic topology   57. Manifolds   58. Global analysis   60. Probability theory   62. Statistics   65. Numerical analysis   68. Computer science   70. Mechanics   74. Mechanics of deformable solids   76. Fluid mechanics   78. Optics   80. Classical thermodynamics   81. Quantum theory   82. Statistical mechanics   83. Relativity   85. Astronomy and astrophysics   86. Geophysics   90. Operations research   91. Game theory   92. Biology   93. Systems theory   94. Information and communication   97. Mathematics education.[a]

---

[a]http://www.ams.org/msc/msc2010.html.

Mathematics stretches back into the past and is also something that is being vigorously pursued today. This raises the question of just what the subject matter of modern mathematics is. So here is what might be called the mathematical panorama. The official *Mathematics Subject Classification* currently divides mathematics into 64 broad areas. You can see what they are in Box 1.1. The missing numbers, by the way, are deliberate. To get some idea of scale, a mathematician could work their entire professional life in just one of these topics. In addition, each topic is subdivided into a large number of smaller areas, any one of which could be the subject of a

---

[5]There is a BBC documentary *Fermat's last theorem* about Andrew Wiles directed by Simon Singh made for the BBC's Horizon series first broadcast on 15th January 1996. It is an exemplary example of how to portray complex mathematics in an accessible way and is highly recommended.

PhD thesis. This can seem overwhelming at first, but take heart: the individual topics share a common language of ideas and techniques. In fact, one of the goals of an undergraduate degree in mathematics is to provide this common language.

1.4 WHAT THEY (PROBABLY) DIDN'T TELL YOU IN SCHOOL

University mathematics is not just school mathematics with harder sums and fancier notation. It is, rather, fundamentally different from what you did at school. The failure to understand those differences can cause problems. To be successful in university mathematics, you have to learn to think in new ways. Here are two key differences:

1. *In much of school mathematics, you learn methods for solving specific problems, and often you just learn formulae.* A method for solving a problem that requires little thought in its application is called an *algorithm*. Computer programs are the supreme examples of algorithms. It is certainly true that finding algorithms for solving specific problems is an important part of mathematics, but crucially it is not the only part. Problems do not come neatly labelled with the methods needed for their solution. A new problem might be solvable using old methods or it might require those methods to be adapted. In extreme cases, completely new methods may be needed requiring new ideas. In other words, mathematics is a creative enterprise.

2. *Mathematics at school is often taught without reasons being given for why the methods work.* This is the fundamental difference between school mathematics and university mathematics. A reason why something works is called a proof. Mathematics without proofs is an oxymoron. Proofs are the subject of Chapter 2.

Box 1.2: The Millennium Problems

Mathematics is hard but intellectually rewarding. Just how hard can be gauged by the following. The Millennium Problems[a] is a list of seven outstanding problems posed by the Clay Institute in the year 2000. A correct solution to any one of them carries a one million dollar prize. To date, only one has been solved, the Poincaré conjecture, by Grigori Perelman in 2010, who declined to take the prize money. The point is that no one offers a million dollars for something trivial. An accessible account of these problems can be found in [36].

It should also be mentioned here that there is no Nobel Prize for mathematics, but there are two mathematics prizes that are as highly esteemed: the *Fields Medal*[b] and the *Abel Prize*.[c] The first is only awarded every four years and the recipients have to be under 40, so that it is both a recognition of achievement and an encouragement to greater things, whereas the second is awarded to outstanding mathematicians of any age who have reshaped the subject.

---

[a]http://www.claymath.org/millennium-problems.
[b]http://www.fields.utoronto.ca/aboutus/jcfields/fields_medal.html.
[c]http://www.abelprize.no/.

1.5   FURTHER READING

To pursue the topics introduced in this book in more depth, there are a wealth of possibilities both printed and online. In particular, the books in the bibliography were all the ones I consulted whilst writing this one and all have something to offer.

For general algebra, [5] and [25] are good places to start because they take complementary viewpoints: namely, conceptual and computational, respectively. For a linear algebra book that asks more of the reader than to perform endless elementary row operations, there is the peerless Halmos [54]. The way the concept of number has developed is a fascinating story that lies at the heart of algebra. See [40] for a detailed survey of a whole range of different kinds of number. For geometry, the choices are less clear-cut but [18] and [60] are both good books that take different approaches. My book is not about analysis, but algebra, geometry and analysis interact at every level. Spivak [114] was my first analysis book and a favourite of many. For the historical development of analysis, see [53].

A book that is in tune with the goals of this chapter is [33]. It is one of those that can be dipped into at random and something interesting will be learnt but, most importantly, it will expand your understanding of what mathematics is, as it did mine.

A good source for the history of mathematics from a chronological perspective is [15], which I referred to constantly when writing this chapter, whereas [117] is written from a more thematic point of view. The MacTutor[6] history of mathematics archive is an invaluable resource. There is a danger, however, of viewing the history of mathematics as a catalogue of names and achievements. This is unavoidable at the beginning when you are trying to get your bearings, but should not be mistaken for true history. To get a sense of the issues involved, I would recommend Robson's articles on Plimpton 322 [101, 102] and Fowler's book on the nature of Greek mathematics [44] as antidotes to my naïve expositions.

If you have never looked into Euclid's book the *Elements*, then you should.[7] There is an online version that you can access via David E. Joyce's website at Clark University.[8] A handsome printed version, edited by Dana Densmore, is [35]. If you have ever wondered what geometry would have been like if Euclid had been Japanese read [3, 23]. This is less whimsical than it sounds.

There is now a plethora of popular mathematics books, and if you pick up any of the books by *Ian* Stewart[9] and Peter Higgins then you will find something interesting. Sir (William) Timothy Gowers won a Fields Medal in 1998 and so can be assumed to know what he is talking about [50]. It is worth checking out his homepage for some interesting links and his blog. The Web is serving to humanize mathematicians: their ivory towers now all have wi-fi. A classic popular, but serious, mathematics book is [31] which can also be viewed as an introduction to university-level mathematics.

---

[6]http://www-history.mcs.st-and.ac.uk/.

[7]Whenever I refer to Euclid, it will always be to this book. It consists of thirteen chapters, themselves called 'books', which are numbered in the Roman fashion I–XIII.

[8]http://aleph0.clarku.edu/~djoyce/.

[9]If the book appears to be rather more about volcanoes than is seemly then it is by *Iain* Stewart.

The books of W. W. Sawyer, such as [109], are pedagogical classics which sparked the interest of many people in mathematics.

Mathematicians themselves have written books which take a broader view. Two classics are by G. H. Hardy [57] and J. E. Littlewood [85]. Though they were close collaborators their books could not be more different. Cédric Villani won the Fields Medal in 2010 and his book [124], described by Alexander Masters in *The Spectator* as 'impenetrable from page four', was serialized in March 2015 on BBC Radio 4 as their book of the week. If you want to get an idea of what mathematicians write for each other, go to the arXiv at http://arxiv.org/

A list of further reading would be incomplete without mentioning the wonderful books of Martin Gardner. For a quarter of a century, he wrote a monthly column on recreational mathematics for the *Scientific American* which inspired amateurs and professionals alike. I would start with [46] and follow your interests.

CHAPTER 2

Proofs

"Mathematics may be defined as the subject in which we never know what we are talking about, nor whether what we are saying is true." – Bertrand Russell

The goal of this chapter is to introduce you to proofs in mathematics. We start with what we mean by truth.

2.1 MATHEMATICAL TRUTH

Humans can believe things for purely emotional reasons.[1] An advance on this is believing something because somebody in authority told you. Most of what we take to be true, certainly as children, is based on such *appeals to authority*. Although this is not automatically a bad thing, after all we cannot be experts in everything, it has also been at the root of humanity's darkest episodes. There must be a better way of getting at the truth. And there is. The scientific method with the *appeal to experiment* at its centre, combined with a culture of debate and criticism, is the only basis for any rational attempt to understand our world. This makes the following all the more surprising: in mathematics, the appeal to experiment, let alone the appeal to authority or the appeal to psychological well-being, plays no rôle whatsoever in deciding mathematical truth.

- Results are not true in mathematics 'just because they are'.

- Results are not true in mathematics because I say so, or because someone a long time ago said they were true.

- Results are not true in mathematics because I have carried out experiments and always get the same answer.

How then can we determine whether something in mathematics is true?

- Results are true in mathematics *only* because they have been *proved* to be true.

---

[1] Unlike Vulcans.

- A *proof* shows that a result is true.

- A proof is something that you can follow for yourself and at the end you will *see* the truth of what has been proved.

- A result that has been proved to be true is called a *theorem*.

The remainder of this chapter is devoted to an introductory answer to the question of what a proof is.

2.2   FUNDAMENTAL ASSUMPTIONS OF LOGIC

In order to understand how proofs work, three simple assumptions are needed.

1. *Mathematics only deals in statements that are capable of being either true or false.* Mathematics does not deal in sentences which are 'almost true' or 'mostly false'. There are no approximations to the truth in mathematics and no grey areas. Either a statement is true or a statement is false, though we might not know which. This is quite different from everyday life, where we often say things which contain a grain of truth or for purely rhetorical reasons. In addition, mathematics does not deal in sentences that are neither true nor false such as exclamations 'Out damned spot!' or questions 'To be or not to be?'.

2. *If a statement is true then its negation is false, and if a statement is false then its negation is true.* In English, the negation of 'it is raining' is 'it is not raining'. Different languages have different ways of negating statements. To avoid grammatical idiosyncrasies, we can use the formal phrase 'it is not the case that' and place it in front of any sentence to negate it. So, 'it is not the case that it is raining' is the negation of 'it is raining'.[2]

3. *Mathematics is free of contradictions.* A *contradiction* is where both a statement and its negation are true. This is impossible by (2) above. This assumption plays a vital rôle in proofs as we shall see later.

2.3   FIVE EASY PROOFS

"An argument is a connected series of statements intended to establish a proposition." – Monty Python

Armed with the three assumptions above, we shall work through five proofs of five results, three of them being major theorems. Before we do so, we describe in general terms how proofs are put together.

Irrespective of their length or complexity, proofs tend to follow the same general

---

[2]In some languages and in older forms of English adding negatives is used for emphasis. In formal English, however, we are taught that two negatives make a positive, but this is in fact a made-up rule based on mathematics. In reality, negating negatives in natural languages is more complex than this. For example, 'not unhappy' is not quite the same as being 'happy'. There are no such subtleties in mathematics.

pattern. First, there is a statement of what is going to be proved. This usually has the form: if some things are assumed true then something else must also be true. If the things assumed true are lumped together as *A*, for *assumptions*, and the thing to be proved true is labelled *B*, for *conclusion*, then a statement to be proved usually has the shape 'if *A* then *B*' or '*A* implies *B*' or, symbolically, '*A* ⇒ *B*'.

The proof itself should be thought of as an argument[3] between two protagonists whom we shall call *Alice* and *Bob*. Alice wants to prove that *B* must be true if *A* is assumed true. To do so, she has to obey certain rules: she can use any of the assumptions *A*; any previously proved theorems; the rules of logic;[4] and definitions. Bob's rôle is to act like a judge challenging Alice to justify each statement she makes to ensure she follows the rules. At the end of the process, Alice can say ' ... and so *B* is proved' and Bob is forced to agree with her conclusion.

Proofs are difficult because it is usually far from obvious how to reach the conclusions from the assumptions. In particular, we are allowed to assume anything that has previously been proved, which is daunting given the scale of the subject. We say more about how to construct proofs later, but without more ado we describe some examples of proofs.

Proof 1

The square of an even number is even.

Before we can prove this statement, we need to understand what it is actually saying. The terms *odd* and *even* are only used of whole numbers such as $0, 1, 2, 3, 4, \ldots$. These numbers are called the *natural numbers*, the first kinds of numbers we learn about as children. Thus we are being asked to prove a statement about natural numbers. The terms 'odd' and 'even' might seem obvious, but we need to be clear about how they are used in mathematics. By definition, a natural number *n* is *even* if it is exactly divisible by 2, otherwise it is said to be *odd*. In mathematics, we just say *divisible* rather than *exactly divisible*. This definition of divisibility only makes sense when talking about whole numbers. For example, it is pointless to ask whether one fraction divides another because any non-zero fraction always divides another. By the way, observe that 0 is an even number because $0 = 2 \times 0$. In other words, 0 is divisible by 2. Remember, you cannot *divide by* 0 but you can certainly *divide into* 0. You might have been told that a number is even if its last digit is one of the digits $0, 2, 4, 6, 8$. In fact, this is a consequence of our definition rather than a definition itself. We do not take this as the definition because it is based on the particular number system we use which is in some sense arbitrary.

That deals with the definition of even. What about the definition of odd? A number is *odd* if it is not even. This is not very helpful since it is defined in terms of what it is not rather than what it is. We want a more positive characterization. So we shall describe a better one. If you attempt to divide a number by 2 then there are

---

[3] A rational one not the automatic gainsaying of what the other person says.

[4] Some books spend quite a bit of time on these. I won't. If you can understand [88], you can understand the rules of logic used in this book.

two possibilities: either it goes exactly, in which case the number is even, or it goes so many times plus a remainder of 1, in which case the number is odd. It follows that a better way of defining an odd number n is one that can be written $n = 2m + 1$ for some natural number m. So, the even numbers are those natural numbers that are divisible by 2, that is the numbers of the form $2m$ for some m, and the odd numbers are those that leave the remainder 1 when divided by 2, that is the numbers of the form $2m + 1$ for some m. Every number is either odd or even but not both. There is an important moral to be drawn from this discussion.[5]

> *Every time you are asked to prove a statement, you must ensure that you understand what that statement is saying. This means, in particular, checking that you understand what all the words in the statement mean.*

The next point is that we are making a claim about all even numbers. If you pick a few even numbers at random and square them then you will find in every case that the result is even but this does not prove our claim. Even if you checked a trillion even numbers and squared them and the results were all even it would not prove the claim. Mathematics, remember, is not an experimental science. There are plenty of examples in mathematics of statements that look true and are true for umpteen cases but are in fact false in general.

Example 2.3.1. In Chapter 5, we shall study prime numbers. These are the natural numbers n, excluding 0 and 1, which are divisible only by 1 and n. You can check that the polynomial

$$p(n) = n^2 - n + 41$$

has the property that its value for $n = 1, 2, 3, 4, \ldots, 40$ is always prime. Even the first few cases might be enough to lead you into believing that it always took prime values. It does not, of course, since when $n = 41$ it is clearly not prime.

All this means that, in effect, we have to prove an infinite number of statements: 0^2 is even, and 2^2 is even, and 4^2 is even .... We cannot therefore prove the claim by picking a specific even number, like 12, and checking that its square is even. This does no more than verify one of the infinitely many statements above. As a result, the starting point for a proof cannot be a specific even number. We are now in a position to prove our claim

1. Let n be an even number. Observe that n is not a specific even number. We want to prove something for all even numbers so we cannot argue with a specific one.

2. Then $n = 2m$ for some natural number m. Here we are using the definition of what it means to be an even number.

3. Square both sides of the equation in (2) to get $n^2 = 4m^2$. To do this correctly, you need to know and apply the appropriate rules of algebra.

---

[5] I cannot emphasize enough how important this moral is. It's *really important*.

4. Now rewrite equation (3) as $n^2 = 2(2m^2)$. This uses more rules of algebra.

5. Since $2m^2$ is a natural number, it follows that n^2 is even using our definition of an even number. This proves our claim.

Now that we have proved our statement, we can call it a theorem. There is an obvious companion statement.

The square of an odd number is odd.

Here is a proof.

1. Let n be an odd number.

2. By definition $n = 2m + 1$ for some natural number m.

3. Square both sides of the equation in (2) to get $n^2 = 4m^2 + 4m + 1$.

4. Now rewrite the equation in (3) as $n^2 = 2(2m^2 + 2m) + 1$.

5. Since $2m^2 + 2m$ is a natural number, it follows that n^2 is odd using our definition of an odd number. This proves our claim.

Proof 2

If the square of a number is even then that number is even.

At first reading, you might think this is simply a repeat of what was proved above. But in Proof 1, we proved

if n is even then n^2 is even

whereas now we want to prove

if n^2 is even then n is even.

The assumptions in each case are different as are the conclusions. It is therefore important to distinguish between $A \Rightarrow B$ and $B \Rightarrow A$. The statement $B \Rightarrow A$ is called the *converse* of the statement $A \Rightarrow B$. Experience shows that it is easy to swap assumptions and conclusions without being aware of it. Here is a proof of the statement.

1. Suppose that n^2 is even.

2. Now it is very tempting to try and use the definition of even here, just as we did in Proof 1, and write $n^2 = 2m$ for some natural number m. This turns out to be a dead-end. Just like playing a game such as chess, not every possible move is a good one. Choosing the right move comes with experience and sometimes trial and error.

3. So we make a different move. We know that n is either odd or even. Our goal is to prove that it must be even. We shall do this by ruling out the possibility that n could be odd. We are applying here a simple rule of logic.

4. Could n be odd? The answer is no, because as we showed in Proof 1, if n is odd then n^2 is odd and we are given that n^2 is even.

5. Therefore n is not odd.

6. But a number that is not odd must be even. It follows that n is even.

We may use a similar strategy to prove the following statement.

If the square of a number is odd then that number is odd.

The proofs here were more subtle and less direct than in our first example and they employed the following important rule of logic.

If there are two possibilities, exactly one of which is true, and we rule out one of those possibilities then we can deduce that the other possibility must be true.[6]

If $A \Rightarrow B$ and $B \Rightarrow A$ then we write $A \Leftrightarrow B$ and say A *if and only if B*. You will sometimes meet the abbreviation *iff* to mean 'if and only if'. It is important to remember that the statement $A \Leftrightarrow B$ is in fact two statements combined into one. It means *both $A \Rightarrow B$ and $B \Rightarrow A$*. Thus a proof of such a statement will require two proofs: one for each direction. There is also some traditional terminology that you will meet in this setting. Suppose that $A \Rightarrow B$. Then we say that *A is a sufficient condition for B* and that *B is a necessary condition for A*. The first statement is reasonably clear, because if A is true then B must be true as well. The second statement can be explained by observing that if B is not true then A cannot be true. As a result, the truth of *A if and only if B* is often expressed by saying that *A is a necessary and sufficient condition for B*. It is worth mentioning here that we often have to prove that a number of different statements are equivalent. For example, that $A \Leftrightarrow B$ and $B \Leftrightarrow C$ and $C \Leftrightarrow D$. In fact in this case, it is enough to prove that $A \Rightarrow B$ and $B \Rightarrow C$ and $C \Rightarrow D$ and $D \Rightarrow A$.

If we combine Proofs 1 and 2, we have proved both of the following two statements for natural numbers n.

n is even if and only if n^2 is even and n is odd if and only if n^2 is odd.

Proof 3

$\sqrt{2}$ *cannot be written as an exact fraction.*

This is our first interesting theorem and one that takes us back to the very origins of proof in Greek mathematics. If each one of the following fractions is squared in turn

$$\frac{3}{2}, \frac{7}{5}, \frac{17}{12}, \frac{41}{29}, \ldots$$

---

[6]This might be called the *Sherlock Holmes method*. "How often have I said to you that when you have eliminated the impossible, whatever remains, however improbable, must be the truth?" *The Sign of Four*, 1890.

you will find that you get closer and closer to 2 and so each of these numbers is an approximation to the square root of 2. This raises the question: is it possible to find a fraction $\frac{x}{y}$ whose square is *exactly 2?* In fact, it is not, but that is not proved because the attempts above failed. Perhaps we have just not looked hard enough. So we have to prove that it is impossible. To prove that $\sqrt{2}$ is not an exact fraction, we assume that it is and see what follows.

1. Suppose that $\sqrt{2} = \frac{x}{y}$ where x and y are positive whole numbers where $y \neq 0$.

2. We can assume that $\frac{x}{y}$ is a fraction in its lowest terms so that the only natural number that divides both x and y is 1. Keep your eye on this assumption because it will return to sting us.

3. Square both sides of the equation in (2) to get $2 = \frac{x^2}{y^2}$.

4. Multiply both sides of the equation in (3) by y^2.

5. We therefore obtain the equation $2y^2 = x^2$ and have eliminated the square root.

6. Since 2 divides the lefthand side of the equation in (5), it must divide the righthand side. This means that x^2 is even.

7. We now use Proof 2 to deduce that x *is even.*

8. We can therefore write $x = 2u$ for some natural number u.

9. Substitute this value for x we have found into (5) to get $2y^2 = 4u^2$.

10. Divide both sides of the equation in (9) by 2 to get $y^2 = 2u^2$.

11. Since the righthand side of the equation in (10) is even so is the lefthand side. Thus y^2 is even.

12. Since y^2 is even, it follows by Proof 2 that y *is even.*

13. If (1) is true then we are led to the following two conclusions. From (2), *the only natural number to divide both x and y is 1.* From (7) and (12), *2 divides both x and y.* This is a contradiction. Thus (1) cannot be true. Hence $\sqrt{2}$ cannot be written as an exact fraction.

This result is phenomenal. It says that no matter how much money you spend on a computer it will never be able to calculate the exact value of $\sqrt{2}$, just a good approximation. We now make an important definition. A real number that is not rational is called *irrational.* We have therefore proved that $\sqrt{2}$ is irrational.

Proof 4

The sum of the angles in a triangle add up to 180°.

Everyone knows this famous result. You might have learnt about it in school by drawing lots of triangles and measuring their angles but as we have explained mathematics is not an experimental science and so this proves nothing. The proof we give is a slight variant of the one to be found in Euclid's book the *Elements*: specifically, Proposition I.32. Draw a triangle, and label its three angles α, β and γ respectively.

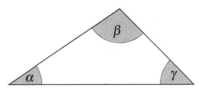

Our goal is to prove that $\alpha + \beta + \gamma = 180°$. In fact, we show that the three angles add up to a straight line which is the same thing. Draw a line through the point P as shown parallel to the base of the triangle.

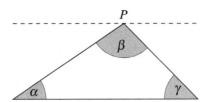

Then extend the two sides of the triangle that meet at the point P as shown.

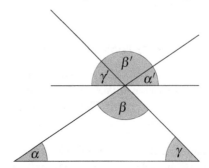

As a result, we get three angles called α', β' and γ', respectively. We claim the following.

- $\beta' = \beta$ because the angles are opposite each other in a pair of intersecting straight line.

- $\alpha' = \alpha$ because these two angles are formed from a straight line cutting two parallel lines.

- $\gamma' = \gamma$ for the same reason as above.

Since α', β' and γ' add up to give a straight line, so too do α, β and γ, and the claim is proved.

Now this is all well and good, but we have proved our result on the basis of three other results currently unproved:

1. That given a line l and a point P not on that line we can draw a line through the point P and parallel to l.

2. If two lines intersect, then opposite angles are equal.

3. If a line l cuts two parallel lines l_1 and l_2 the angle l makes with l_1 is the same as the angle it makes with l_2.

How do we know they are true? Result (2) can readily be proved. We use the diagram below.

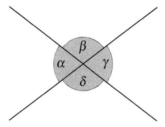

The proof that $\alpha = \gamma$ follows from the simple observation that $\alpha + \beta = \beta + \gamma$. This still leaves (1) and (3). We shall say more about them later when we talk about axioms in Section 2.4.

Proof 5

We have left to last the most famous elementary theorem of them all attributed to Pythagoras and proved as Proposition I.47 of Euclid. We are given a right-angled triangle.

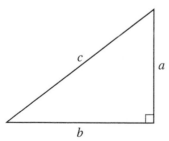

We are required to prove, of course, that $a^2 + b^2 = c^2$. Consider the picture below. It has been constructed from four copies of our triangle and two squares of areas a^2 and b^2, respectively. We claim that this shape is itself actually a square. First, the sides all have the same length $a + b$. Second, the angles at the corners are right angles by Proof 4.

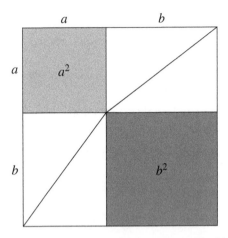

Now look at the following picture. This is also a square with sides $a+b$ so it has the same area as the first square. Using Proof 4, the shape in the middle really is a square with area c^2.

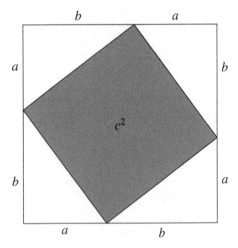

If we subtract the four copies of the original triangle from both squares, the shapes that remain must have the same areas, and we have proved the claim.

Exercises 2.3

1. Raymond Smullyan is both a mathematician and a magician. Here are two of his puzzles. On an island there are two kinds of people: *knights* who always tell the truth and *knaves* who always lie. They are indistinguishable.

   (a) You meet three such inhabitants A, B and C. You ask A whether he is a knight or knave. He replies so softly that you cannot make out what he said. You ask B what A said and B says 'he said he is a knave'. At which point C interjects and says 'that's a lie!'. Was C a knight or a knave?

(b) You encounter three inhabitants: A, B and C.

A says 'exactly one of us is a knave'.

B says 'exactly two of us are knaves'.

C says: 'all of us are knaves'.

What type is each?

2. This question is a variation of one that has appeared in the puzzle sections of many magazines. I first came across it in a copy of *Reader's Digest* as a child. There are five houses, from left to right, each of which is painted a different colour, their inhabitants are called Sarah, Charles, Tina, Sam and Mary, but not necessarily in that order, who own different pets, drink different drinks and drive different cars.

(a) Sarah lives in the red house.

(b) Charles owns the dog.

(c) Coffee is drunk in the green house.

(d) Tina drinks tea.

(e) The green house is immediately to the right (that is: your right) of the white house.

(f) The Oldsmobile driver owns snails.

(g) The Bentley owner lives in the yellow house.

(h) Milk is drunk in the middle house.

(i) Sam lives in the first house.

(j) The person who drives the Chevy lives in the house next to the person with the fox.

(k) The Bentley owner lives in a house next to the house where the horse is kept.

(l) The Lotus owner drinks orange juice.

(m) Mary drives the Porsche.

(n) Sam lives next to the blue house.

There are two questions: who drinks water and who owns the aardvark?

3. Prove that the sum of any two even numbers is even, that the sum of any two odd numbers is even and that the sum of an odd and an even number is odd.

4. Prove that the sum of the interior angles in any quadrilateral is equal to $360°$.

5. (a) A rectangular box has sides of length 2, 3 and 7 units. What is the length of the longest diagonal?

(b) Draw a square. Without measuring any lengths, construct a square that has exactly twice the area.

(c) A right-angled triangle has sides with lengths x, y and hypotenuse z. Prove that if the area of the triangle is $\frac{z^2}{4}$ then the triangle is isosceles.

6. (a) Prove that the last digit in the square of a positive whole number must be one of 0,1,4,5,6 or 9. Is the converse true?

   (b) Prove that a natural number is even if and only if its last digit is even.

   (c) Prove that a natural number is exactly divisible by 9 if and only if the sum of its digits is divisible by 9.

7. Prove that $\sqrt{3}$ cannot be written as an exact fraction.

8. *The goal of this question is to prove *Ptolomy's theorem*.[7] This deals with *cyclic quadrilaterals*, that is those quadrilaterals whose vertices lie on a circle. With reference to the diagram below,

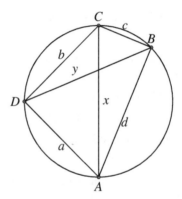

this theorem states that $xy = ac + bd$. Hint. Show that on the line $\overline{BD}$ there is a point X such that the angle $X\hat{A}D$ is equal to the angle $B\hat{A}C$. Deduce that the triangles AXD and ABC are similar, and that the triangles AXB and ACD are similar. Let the distance between D and X be e. Show that

$$\frac{e}{a} = \frac{c}{x} \text{ and that } \frac{y-e}{d} = \frac{b}{x}.$$

From this, the result follows by simple algebra. To help you show that the triangles are similar, you will need to use Proposition III.21 from Euclid which is illustrated by the following diagram.

---

[7]Claudius Ptolomeus was a Greek mathematician and astronomer who flourished around 150 CE in the city of Alexandria.

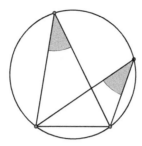

9. *The goal of this question is to find all rational points on the unit circle with centre at the origin. This means finding all those points (x, y) where $x^2 + y^2 = 1$ such that both x and y are rational numbers. We do this using geometry and will refer to the diagram below.

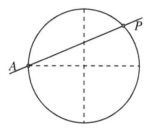

We have drawn a unit circle centre the origin. Denote the point $(-1, 0)$ by A. This is an obvious rational point on the unit circle. We shall use it to find all the others. Denote by P an arbitrary point on the unit circle different from A.

(a) Show that any line passing through the point A and intersecting the circle at a point P different from A has the equation $y = t(x + 1)$ where t is any real number.

(b) Deduce that such a point P has coordinates given by

$$(x, y) = \left(\frac{1 - t^2}{1 + t^2}, \frac{2t}{1 + t^2} \right)$$

where t is any real number.

(c) Deduce that the rational points on the circle correspond to the values of t which are rational.

(d) Describe the rational points on the unit circle.

10. *Inscribe a regular polygon with 2^{m+1} sides inside the circle with radius 1. That is, the vertices of the polygon lie on the circumference of the circle. Show that

$$2^m \underbrace{\sqrt{2 - \sqrt{2 + \sqrt{2 + \ldots}}}}_{m \text{ times}}$$

is an approximation to π.

11. *Take any positive natural number $n > 1$. So $n = 2, 3, \ldots$. If n is even, divide it by 2 to get $\frac{n}{2}$; if n is odd, multiply it by 3 and add 1 to obtain $3n + 1$. Now repeat this process and stop only if you get 1. For example, if $n = 6$ you get $6, 3, 10, 5, 16, 8, 4, 2, 1$. Prove that no matter what number you start with, you will always eventually reach 1. A good reference for this question is [81].

2.4 AXIOMS

We did not complete Proof 4 because to do so required a new ingredient in understanding proofs. The goal of this section is to explain what that ingredient is. Suppose we are trying to prove the statement S. Then we are done if we can find a theorem S_1 so that $S_1 \Rightarrow S$. But this raises the question of how we know that S_1 is a theorem. This can only be because we can find a theorem S_2 such that $S_2 \Rightarrow S_1$. There are now three possibilities.

1. At some point, we find a theorem S_n such that $S \Rightarrow S_n$. This is clearly a bad thing. In trying to prove S, we have in fact used S and so have not proved anything at all. This is an example of *circular reasoning* and has to be avoided. We can do this by organizing what we know into a hierarchy, so that to prove a result we are only allowed to use those theorems already proved.

2. Assuming we have avoided the above pitfall, the next possibility is that we get an infinite sequence of implications

$$\ldots \Rightarrow S_2 \Rightarrow S_1 \Rightarrow S.$$

We never actually know that S is a theorem because it is always proved in terms of something else without end. This is also clearly a bad thing. We establish relative truth: a statement is true if another is true, but not absolute truth. We clearly do not want this to happen. We are therefore led inexorably to the third possibility.

3. To prove S, we only have to prove a finite number of implications

$$S_n \Rightarrow S_{n-1} \Rightarrow \ldots \Rightarrow S_1 \Rightarrow S.$$

If S_n is supposed to be a theorem, however, then how do we know it is true if not in terms of something else, contradicting the assumption that this was supposed to be a complete argument?

We now examine case (3) above in more detail, since resolving it will lead to an important insight. Mathematics is supposed to be about proving theorems but the analysis above has led us to the uncomfortable possibility that some things have to be accepted as true 'because they are' undermining our homily on mathematical truth in Section 2.1. Before we explain the way out of this conundrum, we first consider an example from an apparently different enterprise: playing a game.

Example 2.4.1. To be concrete, take the game of chess. This consists of a board and some pieces. The pieces are of different types, such as kings, queens, knights and so on, each of which can be moved in different ways. To play chess means to accept the rules of chess and to move the pieces in accordance with those rules. It is clearly meaningless to ask whether the rules of chess are true, but a move in chess is *valid* if it is made according to those rules.

This example provides a way of understanding how mathematics works, at least as a starting point. Mathematics should be viewed as a collection of different mathematical domains each described by its own 'rules of the game' which in mathematics are termed *axioms*. These axioms are the basic assumptions on which the theory is built and are the building blocks of all proofs within that mathematical domain. The goal is to prove interesting theorems from those axioms. Thus to ask whether a statement in a particular mathematical domain is true is actually to ask whether it can be derived from the axioms of that domain. This is unsettling, so we shall expand on this point a little in Section 2.5. In the mean time, here is an example.

Example 2.4.2. The Greeks themselves attributed the discovery of geometry to the Egyptians who needed it in recalculating land boundaries for tax purposes after the yearly flooding of the Nile. Thus geometry probably first existed as a collection of geometrical methods that worked: the tax was calculated, the pyramids were built and everybody was happy. The Greeks transformed these methods into a mathematical science and a model for what could be achieved in mathematics. Euclid's book the *Elements* codified what was known about geometry into a handful of axioms and then showed that all of geometry could be deduced from those axioms by the use of mathematical proof.[8]

Axioms for plane geometry

(E1) Two distinct points determine a unique straight line.

(E2) A line segment can be extended infinitely in either direction.

(E3) Circles can be drawn with any centre and any radius.

(E4) Any two right angles are equal to each other.

(E5) Suppose that a straight line cuts two lines l_1 and l_2. If the interior angles on the same side add up to strictly less than $180°$, then if l_1 and l_2 are extended on that side they will eventually meet.

---

[8]Euclid's *Elements* is one of the most important books ever written and is in the same league as Homer's *Illiad* and Darwin's *Origin of Species*.

The last axiom needs a picture to illustrate what is going on.

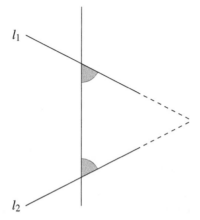

In principle, all the results you learnt in school about triangles and circles can be proved from these axioms. There were a few bugs in Euclid's work but they were later fixed by a number of mathematicians, most notably David Hilbert (1862–1943). This should not detract from what an enormous achievement Euclid's book was and is. With this understanding, we can now finish off Proof 4: claim (1) is proved in Book I, Proposition 31, and claim (3) is proved in Book I, Proposition 29. Thus the proof is completed by showing that it can be proved from the axioms of Euclidean geometry.

One way of learning mathematics is to begin with a list of axioms for some mathematical domain and then start proving theorems. This approach is logically hygienic, because it leaves you in no doubt what you are allowed to assume and where to start, and it has an austere kind of intellectual beauty to it. On the other hand, it takes a long time to get to anywhere remotely interesting and it also leaves out intuition which is vital if the mathematics is to mean anything and not become merely an empty game. In this book, I have erred on the side of developing intuition with gestures towards axioms.

2.5 UN PETIT PEU DE PHILOSOPHIE

There is unfinished business. What we have said so far might lead you to think that mathematics is nothing more than a game and that it is enough to write down some axioms, randomly if you like, and then start proving theorems. The truth is that this is unlikely to result in anything remotely interesting if not actually self-contradictory. The axiom systems that underpin different mathematical domains were in fact the result of a long process of analysis. They are designed to encode what our intuition tells us is important. If they were not satisfactory, they would have been replaced.

There are interesting questions on the nature of the relationship between mathematics and the real world but unlike physicists, who often discuss the philosophical meaning of their work, mathematicians tend to be impatient with the subject which I think is to be regretted. I shall say no more about this here but recommend [11, 13, 29, 33, 52, 62, 82] for further reading, and leave you with the following intriguing quote from NASA on the findings of the Wilkinson Microwave Anisotropy Probe (WMAP) which relates directly to what is often called the *unreasonable effectiveness of mathematics*.

> "WMAP also confirms the predictions that the amplitude of the variations in the density of the universe on big scales should be slightly larger than smaller scales, and that the universe should obey the rules of Euclidean geometry so the sum of the interior angles of a triangle add to 180 degrees".[9]

Box 2.1: Non-Euclidean Geometry

Axiom (E5) of Euclidean is notably more complex than the other axioms, and from earliest times mathematicians have handled it as if it were mildly toxic. Euclid himself avoids using it until about halfway through Book I. Subsequently, attempts were made to derive it from the other axioms. All failed. Axiom (E5) is actually equivalent, in the presence of the other axioms, to the statement that given a line l and a point P not on that line there is exactly one line that contains P and is parallel to l. One way of investigating the status of this axiom is to consider axioms (E1)–(E4) together with a new axiom: that there are at least two lines through P parallel to l. Remarkably there are no contradictions. Instead, a new kind of geometry arises, called *hyperbolic geometry*, the first example of what are called *non-Euclidean geometries*, and notice that the word geometry is in the plural. Hyperbolic geometry was discovered independently by Carl Friedrich Gauss (1777–1855), János Bolyai (1802–1860) and Nikolai Lobachevsky (1793–1856). There are now two geometries: so which is the 'real' geometry? In hyperbolic geometry, the sum of the angles in a triangle is always strictly less than 180° although for smaller and smaller triangles, the discrepancy between the sum of the angles and 180° decreases. Thus we could be living in a hyperbolic universe and not be aware of it because the discrepancies could be below experimental error. The important point is that Euclidean geometry and hyperbolic geometry cannot both be models of the real world. Thus geometry, which seemed so natural, now becomes problematical. The real question is the following: what is the relationship between mathematics and the real, that is physical, world?

2.6 MATHEMATICAL CREATIVITY

One of the themes of this book is that it is important to know the rules that govern different algebraic systems. We shall explain this in more detail in Chapter 4. We have also emphasized the central importance of proofs, and that in mathematics we accept nothing without proof. All this talk of rules and proofs makes mathematics seem like an intellectual straightjacket. This is because although everything we have said is true, it is not the whole story *because absolutely nothing we have said so far explains how theorems are discovered nor how proofs are constructed.* What we have omitted is this: mathematics is a creative endeavour. Creativity eludes simple description and so it is also the hardest aspect of mathematics to convey except by experiencing it. Creativity has few rules as the following two examples illustrate.

---

[9]http://map.gsfc.nasa.gov/news/index.html.

Example 2.6.1. A deep connection was discovered between an algebraic object called the Monster group and a specific analytic object called the elliptic modular form on the basis of the following observation: the number 196883 arises naturally from the Monster group, the number 196884 arises naturally from the elliptic modular form and $196884 = 196883 + 1$. The theory that developed to explain this observation was called *Moonshine* [104]. This is frankly kabbalistic.

Example 2.6.2. The Indian mathematician Srinivasa Ramanujan (1887–1920) attributed his extraordinary creativity to the influence of a goddess from the Hindu pantheon which is reminiscent of the way that poets in Greece would invoke the Muse.

These examples are too esoteric to provide general guidance on creativity, though a mathematical muse would come in handy occasionally. On a more practical level, new mathematics is often discovered by pushing known mathematics beyond its design limits which is something physicists do all the time. This is how the complex numbers were discovered, for example, and Section 6.3 is a further illustration of this idea. One of the most common devices in creating new mathematics is analogy. In this book, the analogy between integers and polynomials is used to motivate the development of the theory of polynomials in Chapter 7 along the lines of the theory of integers developed in Chapter 5. In fact, this informal analogy can be made mathematically precise and leads to a general theory of Euclid's algorithm. In Section 9.7, the quaternions are introduced. These first arose by analogy with the complex numbers, when Hamilton wondered if the two-dimensional complex numbers could be generalized to three. Although creativity has no rules, and what works works, we ultimately ground our speculations by proving them. This is the distinguishing feature of mathematics.

2.7 PROVING SOMETHING FALSE

'Proving a statement true' and 'proving a statement false' sound similar but there is an asymmetry between them. 'Proving a statement false' usually requires less work than 'proving a statement true'. To prove a statement false all you need do is find a single *counterexample*. For example, consider the following statement: every odd number bigger than 1 is a prime. This is false. The reason is that 9 is odd, bigger than 1 and not prime. Thus 9 is a counterexample. Once you have found the counterexample you are done.

2.8 TERMINOLOGY

There is some terminology worth mentioning here. If we believe something might be true, but there is not yet a proof, we say that it is a *conjecture*. The things we do prove fall, roughly, into the following categories: a *theorem* is a major result, worthy of note; a *proposition* is a result, and a *lemma* is an auxiliary result, a tool, useful in many different places; a *corollary* is a result we can deduce with little or no effort from a proposition or theorem. The conclusion of a proof is marked using the symbol □ which replaces the older use of QED.

2.9   ADVICE ON PROOFS

- One of the goals of this book is to introduce you to proofs. This does not mean you will afterwards be able to do proofs. That takes time and practice.

- Initially, you should aim to understand proofs. This means seeing why a proof is true. A good test of whether you really understand a proof is whether you can explain it to someone else.

- It is much easier to check that a proof is correct then it is to invent the proof in the first place. Nevertheless, be warned: it can also take a long time just to understand a proof.

- If you are asked to construct a proof, do not expect to find it in a few minutes. Constructing proofs is usually not trivial.

- If you do not understand the words used in a statement you are asked to prove then you are not going to be able to prove that statement. Definitions are vitally important in mathematics.

- Every statement that you make in a proof must be justified: if it is a definition, say that it is a definition; if it is a theorem, say that it is a theorem; if it is one of the assumptions, say that it is one of the assumptions; if it is an axiom, say that it is an axiom; if it is a rule of logic, say which rule it is.

CHAPTER 3

Foundations

"No one shall drive us out of the paradise that Cantor has created." – David Hilbert

This is a lengthy and diverse chapter.

You should think of it, in the first instance, as a reference chapter. I would recommend reading Section 3.1 and Section 3.8 and then enough of Section 3.4 to understand what a bijective function is and then move on to Chapter 4, returning as and when needed.

The goal of this chapter is to introduce certain concepts that play a rôle in virtually every branch of mathematics. Although there are not very many, basically, sets, relations and functions, learning them for the first time can be difficult: partly because they are abstract, and partly because they are unmotivated. In fact, it took mathematicians themselves a long time to recognize that they were important. It was Georg Cantor (1845–1918) in the last quarter of the nineteenth century who was instrumental in setting up set theory, and the vindication of his work is that subsequent developments in mathematics would have been impossible without it. Whilst foundations are important they do not sound inspiring, so the chapter closes with something that I think is truly extraordinary. This is Cantor's vertiginous discovery that there is not one infinity but an infinity of different infinities.

3.1 SETS

Set theory begins with two deceptively simple definitions on which everything is based.

1. A *set* is a collection of objects, called *elements*, which we wish to regard as a whole.[1]

2. Two sets are *equal* precisely when they contain the same elements.

---

[1] This should be viewed as conveying intuition rather than providing a precise definition.

That is it. It is customary to use capital letters to name sets such as $A, B, C \ldots$ or fancy capital letters such as $\mathbb{N}, \mathbb{Z} \ldots$ with the elements of a set usually being denoted by lower case letters. If *x is an element of* the set A then we write $x \in A$ and if *x is not an element of* the set A then we write $x \notin A$. To indicate that some things are to be regarded as a set rather than just as isolated individuals, we enclose them in 'curly brackets' $\{$ and $\}$ formally called braces. Thus the set of suits in a pack of cards is $\{\clubsuit, \diamondsuit, \heartsuit, \spadesuit\}$. A set should be regarded as a bag of elements, and so the order of the elements within the set is not important. Thus $\{a, b\} = \{b, a\}$. Perhaps more surprisingly, repetition of elements is ignored. Thus $\{a, b\} = \{a, a, a, b, b, a\}$. This latter feature of sets can sometimes be annoying. For example, the set of roots of the equation $(x - 1)^2 = 0$ is $\{1\}$ but we would usually want to say that this root occurs twice. However, the set $\{1, 1\}$ is equal to the set $\{1\}$ and so cannot be used to record this information. Sets can be generalized to what are called *multisets* where repetition is recorded. Although used in computer science, this natural notion is almost never used by mathematicians. So if you need to record repetitions you will have to find circumlocutions. The set $\{\}$ is empty and is called the *empty set*. It is given a special symbol $\emptyset$, which is not the Greek letter ϕ or Φ, but is allegedly the first letter of a Danish word meaning 'desolate'.[2] The symbol $\emptyset$ means exactly the same thing as $\{\}$. Observe that $\emptyset \neq \{\emptyset\}$ since the empty set contains no elements whereas the set $\{\emptyset\}$ contains one element. In addition to the word set, words like *family*, *class* and *collection* are also used to add a bit of variety to the English. This is fine most of the time but be aware that the word 'class' in particular can also be used with a precise technical meaning.

The number of elements a set contains is called its *cardinality* denoted by $|X|$. A set is *finite* if it only has a finite number of elements; otherwise, it is *infinite*. A set with exactly one element is called a *singleton set*. We shall explore cardinality in more depth in Sections 3.9 and 3.10. We can sometimes define infinite sets by using curly brackets but then, because we cannot list all elements in an infinite set, we use '...' to mean 'and so on in the obvious way'. This can also be used to define big finite sets where there is an obvious pattern. The most common way of describing a set is to say what properties an element must have to belong to it. By a *property* we mean a sentence containing a variable such as x so that the sentence becomes true or false depending on what we substitute for x. For example, the sentence 'x is an even natural number' is true when x is replaced by 2 and false when x is replaced by 3. If we abbreviate 'x is an even natural number' by $E(x)$ then the set of even natural numbers is the set of all natural numbers n such that $E(n)$ is true. This set is written $\{x: E(x)\}$ or $\{x \mid E(x)\}$. More generally, if $P(x)$ is any property then $\{x: P(x)\}$ means 'the set of all things x that satisfy the condition P'. Here are some examples of sets defined in various ways.

Examples 3.1.1.

1. $D = \{$ Monday, Tuesday, Wednesday, Thursday, Friday, Saturday, Sunday $\}$, the set of the days of the week. This is a small finite set and so we can conveniently list all its elements.

---

[2]Mathematical Nordic noir.

2. $M = \{$ January, February, March, $\ldots$, November, December $\}$, the set of the months of the year. This is a finite set but we did not want to write down all the elements explicitly so we wrote '$\ldots$' instead.

3. $A = \{x \colon x$ is a prime natural number$\}$. We here define a set by describing the properties that the elements of the set must have. In this case $P(x)$ is the statement 'x is a prime natural number' and those natural numbers x are admitted membership to the set when they are indeed prime.

In this book, the following sets of numbers play a special rôle. We shall use this notation throughout and so it is worth getting used to it.

Examples 3.1.2.

1. The set $\mathbb{N} = \{0, 1, 2, 3, \ldots\}$ of all *natural numbers*. Caution is required here since some books eccentrically do not regard 0 as a natural number.

2. The set $\mathbb{Z} = \{\ldots, -3, -2, -1, 0, 1, 2, 3, \ldots\}$ of all *integers*. The reason $\mathbb{Z}$ is used to designate this set is because 'Z' is the first letter of the word 'Zahl', the German for number.[3]

3. The set $\mathbb{Q}$ of all *rational numbers*. That is those numbers that can be written as quotients of integers with non-zero denominators.

4. The set $\mathbb{R}$ of all *real numbers*. That is all numbers which can be represented by decimals with potentially infinitely many digits after the decimal point.

5. The set $\mathbb{C}$ of all *complex numbers*, introduced in Chapter 6.

Given a set A, a new set B can be formed by *choosing* elements from A to put into B. We say that B is a *subset* of A, denoted by $B \subseteq A$. In mathematics, the word 'choose', unlike in polite society, also includes the possibility of *choosing nothing* and the possibility of *choosing everything*. In addition, there does not need to be any rhyme or reason to your choices: you can pick elements 'at random' if you want. If $A \subseteq B$ and $A \neq B$ then we say that A is a *proper subset* of B.

Examples 3.1.3.

1. $\emptyset \subseteq A$ for every set A, where we choose no elements from A.

2. $A \subseteq A$ for every set A, where we choose all the elements from A.

3. $\mathbb{N} \subseteq \mathbb{Z} \subseteq \mathbb{Q} \subseteq \mathbb{R} \subseteq \mathbb{C}$. Observe that $\mathbb{Z} \subseteq \mathbb{Q}$ because an integer n is equal to the rational number $\frac{n}{1}$.

4. $\mathbb{E}$, the set of even natural numbers, is a subset of $\mathbb{N}$.

5. $\mathbb{O}$, the set of odd natural numbers, is a subset of $\mathbb{N}$.

---

[3]German mathematics was predominant in the nineteenth and early twentieth centuries and so has bequeathed to English some important German mathematical terms and the odd piece of notation.

6. $\mathbb{P} = \{2, 3, 5, 7, 11, 13, 17, 19, 23, \ldots\}$, the set of *primes*, is a subset of $\mathbb{N}$.

7. $A = \{x : x \in \mathbb{R}$ and $x^2 = 4\}$ which is equal to the set $\{-2, 2\}$. This example demonstrates that you may have to do some work to actually produce specific elements of a set defined by a property.

The set whose elements are all the subsets of X is called the *power set* of X and is denoted by $\mathsf{P}(X)$ in this book.[4] It is important to remember that the power set of a set X contains both $\emptyset$ and X as elements. This example illustrates something important about sets: the elements of a set may themselves be sets. It is this ability to nest sets, like Russian dolls one inside the other, which leads to the notion of set being more powerful than it first appears.

Example 3.1.4. We find all the subsets of the set $X = \{a, b, c\}$. First there is the subset with no elements, the empty set. Then there are the subsets that contain exactly one element: $\{a\}, \{b\}, \{c\}$. Then the subsets containing exactly two elements: $\{a, b\}, \{a, c\}, \{b, c\}$. Finally, there is the whole set X. It follows that X has 8 subsets and so

$$\mathsf{P}(X) = \{\emptyset, \{a\}, \{b\}, \{c\}, \{a, b\}, \{a, c\}, \{b, c\}, X\}.$$

The idea that sets are defined by properties is a natural one, but there are murky logical depths which the next box describes.

Box 3.1: Russell's Paradox

It seems obvious that given a property $P(x)$, there is a corresponding set $\{x : P(x)\}$ of all those things that have that property. We shall now describe a famous result in the history of mathematics called *Russell's Paradox*, named after Bertrand Russell (1872–1970),[a] which shows that just because something is obvious does not make it true. Define $\mathscr{R} = \{x : x \notin x\}$. In other words: the set of all sets that do not contain themselves as an element. For example, $\emptyset \in \mathscr{R}$. We now ask the question: is $\mathscr{R} \in \mathscr{R}$? There are only two possible answers and we investigate them both. Suppose that $\mathscr{R} \in \mathscr{R}$. This means that $\mathscr{R}$ must satisfy the entry requirements to belong to $\mathscr{R}$ which it can only do if $\mathscr{R} \notin \mathscr{R}$. Suppose that $\mathscr{R} \notin \mathscr{R}$. Then it satisfies the entry requirement to belong to $\mathscr{R}$ and so $\mathscr{R} \in \mathscr{R}$. Thus exactly one of $\mathscr{R} \in \mathscr{R}$ and $\mathscr{R} \notin \mathscr{R}$ must be true but assuming one implies the other. We therefore have an honest-to-goodness contradiction. Our only way out is to conclude that, whatever $\mathscr{R}$ might be, it is not a set. This contradicts the obvious statement we began with. If you want to understand how to escape this predicament, you will have to study *set theory*. Disconcerting as this might be, imagine how much more so it was to the mathematician Gottlob Frege (1848–1925). He was working on a book which based the development of mathematics on sets when he received a letter from Russell describing this paradox thereby undermining what Frege was attempting to achieve.

---

[a]Bertrand Russell is an interesting figure not merely for mathematicians but in twentieth century cultural and political history. He was an Anglo-Welsh philosopher born when Queen Victoria still had another thirty years on the throne as 'Queen empress' and died only a few months after Neil Armstrong stepped onto the moon. See [52] for more on Russell's extraordinary life and work.

Consider the set $\{a, b\}$. We have explained that order does not matter and so this is the same as the set $\{b, a\}$ but there are many occasions where we do want order

---

[4]Many books use $\wp(X)$ which I find ugly.

to matter. For example, in giving the coordinates of a point in the plane we have to know what the x- and y-coordinates are. So we need a new notion where order does matter. It is called an *ordered pair* and is written (a,b), where a is called the *first component* and b is called the *second component*.[5] The key feature of this new object is that $(a,b) = (c,d)$ if and only if $a = c$ and $b = d$. At first blush, set theory seems inadequate to define ordered pairs. In fact it can. The details have been put in a box, and do not need to be read now. They illustrate something which in many ways is astonishing given the paucity of ingredients: all of mathematics can be developed using only set theory.

Box 3.2: Ordered Pairs

We are going to show you how sets, which do not encode order directly, can nevertheless be used to define ordered pairs. It is an idea due to Kazimierz Kuratowski (1896–1980). Define

$$(a,b) = \{\{a\},\{a,b\}\}.$$

We have to prove, using only this definition, that $(a,b) = (c,d)$ if and only if $a = c$ and $b = d$. The proof is essentially an exercise in special cases. We shall prove the hard direction. Suppose that

$$\{\{a\},\{a,b\}\} = \{\{c\},\{c,d\}\}.$$

Since $\{a\}$ is an element of the lefthand side it must also be an element of the righthand side. So $\{a\} \in \{\{c\},\{c,d\}\}$. There are now two possibilities. Either $\{a\} = \{c\}$ or $\{a\} = \{c,d\}$. The first case gives us $a = c$, and the second case gives us $a = c = d$. Since $\{a,b\}$ is an element of the lefthand side it must also be an element of the righthand side. So $\{a,b\} \in \{\{c\},\{c,d\}\}$. There are again two possibilities. Either $\{a,b\} = \{c\}$ or $\{a,b\} = \{c,d\}$. The first case gives us $a = b = c$, and the second case gives us ($a = c$ and $b = d$) or ($a = d$ and $b = c$). We therefore have the following possibilities:

- $a = b = c$. But then $\{\{a\},\{a,b\}\} = \{\{a\}\}$. It follows that $c = d$ and so $a = b = c = d$ and, in particular, $a = c$ and $b = d$.

- $a = c$ and $b = d$.

- In all remaining cases, $a = b = c = d$ and so, in particular, $a = c$ and $b = d$.

Taking the notion of ordered pair for granted, we can now build sets of ordered pairs. Define $A \times B$, the *product* of the sets A and B, to be the set

$$A \times B = \{(a,b) : a \in A \text{ and } b \in B\}.$$

Example 3.1.5. Let $A = \{1,2,3\}$ and let $B = \{a,b\}$. Then

$$A \times B = \{(1,a),(1,b),(2,a),(2,b),(3,a),(3,b)\}$$

and

$$B \times A = \{(a,1),(a,2),(a,3),(b,1),(b,2),(b,3)\}.$$

In particular, $A \times B \neq B \times A$, in general.

---

[5] This notation should not be confused with the notation for *open intervals* where (a,b) denotes the set $\{r : r \in \mathbb{R} \text{ and } a < r < b\}$, nor with the use of brackets in clarifying the meaning of algebraic expressions. The context should make clear what meaning is intended. This is an opportune moment to remind you that $[a,b]$ denotes the *closed interval*.

If $A = B$ it is natural to abbreviate $A \times A$ as A^2. This now agrees with the notation $\mathbb{R}^2$ which is the set of all ordered pairs of real numbers and, geometrically, can be regarded as the real plane with its distinguished origin. We have defined ordered pairs but there is no reason to stop there. We can also define *ordered triples*. This can be done by defining

$$(x, y, z) = ((x, y), z).$$

The key property of ordered triples is that $(a, b, c) = (d, e, f)$ if and only if $a = d$, $b = e$ and $c = f$. Given three sets A, B and C we can define their *product* $A \times B \times C$ to be the set of all ordered triples (a, b, c) where $a \in A$, $b \in B$ and $c \in C$.

Example 3.1.6. A good example of an ordered triple in everyday life is a date that consists of a day, a month and a year. Thus the 16th June 1904 is really an ordered triple $(16, \text{June}, 1904)$ where we specify day, month and year in that order. Order here is important since dates are written in different orders in the US as opposed to the UK, for example.

If $A = B = C$ then we write A^3 rather than $A \times A \times A$. Thus the set $\mathbb{R}^3$ consists of all cartesian coordinates (x, y, z), and, geometrically, can be regarded as real space with its distinguished origin. In general, we can define *ordered n-tuples*, $(x_1, \ldots, x_n)$, and products of n sets $A_1 \times \ldots \times A_n$. If $A_1 = \ldots = A_n = A$ then we write A^n for their n-fold product. Depending on the context, an ordered n-tuple might also be called a *list* or a *sequence*.

A related concept here is that of a *string*. Let A be a non-empty set called, in this context, an *alphabet*. This change in terminology reflects the way that we wish to regard the set A: it is a psychological crutch rather than a mathematical necessity. The elements of A are called *letters* or *symbols*. Then a *string over* A is simply a finite list of elements of A. We also allow the *empty string* denoted ε. In this context, we would write a string as $a_1 \ldots a_n$ rather than $(a_1, \ldots, a_n)$. For example, the English word 'aardvark' is a string but it is also an ordered 8-tuple (a,a,r,d,v,a,r,k). The *length* of a string is the total number of symbols that occur in it counting repetitions. The set of all strings over A is denoted by A^* and the set of all strings of length n is denoted by A^n.

Examples 3.1.7. Here are a few examples of alphabets. You will meet many more throughout this book.

1. The alphabet $A = \{0, 1\}$ of bits. Strings over this are called *binary strings*. We have

$$\{0, 1\}^* = \{\varepsilon, 0, 1, 00, 01, 10, 11, \ldots\}$$

and

$$\{0, 1\}^2 = \{00, 01, 10, 11\}.$$

2. The alphabet $A = \{A, G, T, C\}$ represents the four building blocks of DNA. Strings over this alphabet describe everything from aardvarks to you.

3. The alphabet $A = \{0, 1, 2, \ldots, 9\}$ of decimal digits. Strings over this alphabet describe the natural numbers in base 10.

The following three operations defined on sets are called *Boolean operations*, named after George Boole (1815–1864). Let A and B be sets. Define a set, called the *intersection* of A and B, denoted by $A \cap B$, whose elements consist of all those elements that belong to A **and** B.

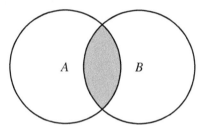

Define a set, called the *union* of A and B, denoted by $A \cup B$, whose elements consist of all those elements that belong to A **or** B. The word *or* in mathematics does not mean quite the same as it does in everyday life. Thus X or Y means X or Y or both. It is therefore *inclusive or*.[6]

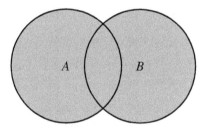

Define a set, called the *difference* or *relative complement* of A and B, denoted by $A \setminus B$,[7] whose elements consist of all those elements that belong to A **and not** to B.

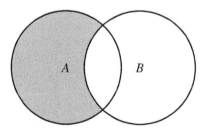

The diagrams used to illustrate the above definitions are called *Venn diagrams* where a set is represented by a region in the plane.

Example 3.1.8. Let $A = \{1,2,3,4\}$ and $B = \{3,4,5,6\}$. Determine $A \cap B$, $A \cup B$, $A \setminus B$ and $B \setminus A$.

---

[6] In everyday life, we might, on occasion, assume that *or* means *exclusive or*. That is, X or Y and not both.

[7] Sometimes denoted by $A - B$.

- $A \cap B$. We have to find the elements that belong to both A and B. We start with the elements in A and work left-to-right: 1 is not an element of B; 2 is not an element of B; 3 and 4 are elements of B. Thus $A \cap B = \{3,4\}$.

- $A \cup B$. We join the two sets together $\{1,2,3,4,3,4,5,6\}$ and then read from left-to-right weeding out repetitions to get $A \cup B = \{1,2,3,4,5,6\}$.

- $A \setminus B$. We have to find the elements of A that do not belong to B. Read the elements of A from left-to-right comparing them with the elements of B: 1 does not belong to B; 2 does not belong to B: but 3 and 4 do belong to B. It follows that $A \setminus B = \{1,2\}$.

- To calculate $B \setminus A$ we have to find the set of elements of B that do not belong to A. This set is equal to $\{5,6\}$.

Examples 3.1.9.

1. $\mathbb{E} \cap \mathbb{O} = \emptyset$. This says that there is no number which is both odd and even.

2. $\mathbb{P} \cap \mathbb{E} = \{2\}$. This says that the only even prime number is 2.

3. $\mathbb{E} \cup \mathbb{O} = \mathbb{N}$. This says that every natural number is either odd or even.

Sets A and B are said to be *disjoint* if $A \cap B = \emptyset$. If A and B are disjoint then their union is called a *disjoint union*.

Example 3.1.10. Probability theory uses the language of set theory. Here is an example. Throw two dice. The set of possible outcomes is the set

$$\Omega = \{1,2,3,4,5,6\} \times \{1,2,3,4,5,6\}$$

called in this context the *sample space*. Subsets of Ω are called *events*. Thus the singleton subset $\{(3,4)\}$ records the event that the two dice are thrown and we obtain a 3 on the first dice and a 4 on the second, whereas the event that both dice show an even number is the subset $\{2,4,6\} \times \{2,4,6\}$. All subsets containing one element are assumed to be *equally likely* at least when the dice are not loaded. Since the set Ω has 36 elements, the probability of any one-element event is defined to be $\frac{1}{36}$. Probability theory is not developed in this book, but it starts with counting because the cardinality of an event that we might be interested in divided by the cardinality of the finite sample space is defined to be the probability of that event, assuming all elementary events to be equally likely.

Exercises 3.1

1. Let $A = \{\clubsuit, \diamondsuit, \heartsuit, \spadesuit\}$, $B = \{\spadesuit, \diamondsuit, \clubsuit, \heartsuit\}$ and $C = \{\spadesuit, \diamondsuit, \clubsuit, \heartsuit, \clubsuit, \diamondsuit, \heartsuit, \spadesuit\}$. Is it true or false that $A = B$ and $B = C$? Explain.

2. Let $X = \{1,2,3,4,5,6,7,8,9,10\}$. Write down the following subsets of X.

(a) The subset A of even elements of X.

(b) The subset B of odd elements of X.

(c) $C = \{x \colon x \in X \text{ and } x \geq 6\}$.

(d) $D = \{x \colon x \in X \text{ and } x > 10\}$.

(e) $E = \{x \colon x \in X \text{ and } x \text{ is prime}\}$.

(f) $F = \{x \colon x \in X \text{ and } (x \leq 4 \text{ or } x \geq 7)\}$.

3.   (a) Find all subsets of $\{a,b\}$. How many are there? Write down also the number of subsets with, respectively, 0, 1 and 2 elements.

    (b) Find all subsets of $\{a,b,c\}$. How many are there? Write down also the number of subsets with, respectively, 0, 1, 2 and 3 elements.

    (c) Find all subsets of the set $\{a,b,c,d\}$. How many are there? Write down also the number of subsets with, respectively, 0, 1, 2, 3 and 4 elements.

    (d) What patterns do you notice arising from these calculations?

4. Write down the elements of the set $\{A,B,C\} \times \{a,b\}$.

5. If the set A has m elements and the set B has n elements how many elements does the set $A \times B$ have?

6. If A has m elements, how many elements does the set A^n have?

7. Let $S = \{4,7,8,10,23\}$, $T = \{5,7,10,14,20,25\}$ and $V = \{2,5,10,20,30,36\}$. Determine the following.

    (a) $S \cup (T \cap V)$.

    (b) $S \setminus (T \cap V)$.

    (c) $(S \cap T) \setminus V$.

8. Let $A = \{a,b,c,d,e,f\}$ and $B = \{g,h,k,d,e,f\}$. What are the elements of the set $A \setminus ((A \cup B) \setminus (A \cap B))$?

9. Let $A = \{1,2,3\}$ and $B = \{a,b,c\}$. What is the set

$$(A \times B) \setminus ((\{1\} \times B) \cup (A \times \{c\}))?$$

10. Prove that two sets A and B are equal if and only if $A \subseteq B$ and $B \subseteq A$.

11. Given a set A define a new set $A^+ = A \cup \{A\}$. Calculate in succession the sets

$$\emptyset^+, \emptyset^{++}, \emptyset^{+++}$$

which are obtained by repeated application of the operation $+$. Write down the cardinalities of these sets.

3.2 BOOLEAN OPERATIONS

The Boolean operations introduced in Section 3.1 are ubiquitous, being particularly important in probability theory. It is, therefore, important to know how to compute with them effectively and this means knowing the *properties* these operations have. In the theorem below, these properties and the names they are known by are listed. We discuss how to prove them afterwards. There is more to be said about these properties, and it is said in Section 4.5.

Theorem 3.2.1 (Properties of Boolean operations). *Let A, B and C be any sets.*

1. $A \cap (B \cap C) = (A \cap B) \cap C$. *Intersection is associative.*

2. $A \cap B = B \cap A$. *Intersection is commutative.*

3. $A \cap \emptyset = \emptyset = \emptyset \cap A$. *The empty set is the zero for intersection.*

4. $A \cup (B \cup C) = (A \cup B) \cup C$. *Union is associative.*

5. $A \cup B = B \cup A$. *Union is commutative.*

6. $A \cup \emptyset = A = \emptyset \cup A$. *The empty set is the identity for union.*

7. $A \cap (B \cup C) = (A \cap B) \cup (A \cap C)$. *Intersection distributes over union.*

8. $A \cup (B \cap C) = (A \cup B) \cap (A \cup C)$. *Union distributes over intersection.*

9. $A \setminus (B \cup C) = (A \setminus B) \cap (A \setminus C)$. *De Morgan's law part one.*

10. $A \setminus (B \cap C) = (A \setminus B) \cup (A \setminus C)$. *De Morgan's law part two.*

11. $A \cap A = A$. *Intersection is idempotent.*

12. $A \cup A = A$. *Union is idempotent.*

To *illustrate* these properties, we can use Venn diagrams.

Example 3.2.2. We illustrate property (7). The Venn diagram for $(A \cap B) \cup (A \cap C)$ is given below.

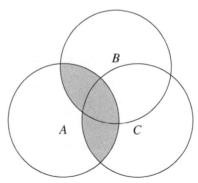

This is exactly the same as the Venn diagram for $A \cap (B \cup C)$. It follows that $A \cap (B \cup C) = (A \cap B) \cup (A \cap C)$ at least as far as Venn diagrams are concerned.

To *prove* these properties hold, we have to proceed more formally and introduce some of the basic ideas of *propositional logic*.

In Don Siegel's classic 1956 science fiction movie *Invasion of the Body Snatchers* people are replaced by pod-people who outwardly look like their replacements but who inside are empty and devoid of all human emotion. Mathematicians do a similar job on the English language. The words they use look like ordinary English words but have had all the juices sucked out of them. In particular, the words *and, or* and *not* are not what they appear to be. They are empty shells that contain exactly one technical meaning apiece and should not be confused with the full-bodied English words they resemble. To explain what these meanings are, we replace them in this section by mathematical symbols so that there is no confusion. In the rest of this book we shall use the ordinary English words, but remember: they are aliens. We use the following dictionary.

and	or	it is not the case that
$\wedge$	$\vee$	$\neg$

These symbols are called *propositional connectives*. We can now write down specific statements replacing occurrences of and, or, not by $\wedge$, $\vee$, $\neg$, respectively.

Examples 3.2.3.

1. $\neg$(Frege is happy)$\wedge$(Frege's postman is happy).

2. $\neg$(There is a rhino under the table)$\vee$(There is a rhino under the table).

3. (Popper threatened Wittgenstein with a poker)$\vee$(Wittgenstein threatened Popper with a poker).

We now go one step further and replace specific statements by *statement variables* or *atoms* using letters like p, q, r. This is because the only feature of mathematical statements that is interesting to us is whether they are true or false, and not whether they are witty or wise. Thus the atoms can be assigned one of two *truth-values*: either true (T) or false (F). A statement constructed out of atoms using propositional connectives is called a *formula*. For example, $(p \wedge q) \vee r$ is a formula. To work out whether a formula is true or false, we need to know the truth values assigned to the atoms, and the behaviour of the propositional connectives with respect to truth-values. Their behaviour is *defined* as follows.

1. $\neg p$ has the opposite truth-value to p. This operation is called *negation*.

2. $p \wedge q$ is true when both p and q are true, and false otherwise. This operation is called *conjunction*.

3. $p \vee q$ is true when at least one of p or q is true, and false otherwise. This operation is called *disjunction*.

The above definitions can also be made using *truth tables*. For example, here is the truth table for disjunction.

p	q	$p \lor q$
T	T	T
T	F	T
F	T	T
F	F	F

Example 3.2.4. Consider the compound statement $A = (p \land q) \lor r$. Suppose that p is true, q is false and r is true. Then you can calculate that A is true.

Formulae A and B are *logically equivalent*, denoted $A \equiv B$, if they have the same truth tables. We come now to a fundamental point about the meaning of mathematical statements: logically equivalent formulae mean the same thing.[8] We now have all the tools needed to prove Theorem 3.2.1.

Example 3.2.5. We prove part (7) of Theorem 3.2.1 to demonstrate the method. The first step is to prove that the formula $p \land (q \lor r)$ is logically equivalent to the formula $(p \land q) \lor (p \land r)$. We can do this easily by constructing truth tables for each formula and then observing that these truth tables are the same.

p	q	r	$p \land (q \lor r)$	p	q	r	$(p \land q) \lor (p \land r)$
T	T	T	T	T	T	T	T
T	T	F	T	T	T	F	T
T	F	T	T	T	F	T	T
T	F	F	F	T	F	F	F
F	T	T	F	F	T	T	F
F	T	F	F	F	T	F	F
F	F	T	F	F	F	T	F
F	F	F	F	F	F	F	F

We can now go to the second step. Our goal is to prove that

$$A \cap (B \cup C) = (A \cap B) \cup (A \cap C).$$

To do this, we have to prove that the set of elements belonging to the lefthand side is the same as the set of elements belonging to the righthand side. An element x either belongs to A or it does not. Similarly, it either belongs to B or it does not, and it either belongs to C or it does not. Define p to be the statement '$x \in A$'. Define q to be the statement '$x \in B$'. Define r to be the statement '$x \in C$'. If p is true then x is an element of A, and if p is false then x is not an element of A. Using now the definitions of the Boolean operations, it follows that $x \in A \cap (B \cup C)$ precisely when the statement $p \land (q \lor r)$ is true. Similarly, $x \in (A \cap B) \cup (A \cap C)$ precisely

---

[8]For a mathematician, the statements 'The hounds of spring are on winter's traces' and 'Spring's coming' mean the same thing. It is not that mathematicians lack poetry; it is that they put it into their ideas rather than into their words.

when the statement $(p \wedge q) \vee (p \wedge r)$ is true. But these two statements have the same truth-tables. It follows that an element belongs to the lefthand side precisely when it belongs to the righthand side. Consequently, the two sets are equal.

Box 3.3: $P = NP$?

Sometimes the more harmless a problem looks the more treacherous it is. Let $A = A(p_1, \ldots, p_n)$ stand for a formula constructed from the n atoms $p_1, \ldots, p_n$ using the logical connectives. If there is some assignment of truth values to these atoms that makes A true, we say that A is *satisfiable*. To decide whether A is satisfiable or not is therefore straightforward: construct a truth table. This will answer the question in a finite amount of time proportional to the number of rows in the table which will be 2^n. But there's the rub: 2^n increases in size exponentially. This means that even a superfast computer will stall in solving this problem if programmed to construct truth tables. So, is there a better, that is intrinsically faster, way of solving the *satisfiability problem*? This is the essence of the question $P = NP$? and one of the Millennium Problems [36]. You might ask why anyone should care. It turns out that thousands of problems, including many natural and interesting ones, can be encoded via formulae in terms of the satisfiability problem. Thus a fast solution to the satisfiability problem would lead immediately to fast solutions to a host of other problems. The issue is that the truth table method is nothing other than a dressed up version of an exhaustive search. Perhaps there is some mathematical trick that we could perform on a formula which would tell us whether it was satisfiable or not. However, no one knows if there is one.

The fact that $A \cap (B \cap C) = (A \cap B) \cap C$ means that we can just write $A \cap B \cap C$ unambiguously sans brackets. Similarly, we can write $A \cup B \cup C$ unambiguously. This can be extended to any number of unions and any number of intersections. The union[9]

$$A_1 \cup \ldots \cup A_n$$

would usually be written as

$$\bigcup_{i=1}^{n} A_i.$$

There is again no need for brackets since it can be shown that however this big union is bracketed into pairs of unions the same set arises.[10] On one or two occasions we will need infinite unions

$$A_1 \cup A_2 \cup \ldots$$

which we would usually write as

$$\bigcup_{i=1}^{\infty} A_i.$$

Observe that here ∞ does not stand for a number. There is no set A_∞. Instead, it tells us that $i = 1, 2, 3, \ldots$ without limit. More generally, if I is any indexing set, we write

$$\bigcup_{i \in I} A_i.$$

---

[9]The notation introduced here is closely related to the $\sum$-notation introduced in Section 4.2.

[10]This is a special case of a result called generalized associativity proved in Section 4.1 but in this case it is clear from the meaning of union.

There is no need for brackets and no ambiguity in meaning. Of course, what we have said for union applies equally well to intersection.

Exercises 3.2

1. Prove the following logical equivalences using truth-tables.

   (a) $p \wedge (q \wedge r) \equiv (p \wedge q) \wedge r$.

   (b) $p \wedge q \equiv q \wedge p$.

   (c) $p \vee (q \vee r) \equiv (p \vee q) \vee r$.

   (d) $p \vee q \equiv q \vee p$.

   (e) $p \vee (q \wedge r) \equiv (p \vee q) \wedge (p \vee r)$.

   (f) $p \wedge \neg(q \vee r) \equiv (p \wedge \neg q) \wedge (p \wedge \neg r)$.

   (g) $p \wedge \neg(q \wedge r) \equiv (p \wedge \neg q) \vee (p \wedge \neg r)$.

   (h) $p \wedge p \equiv p$.

   (i) $p \vee p \equiv p$.

2. Use Question 1 to prove the unproved assertions of Theorem 3.2.1. Note that parts (3) and (6) can be proved directly.

3. The logical connective disjunction means 'inclusive or'. There is also *exclusive or*, abbreviated *xor*, which is defined by means of the following truth-table.

p	q	p xor q
T	T	F
T	F	T
F	T	T
F	F	F

   Define a new operation on sets by

   $$A \oplus B = \{x : (x \in A) \text{ xor } (x \in B)\}.$$

   (a) Prove that $A \oplus (B \oplus C) = (A \oplus B) \oplus C$.

   (b) Prove that $A \cap (B \oplus C) = (A \cap B) \oplus (A \cap C)$.

4. Prove that the following are equivalent.

   (a) $A \subseteq B$.

   (b) $A \setminus B = \emptyset$.

   (c) $B = A \cup B$.

   (d) $A = A \cap B$.

   Hint. Prove the following implications: (a)⇒(b), (b)⇒(c), (c)⇒(d) and (d)⇒(a).

5. Let $A, B \subseteq X$. Prove that if $A \subseteq B$ then $X \setminus B \subseteq X \setminus A$.

3.3 RELATIONS

Set theory consists of simple ingredients from which everything in mathematics can be constructed. For example, in any piece of mathematics, you will come across all kinds of special symbols. Here is a small list

$$=, \leq, \geq, <, >, \in, \subseteq, \subset, \mid .$$

These are all examples of relations. The word 'relation' is adapted from everyday language and so comes with the health warning we gave earlier about the way mathematicians use that language. The statement 'Darth Vader is the brother of Luke Skywalker' is an instance of the relation 'x is the brother of y' where we have replaced x by Darth Vader and y by Luke Skywalker. In this case, the statement is false, whereas the statement 'Groucho Marx is the brother of Harpo Marx' is true. On the other hand, the statement 'William Pitt the Elder is the father of William Pitt the Younger' is an instance of the relation 'x is the father of y' and happens to be true, whereas the statement 'Luke Skywalker is the father of Darth Vader' is false.[11] To make a mathematical definition of 'one thing being related to another thing' we hollow out the everyday meaning and just look at the ordered pairs where the relation is true.

Let A and B be sets. Any subset $\rho \subseteq A \times B$ is called a *relation from A to B*. We call A the *domain* and B the *codomain* of the relation. It is also possible for $A = B$, in which case we speak of a relation *on A*. In fact, most of the relations considered in this book are defined on a set. We have defined a relation to be a set of ordered pairs. Accordingly, we should write $(a, b) \in \rho$ to mean that the ordered pair (a, b) belongs to the relation ρ. It is more natural, however, to write this as $a \rho b$ because then we can read it as 'a is ρ-related to b'. It then becomes psychologically irresistible to regard relations as having a dynamic quality to them: the relation ρ relates elements in A *to* elements in B. This leads to a natural pictorial representation of relations called *directed graphs* or simply *digraphs*. The elements of A and B, now called *vertices*, are represented by circles or points in the plane, and we draw an *arrow* or *directed edge* from a to b precisely when (a, b) belongs to the relation. An arrow of the form (a, a) is called a *loop*.

Example 3.3.1. Let $A = \{a, b, c\}$ and let $B = \{1, 2, 3, 4\}$. Define the relation ρ from A to B by $\rho = \{(a, 2), (a, 4), (b, 2), (b, 3), (b, 4)\}$. Observe that deliberately none of these ordered pairs has as first component c or as second component 1. This relation is represented by the following digraph.

---

[11] Or the beginning of a rebooted new franchise.

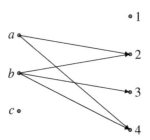

When the relation ρ is defined on the set A, in other words when $\rho \subseteq A \times A$, digraphs are drawn slightly differently. There is now just one set of vertices labelled by the elements of A and an edge is drawn from a to a' precisely when $(a, a') \in \rho$.

Example 3.3.2. Here is a directed graph with set of vertices $A = \{1, 2, 3, 4, 5\}$.

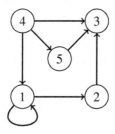

This corresponds to the subset of $A \times A$, and so the relation,

$$\{(1,1), (1,2), (2,3), (4,1), (4,3), (4,5), (5,3)\},$$

on the set A.

Examples 3.3.3.

1. For each set X there is a smallest relation, the *empty relation*, which contains no ordered pairs.

2. For each set X there is a largest relation, the *universal relation*, which contains all ordered pairs.

3. The following $=, \neq, \leq, <, \geq, >$ are all examples of relations on $\mathbb{N}$.

Relations can be as wild as you like: any subset of $A \times B$ defines a relation from A to B. In this book, however, only certain well-behaved relations will actually be important. In Section 3.4, we study the relations called functions, and in Sections 3.5 and 3.6, we study equivalence relations and partial order relations, respectively. To do this we need some more definitions.

Let $\rho \subseteq A \times B$ be a relation.

- ρ is said to be *deterministic* if $(a,b), (a,b') \in \rho$ implies that $b = b'$. In other words, each element in A is at the base of *at most one* arrow.

- ρ is said to be *total* if for each $a \in A$ there exists at least one $b \in B$ such that $(a,b) \in \rho$. In other words, each element of A is at the base of *at least one* arrow.

Let ρ be a relation defined on the set X.

- We say that ρ is *reflexive* if $(x,x) \in \rho$ for all $x \in X$.

- We say that ρ is *symmetric* if $(x,y) \in \rho$ implies that $(y,x) \in \rho$.

- We say that ρ is *antisymmetric* if $(x,y) \in \rho$ and $(y,x) \in \rho$ imply that $x = y$.

- We say that ρ is *transitive* if $(x,y) \in \rho$ and $(y,z) \in \rho$ imply that $(x,z) \in \rho$.

It is worth visualizing these properties in terms of digraphs. A digraph which has a loop at every vertex represents a reflexive relation. A relation is symmetric if whenever there is an arrow from x to y there is also a return arrow from y to x. A relation is antisymmetric if the only way there is an arrow from x to y and an arrow from y to x is if $x = y$ and so the arrow must be a loop. A relation is transitive if whenever there are arrows from x to y and from y to z there is an arrow from x to z.

Examples 3.3.4. We illustrate the above properties by means of some concrete examples. Take as our set all people.

1. The relation is 'x is the ancestor of y'. This is transitive because an ancestor of an ancestor is still an ancestor.

2. The relation is 'x is the same age as y'. This is reflexive, symmetric and transitive.

3. The relation is 'x is at least the same age as y'. This is reflexive and transitive but not symmetric.

Example 3.3.5. If a relation is symmetric, then for every arrow (x,y) there is also an arrow (y,x). So in reality we need only think about the set $\{x,y\}$ rather than the two ordered pairs. When we draw the digraph, we may if we wish dispense with arrows and use instead simply *edges*. The resulting picture is called a *graph*.[12] If there is a directed edge such as (x,x) then we hit the problem that set notation does not record repeats, but then we interpret $\{x\}$ as being a loop based at x. Here is a graph with set of vertices $A = \{1,2,3,4,5\}$.

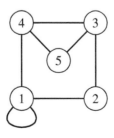

---

[12]Not to be confused with the *graph* of a function.

This corresponds to the following subsets of A

$$\{\{1\},\{1,2\},\{2,3\},\{4,1\},\{4,3\},\{4,5\},\{5,3\}\}.$$

The relations defined in this section are all examples of *binary* relations because they are subsets of products of two sets. It is also possible to define *ternary* relations, which are subsets of $A \times B \times C$ and, more generally, *n-ary* relations, which are subsets of $A_1 \times \ldots \times A_n$. In mathematics, binary relations are the most common but ternary relations are not unknown.

Example 3.3.6. Take as our set the real line $\mathbb{R}$. Consider the ternary relation 'the point x is between the point y and the point z'. The triple $(2,1,3)$ belongs to the relation whereas the triple $(1,2,3)$ does not.

Exercises 3.3

1. This question deals with relations on the set $X = \{1,2,3\}$. For each of the relations below defined on X write down the set of all ordered pairs that defines the relation and the corresponding digraph.

   (a) $=$.

   (b) $\leq$.

   (c) $\geq$.

   (d) $<$.

   (e) $>$.

   (f) $\neq$.

2. For each of the relations in Question 1, determine which are reflexive, symmetric, antisymmetric or transitive.

3. Denote by $B(X)$ the set of all binary relations on the set X. Let $\alpha, \beta \in B(X)$. Define $\alpha \circ \beta$ to be the set of all pairs $(x,z) \in X \times X$ where there exists $y \in X$ such that $(x,y) \in \alpha$ and $(y,z) \in \beta$. Define α^{-1} to be all ordered pairs (y,x) such that $(x,y) \in \alpha$. Define $\Delta = \{(x,x) : x \in X\}$.

   (a) Prove that $\alpha \circ (\beta \circ \gamma) = (\alpha \circ \beta) \circ \gamma$.

   (b) Prove that $\Delta \circ \alpha = \alpha = \alpha \circ \Delta$.

   (c) Prove that $(\alpha \circ \beta)^{-1} = \beta^{-1} \circ \alpha^{-1}$.

3.4 FUNCTIONS

We begin with a definition right off the bat. Let ρ be a relation from A to B. We say that ρ is a *function* if it is total and deterministic. The set A is called the *domain* of the function and B is called the *codomain*. We can also define functions directly without reference to relations. A function from the set A to the set B is determined by a *rule* that associates with each element of $a \in A$ exactly one element of $b \in B$. We think of a as the *input* and b as the corresponding, uniquely determined, *output*. If the function is denoted by f write $b = f(a)$ or $a \mapsto b$ to describe the rule transforming inputs to outputs. The element $f(a)$ is called the *image* of the element a *under* the function f. We also say that a is *mapped* to $f(a)$. A function is a 'black-box' for transforming inputs to outputs. There are no restrictions on how it accomplishes this task. It might do so via a formula but it could also be a program or a look-up table. The definition of a function does not require any information about how the rule is specified.

The *image* of the function f, denoted by $\mathrm{im}(f)$, is the set $\{f(a) : a \in A\}$ and although the image is a subset of the codomain it could well be a proper one. The relation corresponding to the function f is the set of all ordered pairs $(a, f(a)) \in A \times B$ where $a \in A$. We write $f : A \to B$ or $A \xrightarrow{f} B$ to denote a function with domain A and codomain B. It is important to remember that a function is determined by three pieces of information: its domain, its codomain and its rule. The definition of a function looks like overkill but it is one that has proved its worth. The functions $f : A \to B$ and $g : C \to D$ are *equal* if $A = C$ and $B = D$ and $f(a) = g(a)$ for all $a \in A$. Functions have all manner of aliases depending on the context such as *maps, mappings, functionals, operators* and *transformations*. The word 'function' is the neutral all-purpose term.

Box 3.4: Category Theory

All of mathematics can be developed within set theory, and this would suggest that set theory really is the foundation of mathematics. But in 1945, Samuel Eilenberg (1913–1998) and Saunders Mac Lane (1909–2005) introduced a new approach to mathematics, called category theory, in which the notion of function is taken as basic rather than that of set. The essential idea is to focus on the relationships between mathematical objects, and their abstract properties as expressed by functions, rather than the details of their construction as expressed by sets. It is probably fair to say that whereas set theory dominates undergraduate mathematics, category theory is essential in postgraduate mathematics. In Chapter 8 we actually meet a category face-to-face: the set of all $m \times n$ matrices over the real numbers for all non-zero natural numbers m and n forms a category.

Example 3.4.1. Let A be a set of people and B the set of natural numbers. Then a function f is defined when for each $a \in A$ we associate their age $f(a)$. This is a

function, and not merely a relation, because everyone has a uniquely defined age.[13].
On the other hand, if we kept A as before and now let B be the set of nationalities
then we will no longer have a function in general, because some people have dual
nationality. This is a relation rather than a function.

Example 3.4.2. In calculus functions such as $f\colon \mathbb{R}^n \to \mathbb{R}$, where $n \geq 2$, are called
functions of *several variables*.

Example 3.4.3. Functions can be defined between complex sets. For example, let
A be the set of all integrable functions from the closed interval $[0,1]$ to $\mathbb{R}$. Then
$f \mapsto \int_0^1 f$ is a function from A to $\mathbb{R}$.

Example 3.4.4. In algebra, a function from $X \times X$ to X is called a *binary operation*. We could represent such a function by a symbol such as f and so write
$(x,y) \mapsto f(x,y)$, but in algebraic settings we would usually represent it using *infix notation* and so we might write something like $(x,y) \mapsto x * y$. Binary operations
abound. For example in the set $\mathbb{R}$, addition and multiplication are both examples of
binary operations. If the set X is finite and not too big then any binary operation $*$
on it can be described by means of a *Cayley table* similar to the multiplication tables
that children used to make at school. This is an array of elements of X whose rows
and columns are labelled by the elements of X in some specific order. The element of
the array in the row labelled by the element a and the column labelled by the element
b is $a * b$. In algebra, functions from X to itself are usually called *unary operations*,
and a function $X \times X \times X$ to X a *ternary operation*, though the latter are rare. We
shall study binary operations in more detail in Chapter 4. A function from X^n to X is
said to have *arity n*.

Examples 3.4.5. There are some functions that are easy to define and play an important bookkeeping rôle in mathematics.

1. For every set X, the function $1_X\colon X \to X$ defined by $1_X(x) = x$ is called the
identity function on X.

2. Let $X_1, \ldots, X_n$ be sets. For each i, where $1 \leq i \leq n$, the function

$$p_i^n\colon X_1 \times \ldots \times X_n \to X_i,$$

defined by $p_i^n(x_1, \ldots, x_n) = x_i$, is called the *ith projection function*.

3. The function $d\colon X \to X \times X$, defined by $x \mapsto (x,x)$, is called the *diagonal function*.

4. The function $t\colon X \times Y \to Y \times X$ defined by $(x,y) \mapsto (y,x)$ is called the *twist
function*.

5. Let $f_i\colon X \to Y_i$, where $1 \leq i \leq n$, be n functions. The function

$$(f_1, \ldots, f_n)\colon X \to Y_1 \times \ldots \times Y_n$$

---

[13]This also applies to the reigning Queen of England who has two birthdays but only one age.

defined by

$$(f_1,\ldots,f_n)(x) = (f_1(x),\ldots,f_n(x))$$

is called the *product* of the given functions.

The set of all functions from the set X to itself is denoted by $T(X)$ but when $X = \{1,\ldots,n\}$ we write T_n.

Example 3.4.6. The set T_2 consists of four functions. The most convenient way to describe them is using *two-row notation*.

$$f_1 = \begin{pmatrix} 1 & 2 \\ 1 & 2 \end{pmatrix}, f_2 = \begin{pmatrix} 1 & 2 \\ 1 & 1 \end{pmatrix}, f_3 = \begin{pmatrix} 1 & 2 \\ 2 & 2 \end{pmatrix} \text{ and } f_4 = \begin{pmatrix} 1 & 2 \\ 2 & 1 \end{pmatrix}.$$

In this notation, the domain elements are listed along the top and their corresponding images immediately below. Thus $f_2(1) = 1$ and $f_2(2) = 1$. Both f_2 and f_3 are functions whose images are proper subsets of the codomain, whereas f_1 and f_4 are functions whose images are equal to the codomain. The function f_1 is the identity function.

A function $A \xrightarrow{f} B$ is called *surjective* if its image is equal to its codomain. This means that for each $b \in B$ there exists at least one $a \in A$ such that $f(a) = b$. A function is called *injective* if $a \neq a'$ implies that $f(a) \neq f(a')$. Equivalently, if $f(a) = f(a')$ then $a = a'$. In other words, different elements in the domain are mapped to different elements in the codomain. A function is called *bijective* if it is both injective and surjective. Injective, surjective and bijective functions are also referred to as *injections*, *surjections* and *bijections*, respectively.

Example 3.4.7.

1. In Example 3.4.6, the functions f_1 and f_4 are bijective and the functions f_2 and f_3 are neither injective nor surjective.

2. The function $f: \mathbb{N} \to \mathbb{N}$ defined by $n \mapsto n+1$ is injective but not surjective, because zero is not in the image, whereas the function $g: \mathbb{N} \to \mathbb{N}$ defined by $2n \mapsto n$ and $2n+1 \mapsto n$ is surjective but not injective.

Injective functions in general, and bijective ones in particular, are important in coding information in different ways such as error-correcting codes, ciphers and data compression codes.

Example 3.4.8. We have defined functions as special kinds of relations, but we could equally well have defined relations as special kinds of functions. Let $\mathbf{2} = \{0,1\}$. A function $f: X \to \mathbf{2}$ determines a subset $A \subseteq X$ by putting $A = \{x \in X: f(x) = 1\}$. Conversely, given a subset $B \subseteq X$ we can define a function $\chi_B: X \to \mathbf{2}$, called the *characteristic function of B*, by defining $\chi_B(x) = 1$ if $x \in B$ and $\chi_B(x) = 0$ if $x \notin B$. This sets up a bijection between $P(X)$ and the set of all functions from X to $\mathbf{2}$. We can therefore characterize relations from A to B as functions from $A \times B$ to $\mathbf{2}$.

Let $f: X \to Y$ and $g: Y \to Z$ be functions. Let $x \in X$. Then $f(x) \in Y$ and so we may apply g to $f(x)$ to get $g(f(x))$. It therefore makes sense to define a function $g \circ f: X \to Z$ by $x \mapsto g(f(x))$, called the *composition* of g and f.

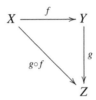

We write gf rather than $g \circ f$ for simplicity. The operation of composing functions enables us to construct new functions from old.[14] The set $T(X)$ together with the binary operation of composition is called the *full transformation monoid* on X.

Example 3.4.9. Let $f: \mathbb{Z} \to \mathbb{Z}$ be the function given by $f(x) = x+1$ and let $g: \mathbb{Z} \to \mathbb{Z}$ be the function given by $g(x) = x^2$. Then

$$(gf)(x) = g(f(x)) = g(x+1) = (x+1)^2$$

whereas

$$(fg)(x) = f(g(x)) = f(x^2) = x^2 + 1.$$

It follows that $fg \neq gf$. This example shows that when composing functions order is important.

Example 3.4.10. We can draw up a Cayley table describing all the ways of composing the functions in T_2. We use the notation from Example 3.4.6.

$\circ$	f_1	f_2	f_3	f_4
f_1	f_1	f_2	f_3	f_4
f_2	f_2	f_2	f_2	f_2
f_3	f_3	f_3	f_3	f_3
f_4	f_4	f_3	f_2	f_1

Bearing in mind the importance of order, observe that the entry in row f_4 and column f_3 is $f_2 = f_4 f_3$.

Lemma 3.4.11 (Category properties).

1. Let $f: X \to Y$ be any function. Then $f 1_X = f$ and $1_Y f = f$.

2. Let $f: U \to V$ and $g: V \to W$ and $h: W \to X$ be functions. Then $h(gf) = (hg)f$.

---

[14]This is the Big-endian definition. I regret to inform you that there is also a Little-endian definition where fg is computed as $((x)f)g$. I cannot stop this sort of thing, but I can at least warn you of the hazards ahead.

Proof. (1) This is left as an exercise.

(2) Both $h(gf)$ and $(hg)f$ have the same domains and codomains, and so it is enough to prove that they do the same thing. Let $u \in U$. Then

$$(h(gf))(u) = h((gf)(u)) = h(g(f(u))),$$

by applying the definition of composition twice, and

$$((hg)f)(u) = (hg)f(u) = h(g(f(u))).$$

$\square$

Part (2) of Lemma 3.4.11 says that composition of functions is *associative*. We shall meet the general notion of associativity in Chapter 4. If $f\colon X \to X$ then we may form the functions $f^2 = ff$ and $f^3 = f^2 f$ and so on. We define $f^0 = 1_X$. An application of the ideas we have so far introduced to circuit design is described in Section 4.5.

A function $f\colon X \to Y$ is said to be *invertible* if there is a function $g\colon Y \to X$ such that $gf = 1_X$ and $fg = 1_Y$. If such a function g exists it is called an *inverse* of f.

Lemma 3.4.12. *If a function has an inverse then that inverse is unique.*

Proof. Let $f\colon X \to Y$ be a function, and suppose that both g and h are both inverses of f. By definition $gf = 1_X$ and $fg = 1_Y$, and $hf = 1_X$ and $fh = 1_Y$. Compose $gf = 1_X$ on the left with h to get $h(gf) = h1_X$. The righthand side simplifies immediately to h using the properties of identity functions. We now look at the lefthand side. By associativity of function composition $h(fg) = (hf)g$. Use the fact that h is an inverse of f to get $(hf)g = 1_X g$ which simplifies to g. Thus $g = h$, as required. $\square$

If the function f is invertible then its unique inverse is denoted by f^{-1}. If a function is invertible then what it does can be undone.

Example 3.4.13. Define the function from the non-negative reals $\mathbb{R}^{\geq 0}$ to itself by $x \mapsto x^2$. This is invertible because every real number $r \geq 0$ has a unique positive (or zero) square root.

Now define the function from $\mathbb{R}$ to $\mathbb{R}^{\geq 0}$ by $x \mapsto x^2$. This is surjective but not invertible because given a real number $r > 0$ there are now two square roots, positive and negative, and so we have no unique way of undoing the squaring function.

Finally, define the function from $\mathbb{R}$ to itself by $x \mapsto x^2$. This is neither surjective nor injective. The negative reals are not the square of any real number.

These examples emphasize the fact that a function is not just what it does, but what it does it to and where it goes. The domain and codomain play an essential rôle in defining exactly what a function is.

Deciding whether a function is invertible is easy in principle.

Lemma 3.4.14. *A function is invertible if and only if it is a bijection.*

Proof. Let $f: X \to Y$ be invertible. We prove that it is a bijection. Suppose that $f(x) = f(x')$. Then $f^{-1}(f(x)) = f^{-1}(f(x'))$ and so $(f^{-1}f)(x) = (f^{-1}f)(x')$. But $f^{-1}f$ is the identity function on X and so $x = x'$. We have proved that f is injective. Let $y \in Y$. Put $x = f^{-1}(y)$. Then $f(x) = f(f^{-1}(y)) = (ff^{-1})(y) = y$, and we have proved that f is surjective.

Let $f: X \to Y$ be a bijection. We prove that it is invertible. Define $g: Y \to X$ by $g(y) = x$ if and only if $f(x) = y$. We need to show that this does indeed define a function. Let $y \in Y$. Then since f is surjective, there is at least one $x \in X$ such that $f(x) = y$. But f is injective so that x is unique. Thus g really is a function. We calculate gf. Let $x \in X$ and $y = f(x)$. Then $(gf)(x) = g(y) = x$. Thus gf is the identity function on X. We may similarly prove that fg is the identity function on Y. □

Example 3.4.15. The needs of computing have led to the consideration of other properties that functions might have. Before the advent of mobile phones, people needed phone-books that listed names, in alphabetical order, together with their phone numbers. For simplicity, we think a phone-book as defining a bijection from a set of names to a set of phone numbers. To find a phone number of someone, you looked up the name and so located their phone number. The alphabetical listing of the names meant that the search could be carried out efficiently. On the other hand, if you only had a telephone number, finding the name that went with it involved a time-consuming exhaustive search.[15] The idea that a function can be easy to compute but hard to 'uncompute', that is find its inverse, is the basis of modern cryptography.

The following lemma will be useful when we discuss cardinalities.

Lemma 3.4.16 (Properties of bijections).

1. *For each set X, the identity function 1_X is bijective.*

2. *If $f: X \to Y$ is bijective, then $f^{-1}: Y \to X$ is bijective, and the inverse of f^{-1} is f.*

3. *If $f: X \to Y$ is bijective and $g: Y \to Z$ is bijective then $gf: X \to Z$ is bijective. The inverse of gf is $f^{-1}g^{-1}$.*

Proof. (1) The function 1_X is both injective and surjective and so it is bijective. For the remaining proofs, we use Lemma 3.4.12 and Lemma 3.4.14 throughout.

(2) Since f is bijective it is invertible. So it has a unique inverse f^{-1} satisfying $f^{-1}f = 1_X$ and $ff^{-1} = 1_Y$. These two equations can also be interpreted as saying that f^{-1} is invertible with inverse f. This implies that f^{-1} is bijective.

(3) Since f and g are both bijective, they are also both invertible. Observe that $gff^{-1}g^{-1} = g1_Yg^{-1} = gg^{-1} = 1_Z$. Similarly $f^{-1}g^{-1}gf = 1_X$. Thus gf is invertible with inverse $f^{-1}g^{-1}$. It follows that gf is bijective. □

---

[15]The police had reverse phone-books for this very reason.

The set of all invertible functions from X to itself is denoted by $S(X)$. Clearly $S(X) \subseteq T(X)$. When $X = \{1, \ldots, n\}$, we write S_n rather than $S(X)$. Invertible functions are, as we have seen, just the bijections. A bijection from a set to itself is often called a *permutation*, particularly when the set is finite. You should be aware that the word 'permutation' can also refer to the effect of a bijection as it does in Section 3.9. We shall explain this point further there. By Lemma 3.4.16, the composition of any two functions in $S(X)$ again belongs to $S(X)$ and the inverse of a function in $S(X)$ is again in $S(X)$. The set $S(X)$ with the binary operation of composition is called the *symmetric group on X*.

There is one piece of terminology you should be aware of that is potentially ambiguous. Permutations are functions and so the way to combine them is to 'compose' them, but it is common to hear the phrase 'product of permutations' instead of 'composition of permutations'. The word 'product' is really being used here to mean 'appropriate binary operation in this context'.

The finite symmetric groups S_n play an important rôle in the history of algebra as we shall explain in Section 7.9, and in the theory of determinants in Section 9.6. For these reasons, it is useful to find an efficient way of representing their elements.

Put $X = \{1, \ldots, n\}$. Let $f \in S_n$ and $i \in X$. If $f(i) = i$ we say that i is *fixed* by f. If $f(i) \neq i$ we say that i is *moved* by f. The identity of S_n, denoted by ι, of course fixes every element of X but every non-identity permutation moves at least one element. We now define a special class of permutations. Let $y_1, \ldots, y_s$ be s distinct elements of X. We define, in this context, the symbol

$$(y_1, \ldots, y_s)$$

to be the permutation of X that does the following

$$y_1 \mapsto y_2, \, y_2 \mapsto y_3, \, \ldots, y_s \mapsto y_1,$$

and which fixes all the elements in $X \setminus \{y_1, \ldots, y_s\}$. We call such a permutation a *cycle (of length s)* or an *s-cycle*. The identity permutation is deemed to be a cycle in a trivial way. Observe that 1-cycles are not usually written down as long as the set being permuted is made clear.

Example 3.4.17. Here is the digraph representing the cycle $(1, 2, 3, 4) \in S_6$.

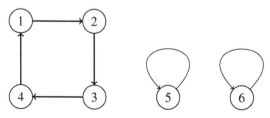

The cycle can also be written as (1234), without commas, as long as no ambiguity arises.

Cycles are easy permutations to work with. For example, the inverse of the cycle $(y_1,\ldots,y_s)$ is $(y_s,\ldots,y_1)$. The only point to watch is that different cycles can represent the same permutation because $(y_1,\ldots,y_s) = (y_2,\ldots,y_s,y_1)$.

Two cycles $(y_1,\ldots,y_s)$ and $(z_1,\ldots,z_t)$ are said to be *disjoint* if the sets $\{y_1,\ldots,y_s\}$ and $\{z_1,\ldots,z_t\}$ are disjoint. If f and g are disjoint cycles then $fg = gf$ because each cycle leaves fixed what the other cycle moves. A finite set of cycles is disjoint if any two distinct cycles in the set are disjoint.

Lemma 3.4.18. *Every element of S_n can be written as a product of disjoint cycles.*

Proof. Let $X = \{1,\ldots,n\}$. We prove the result by describing an algorithm that will write each permutation as a product of disjoint cycles. Let $f \in S_n$ be any non-identity permutation. There is therefore a smallest i where $1 \le i \le n$ such that $f(i) \ne i$. Define

$$Y = \{i, f(i), f^2(i), f^3(i),\ldots\}.$$

By finiteness, there must be natural numbers s and t where $s < t$ such that $f^s(i) = f^t(i)$. The function f is invertible and so in particular injective. Thus $i = f^{t-s}(i)$ by applying the injectivity of f repeatedly. It follows that there is a smallest natural number $r > 0$ such that $i = f^r(i)$. Then $Y = \{i, f(i),\ldots,f^{r-1}(i)\}$ and consists of r elements. Thus

$$g_1 = (i, f(i),\ldots,f^{r-1}(i))$$

is a cycle. If f fixes all other elements of X we are done and $f = g_1$. Otherwise there are elements of X that are moved by f and have not yet been mentioned. Choose the smallest such element. Repeat the above procedure to obtain a cycle g_2 necessarily disjoint from g_1. As before either $f = g_2 g_1$ or there are still elements moved by f that have not yet been mentioned. The set X is finite and so this algorithm terminates in a finite number of steps with $f = g_p \ldots g_1$ for some $p \ge 1$. □

Example 3.4.19. We illustrate the proof of Lemma 3.4.18. Let $f \in S_{12}$ be defined in two-row notation by

$$f = \begin{pmatrix} 1 & 2 & 3 & 4 & 5 & 6 & 7 & 8 & 9 & 10 & 11 & 12 \\ 6 & 5 & 4 & 1 & 3 & 9 & 12 & 11 & 10 & 2 & 8 & 7 \end{pmatrix}.$$

The smallest element moved by f is 1. We now apply f repeatedly to 1 to get

$$1 \mapsto 6 \mapsto 9 \mapsto 10 \mapsto 2 \mapsto 5 \mapsto 3 \mapsto 4 \mapsto 1.$$

This gives rise to the cycle $(1,6,9,10,2,5,3,4)$. The smallest number not yet mentioned is 7. We now apply f repeatedly to 7 to get

$$7 \mapsto 12 \mapsto 7.$$

This gives rise to the cycle $(7,12)$. The smallest number not yet mentioned is 8. We now apply f repeatedly to 8 to get

$$8 \mapsto 11 \mapsto 8.$$

This gives rise to the cycle $(8,11)$. All the numbers between 1 and 12 have now been mentioned and so

$$f = (8,11)(7,12)(1,6,9,10,2,5,3,4).$$

A cycle of length 2 is called a *transposition*. The inverse of a transposition is just itself and so it is *self-inverse*.

Proposition 3.4.20. *Every permutation in S_n can be written as a product of transpositions.*

Proof. Since every permutation can be written as a product of cycles by Lemma 3.4.18, it is enough to prove that every cycle can be written as a product of transpositions which follows from the calculation

$$(y_1,\ldots,y_s) = (y_1,y_s)\ldots(y_1,y_3)(y_1,y_2).$$

$\square$

The above result is intuitively plausible. A line of n objects can be arranged into any order by carrying out a sequence of moves that only interchanges two objects at a time. We shall return to this result in Section 7.9 and again in Section 9.6 since it is more significant than it looks.

Example 3.4.21. The diagram below illustrates Proposition 3.4.20 with reference to the cycle $(1,2,3,4) = (14)(13)(12)$.

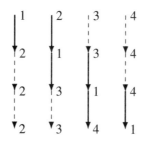

Example 3.4.22. $S_3 = \{\iota, (12), (23), (13), (12)(13), (13)(12)\}$.

Example 3.4.23. When we shuffle a deck of cards, we carry out a sequence of manoeuvres designed to produce a random permutation of the deck. I shuffle cards in the usual clunky way, but the smart way used by casinos is the riffle or dovetail shuffle where the deck is divided into roughly two parts and then the two parts interleaved. Doing this once results in a permutation of the deck. Repeating this a number of times will mix the cards more and more. How many times should you do this to get a random permutation? The answer is 7, a result discovered by Dave Bayer and Persi Diakonis [8] in a detailed analysis of card shuffling. Persi Diakonis is the doyen of mathematical card shuffling.[16]

---

[16]See [68] and the 2012 documentary film *Deceptive practice* about the magician Ricky Jay directed by Molly Bernstein and Alan Edelstein.

A function $f: X \to Y$ determines a function from $\mathsf{P}(X)$ to $\mathsf{P}(Y)$. This is defined as follows. If $A \subseteq X$ then map A to the set $\{f(a): a \in A\}$. In this book, we shall also denote this function by f. Thus we write

$$f(A) = \{f(a): a \in A\}.$$

The function f also determines a function from $\mathsf{P}(Y)$ to $\mathsf{P}(X)$. This is defined as follows. If $B \subseteq Y$ then map B to the set $\{x \in X: f(x) \in B\}$. In this book, this function is denoted by f^{-1}. Thus

$$f^{-1}(B) = \{x \in X: f(x) \in B\}$$

and is called the *inverse image* of B under f. This leads to a notation clash with the inverse of a function defined earlier, so some clarification is needed. The inverse image function f^{-1} is always defined and it maps subsets to subsets. If f^{-1} always maps singleton sets to singleton sets, then f is actually invertible and its inverse function f^{-1} is defined.

There is a weakening of the definition of a function which is widespread and important. A *partial function* f from A to B is defined just like a function except that it may happen that $f(a)$ is not defined for some $a \in A$. Partial functions arise naturally in calculus as partial functions from $\mathbb{R}$ to itself. For example, $x \mapsto \frac{1}{x}$ is a partial function from the reals to itself that is not defined when $x = 0$. In this book, we work only with functions but be aware that in more advanced work it is often more natural to work with partial functions.

Box 3.5: Computable Functions

The popular conception of computers is that they have no limits, but in the early decades of the twentieth century, and before the valves in the first computer had warmed up, mathematicians such as Turing had already figured out that there were things computers cannot do. They focused on the partial functions from $\mathbb{N}^m$ to $\mathbb{N}^n$ and asked which of them could in principle be computed. To make this precise, they had to define what they meant by *computed*. Although many definitions were proposed, they all turned out to be equivalent and the class of computable functions that resulted goes under the name of the *partial recursive functions*. Intuitively, these are functions that can be computed by running a programme on a mathematical model of a computer where no restrictions are placed on its memory capacity. You might think that the choice of programming language would matter but remarkably it does not. As long as it can handle basic arithmetic operations and can loop under control, it can in principle describe anything computable. The differences amongst programming languages are therefore ones of convenience for the user rather than computing capability. Crucially, there are well-defined functions that are not partial recursive, and so not computable.

Finally, let $f: X \to Y$ be a function. If $U \subseteq X$ then we can define a new function $g: U \to Y$ by $g(x) = f(x)$ if $x \in U$. Do not be surprised if the function g is also called f. If V is any subset of Y that contains the image of g then we obtain yet another function $k: U \to V$ where $k(x) = f(x)$ if $x \in U$. This too may well be called f. These functions derived from f are obtained by *restricting its domain*. Notation needs to be clear but the proliferation of notation can lead to confusion.

Exercises 3.4

1. For each of the following functions determine whether they are surjective or injective.

   (a) $f\colon \mathbb{N} \to \mathbb{N}$ defined by $x \mapsto x+1$.

   (b) $g\colon \mathbb{Z} \to \mathbb{Z}$ defined by $x \mapsto x+1$.

   (c) $h\colon \mathbb{Z} \to \mathbb{N}$ defined by $x \mapsto |x|$.

2. Let $f, g, h\colon \mathbb{R} \to \mathbb{R}$ be the following functions $f(x) = x+1$, $g(x) = x^2$ and $h(x) = 3x+2$. Calculate

   (a) f^2.

   (b) fg.

   (c) gf.

   (d) $f(gh)$.

   (e) $(fg)h$.

3. For each of the following permutations, write them first as a product of disjoint cycles and then as a product of transpositions.

   (a) $\begin{pmatrix} 1 & 2 & 3 & 4 & 5 & 6 & 7 & 8 & 9 \\ 9 & 8 & 7 & 6 & 5 & 4 & 3 & 2 & 1 \end{pmatrix}$.

   (b) $\begin{pmatrix} 1 & 2 & 3 & 4 & 5 & 6 & 7 & 8 & 9 \\ 2 & 4 & 6 & 8 & 1 & 3 & 5 & 7 & 9 \end{pmatrix}$.

   (c) $\begin{pmatrix} 1 & 2 & 3 & 4 & 5 & 6 & 7 & 8 & 9 \\ 8 & 6 & 4 & 2 & 9 & 7 & 5 & 3 & 1 \end{pmatrix}$.

4. Let $f\colon X \to Y$. Prove the following where $A, B \subseteq X$.

   (a) $f(\emptyset) = \emptyset$.

   (b) $A \subseteq B$ implies that $f(A) \subseteq f(B)$.

   (c) $f(A \cup B) = f(A) \cup f(B)$.

   (d) $f(A \cap B) \subseteq f(A) \cap f(B)$. Construct an example to show that equality does not always occur.

5. Let $f\colon X \to Y$. Prove the following where $A, B \subseteq Y$.

   (a) $f^{-1}(\emptyset) = \emptyset$.

   (b) $f^{-1}(Y) = X$.

   (c) $A \subseteq B$ implies that $f^{-1}(A) \subseteq f^{-1}(B)$.

   (d) $f^{-1}(A \cup B) = f^{-1}(A) \cup f^{-1}(B)$.

(e) $f^{-1}(A \cap B) = f^{-1}(A) \cap f^{-1}(B)$.

6. Let $f: X \to Y$ be any function. Prove that $f 1_X = f$ and $1_Y f = f$.

7. (a) Prove that if $f: X \to Y$ is injective and $g: Y \to Z$ is injective then $gf: X \to Z$ is injective.

   (b) Prove that if $f: X \to Y$ is surjective and $g: Y \to Z$ is surjective then $gf: X \to Z$ is surjective.

8. Let X be a finite non-empty set. Prove that a function $f: X \to X$ is injective if and only if it is surjective.

9. *This question develops an approach to injectivity and surjectivity that is *category-theoretic* rather than *set-theoretic*.

   (a) Let $f, g: X \to Y$ and $h: Y \to Z$ be functions. Show that if h is injective then $hf = hg$ implies that $f = g$.

   (b) Let $h: Y \to Z$ be a function. Suppose that for *all* functions $f, g: X \to Y$, we have that $hf = hg$ implies that $f = g$. Prove that h is injective.

   (c) Can you state and prove similar results for surjective functions?

Box 3.6: DES and AES

Permutations have become big business in the world of secret codes or more formally *ciphers*. Suppose we want to send a message securely. For simplicity, assume that the message is just a string $\mathbf{x} = x_1 \ldots x_n$, called in this context the *plaintext*, over the alphabet $A = \{a, \ldots, z\}$. Thus $x_i \in A$. So, all spaces and punctuation are ignored as are upper case characters. The simplest way of disguising such a message is to choose a non-identity permutation $f \in S_A$. Rather than transmitting the message $\mathbf{x}$, we would instead transmit the message $\mathbf{y} = f(x_1) \ldots f(x_n) = y_1 \ldots y_n$, called the *ciphertext*. If you receive $\mathbf{y}$ and you know f then you can calculate f^{-1} and retrieve the plaintext as $f^{-1}(y_1) \ldots f^{-1}(y_n)$. Ciphers such as this have a rôle in social contexts but are woefully inadequate in commercial environments. Even if you do not know f, it is often easy to figure out what it is from $\mathbf{y}$ alone. If this is long enough, a statistical analysis of letter frequencies together with knowledge of the language of the plaintext message is enough to discover f and so calculate f^{-1}. Thus this cipher can easily be broken. A better method is as follows. Denote by $B = A^l$ all strings over the alphabet A of length $l \geq 2$. Our message $\mathbf{x} = x_1 \ldots x_n$ should now be viewed as a string over the alphabet B where we pad out our message with random letters if its length is not an exact multiple of l. We choose $g \in S_B$ which is a bigger set of permutations. We now encrypt and decrypt as above using g. Ciphers such as these are called *block ciphers*. However, even these are cryptographically weak. To construct good ciphers we need to combine and iterate these two ideas. This leads to ciphers known as the *Data Encryption Standard (DES)* and the *Advanced Encryption Standard (AES)*. There are two conflicting issues with ciphers such as these: they need to be resistant to attack but they have to be easy to use in commercial environments. What is interesting, or alarming, from a mathematical point of view is that the security of these systems is based on testing and experience rather than actual proof.

3.5 EQUIVALENCE RELATIONS

Equivalence relations are important in mathematics. They have two related rôles: they enable us to classify the elements of a set, and they enable us to identify, that is glue together, the elements of a set to build new sets.

An essential conceptual tool used in understanding the world around us is to classify things according to perceived similarities.[17] At its simplest, we use a binary classification and divide things into two groups. Thus we have an important distinction such as *edible versus inedible*, which has evolutionary importance, and the all-purpose *us versus them* in all its many forms. Binary classifications are pretty crude. Science allows for much more sophisticated systems of classification such as the one used in classifying living things into five big groups.[18] It is in mathematics, however, where this idea is used repeatedly.

The general set-up is as follows. There is a set X whose elements we want to classify using some notion of similarity. Those elements which are similar are put into the same pigeonhole. Similarity is obviously a relation on the set X but not all relations are suitable for classifying elements. So we need to characterize just which relations are. We also need to make precise what we mean by the pigeonholes into which we classify the elements. We now make two fundamental definitions that will deal with both of these issues.

A relation ρ defined on a set X is called an *equivalence relation* if it is reflexive, symmetric and transitive. In practice, equivalence relations are denoted by symbols such as $\cong, \equiv, \sim, \simeq, \ldots$ which serve to remind us of equality because, as we shall see, equivalence relations generalize equality.

Example 3.5.1. Let X be a set of people and define a relation ρ as follows.

1. $x \rho y$ if x and y have the same age.

2. $x \rho y$ if x and y have the same nationality.

3. $x \rho y$ if x and y have the same height.

4. $x \rho y$ if x and y are one and the same person.

5. $x \rho y$ if x and y are people.

You can check that each of these is an equivalence relation, and in each case we define x and y to be ρ-related if they are the same in some respect. There are two extreme examples in list above. In (4) two people are ρ-related if they are one and the same person, whereas in (5) it is enough that they are people in the set X for them to be ρ-related.

---

[17]Otherwise language wouldn't work. We can use the word 'cat' because we can divide the world into cats and not-cats. If we saw each cat as a distinct individual we wouldn't be able to use the word 'cat' at all. This is the curse of *Funes the memorious* in Borges' short story [20]. "To think is to forget a difference, to generalize, to abstract. In the overly replete world of Funes, there were nothing but details".

[18]For the record: monera, protista, fungi, plantae and animalia.

If $\{A_i : i \in I\}$ is a collection of sets such that $A_i \cap A_j = \emptyset$ when $i \neq j$ then we say they are *pairwise disjoint*. Let A be a set. A *partition* $P = \{A_i : i \in I\}$ of A is a set whose elements consist of non-empty subsets A_i of A which are pairwise disjoint and whose union is A. The subsets A_i are called the *blocks* of the partition. To prove that a set $\{A_i : i \in I\}$ of non-empty subsets of a set A is a partition, it is enough to show that $X = \bigcup_{i \in I} A_i$ and that if $A_i \cap A_j \neq \emptyset$ then $i = j$.

Examples 3.5.2.

1. The set
$$P = \{\{a,c\}, \{b\}, \{d,e,f\}\}$$
is a partition of the set $X = \{a,b,c,d,e,f\}$. In this case there are three blocks in the partition.

2. How many partitions does the set $X = \{a,b,c\}$ have?[19] There is 1 partition with 1 block
$$\{\{a,b,c\}\},$$
there are 3 partitions with 2 blocks
$$\{\{a\}, \{b,c\}\}, \ \{\{b\}, \{a,c\}\}, \ \{\{c\}, \{a,b\}\},$$
there is 1 partition with 3 blocks
$$\{\{a\}, \{b\}, \{c\}\}.$$
There are therefore 5 partitions of the set X.

Our next example will prepare us for the main theorem of this section.

Example 3.5.3. It is helpful to picture an equivalence relation ρ on a set X in terms of its associated digraph. Reflexivity means that every vertex has a loop; symmetry means that for every edge out there is a return edge; transitivity means that for every indirect path between two vertices there is a direct edge. The result of these properties is to divide the set of vertices of the digraph into 'islands'. Within each island, any two vertices are joined by an edge, in either direction, but there are no edges linking vertices in different islands.

Theorem 3.5.4 (Fundamental theorem of equivalence relations). *Let X be a set.*

1. *Let ρ be an equivalence relation defined on X. For each element $x \in X$, define*
$$[x]_\rho = \{y \in X : x\rho y\}.$$
The sets $[x]_\rho$ are non-empty since $x \in [x]_\rho$, and if $[x]_\rho \cap [y]_\rho \neq \emptyset$ then $[x]_\rho = [y]_\rho$. Define
$$P_\rho = X/\rho = \{[x]_\rho : x \in X\}.$$
Then P_ρ is a partition of X.

---

[19]The number of partitions of a set with n elements is called the *nth Bell number*.

2. *Let $P = \{A_i : i \in I\}$ be a partition of X. Define a relation ρ_P on X by $x\rho_P y$ precisely when there is a block $A_i \in P$ such that $x, y \in A_i$. Then ρ_P is an equivalence relation on X.*

3. *There is a bijection between the set of equivalence relations on X and the set of partitions on X.*

Proof. (1) The relation ρ is reflexive and so $x\rho x$. Thus $x \in [x]_\rho$ and as a result $[x]_\rho \neq \emptyset$. To show that P_ρ is a partition, we need to prove that the sets $[x]_\rho$ are either disjoint or equal. Suppose that $[x]_\rho \cap [y]_\rho \neq \emptyset$. We prove that $[x]_\rho = [y]_\rho$. Let $z \in [x]_\rho \cap [y]_\rho$. Then $x\rho z$ and $y\rho z$. By symmetry, $z\rho y$ and so by transitivity $x\rho y$. Thus an element of X is ρ-related to x if and only if it is ρ-related to y. It follows that $[x]_\rho = [y]_\rho$. We have therefore proved that X/ρ is a partition of X.

(2) It is left as an exercise to prove that ρ_P is reflexive, symmetric and transitive.

(3) This is the part where we explicitly show that equivalence relations and partitions are two sides of the same coin. We have defined two functions in parts (1) and (2). The first maps equivalence relations to partitions and is given by $\rho \mapsto P_\rho$. The second maps partitions to equivalence relations and is given by $P \mapsto \rho_P$. We prove that these maps are inverses of each other, yielding the required bijection.

Let ρ be an equivalence relation. Its associated partition is P_ρ. Denote the equivalence relation associated with this partition by ρ'. By definition $x\rho' y$ if and only if x and y belong to the same block of P_ρ. Thus $x\rho' y$ precisely when $x, y \in [z]_\rho$ for some z. But $x, y \in [z]_\rho$ for some z precisely when $x\rho y$. We have therefore proved that $\rho = \rho'$.

Let $P = \{A_i : i \in I\}$ be a partition of X. Its associated equivalence relation is ρ_P. Denote the partition of X associated with this equivalence relation by $P' = \{B_j : j \in J\}$. Let $A_i \in P$. The block A_i is non-empty and so we can choose, and fix, $x \in A_i$. Let B_j be the unique block of P' that contains x. If $y \in A_i$ then by definition $x\rho_P y$. Thus $y \in B_j$. It follows that $A_i \subseteq B_j$. Let $z \in B_j$. Then $x\rho_P z$. Thus $z \in A_i$. We have proved that $A_i = B_j$ and so $P \subseteq P'$. The reverse inclusion is proved in a similar way. We have therefore proved that $P = P'$. $\square$

Example 3.5.5. Let $X = \{1, 2, 3, 4, 5, 6\}$. The relation ρ is defined to be the following set of ordered pairs.

$$\{(1,1), \quad (2,2), \quad (3,3), \quad (4,4), \quad (5,5), \quad (6,6), \quad (2,3),$$
$$(3,2), \quad (4,5), \quad (5,4), \quad (4,6), \quad (6,4), \quad (5,6), \quad (6,5)\}.$$

This is an equivalence relation. The associated digraph is

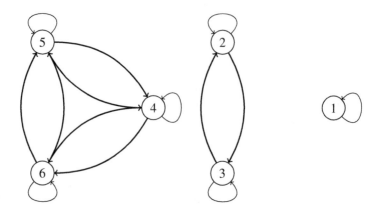

Observe that the vertices are grouped together into 'islands'. These islands lead to the associated partition $\{\{1\}, \{2,3\}, \{4,5,6\}\}$.

Let ρ be an equivalence relation on the set X. The blocks of this partition are called the *ρ-equivalence classes* or *equivalence classes*. The ρ-equivalence class containing the element x is denoted by $[x]_\rho$ or $[x]$. The set of ρ-equivalence classes is denoted by X/ρ. Theorem 3.5.4 tells us that the relations that enable us to classify the elements of a set are precisely the equivalence relations, and that the result of such a classification is to partition the set into equivalence classes. The elements within a block of the partition are to be regarded as being similar as viewed by the equivalence relation.

There is one important feature of equivalence classes that we have used above and that we shall highlight again here. There is equality $[x] = [y]$ if and only if $x \rho y$. This means that there are usually many different ways of describing one and the same equivalence class: that is, they have multiple aliases. This is important when we make a definition that refers to the class $[x]$ using the specific element x. We must ensure that we get the same definition if we choose instead any other element of $[x]$.

If we choose one element from each block of a partition we get what is called a *transversal*. The elements of the transversal are then proxies for the blocks that contain them. A partition will, of course, have many transversals. The elements of a transversal may also, in certain situations, be called *canonical forms*.

Example 3.5.6. Consider the real plane $\mathbb{R}^2$ and the set X of all possible circles in the plane, all possible squares and all possible equilateral triangles. There is an obvious equivalence relation on the set X that has three blocks: the set of all circles C, the set of all squares S and the set of all equilateral triangles T. Let c be the circle of unit radius centre the origin, let s be the square with vertices at $(0,0)$, $(0,1)$, $(1,0)$ and $(1,1)$ and let t be the equilateral triangle with base $(0,0)$ and $(1,0)$ in the first quadrant. Then $c \in C$, and $s \in S$ and $t \in T$. Thus $\{c,s,t\}$ is a transversal of the equivalence relation. It is also a set of canonical forms by definition, but the terminology suggests the additional idea that c is typical of C, and s is typical of S, and t is typical of T.

One important application of equivalence relations to the theory of functions permeates all of mathematics in one form or another.

Theorem 3.5.7. *Let* $f: X \to Y$ *be a function from the set* X *to the set* Y. *Then there is a set* Z, *a surjective function* $g: X \to Z$ *and an injective function* $h: Z \to Y$

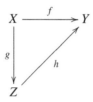

so that $f = hg$. *Thus every function can be factorized into a surjection followed by an injection.*

Proof. Define the relation $\sim$ on the set X by $x \sim y$ precisely when $f(x) = f(y)$. It is left as an exercise to check that $\sim$ is an equivalence relation. Put $Z = X/\sim$, the set of $\sim$-equivalence classes. Write $[x]$ for the $\sim$-equivalence class containing x. Define $g: X \to Z$ by $g(x) = [x]$. This is evidently a surjective function. Define $h: Z \to Y$ by $h([x]) = f(x)$. We have to check that this definition really makes sense. This means that if $[x] = [x']$ then we have to check that $f(x) = f(x')$. Now $[x] = [x']$ implies that $x \sim x'$ and by definition this actually means that $f(x) = f(x')$. Thus h is indeed a function. It remains to check injectivity. Suppose that $h([x]) = h([x'])$. Then $f(x) = f(x')$. By definition, this means that $x \sim x'$ and so $[x] = [x']$, as required. The functions f and hg have the same domains and codomains. Let $x \in X$. Then $(hg)(x) = h([x]) = f(x)$. We have therefore proved that $f = hg$. $\qquad\square$

Example 3.5.8. Let $f: \mathbb{R}^2 \to \mathbb{R}$ be the function defined by $f(x,y) = x$. Then the equivalence relation induced on $\mathbb{R}^2$ by f has blocks consisting of vertical lines perpendicular to the x-axis. The function g, defined in the above theorem, maps a point to the unique vertical line that contains it. The function h, defined in the above theorem, maps a vertical line to its position on the x-axis.

Equivalence relations classify the elements of a set into equivalence classes but the set of equivalence classes is itself a new set. From this perspective, each equivalence class is regarded as an entity in its own right and the elements that make up the class are said to be *identified*. Whether an equivalence relation is being used in its classifying rôle or in its identifying rôle depends on the context. By and large, it is the identifying rôle which is of greatest importance in this book. This is how equivalence relations are used in modular arithmetic in Section 5.4, the modular arithmetic of polynomials in Section 7.11 and the definition of vectors in Section 9.1. On the other hand, it is the classifying rôle of equivalence relations that is to the fore in our discussion of similarity between matrices in Section 8.6.

Example 3.5.9. Take a strip of paper with the following proportions.

We can make two different shapes with this strip. The first is a cylinder which is constructed by just taping (or glueing) the two ends of the strip together. One side of the cylinder can be coloured red and the other blue. The second shape is similar to the first except that the strip of paper is twisted before the ends are taped together. We find this time there is only one side: if you start colouring the strip red you will find that the whole strip is ultimately coloured red. This shape is called a *Möbius band*. The point of the colouring is to show that the two shapes we obtain are quite different from each other. To see what is going on in mathematical terms, we model the strip of paper by the following set $[0,10] \times [0,1]$, a product of two closed intervals. We now define two different partitions on this set.

1. The first partition has the following blocks: all the singleton sets of the elements of the set $[0,10] \times [0,1]$ with the ends $\{0\} \times [0,1]$ and $\{10\} \times [0,1]$ removed together with all sets of the form $\{(0,x),(10,x)\}$, where $x \in [0,1]$. The set of blocks is a cylinder.

2. The second partition is defined slightly differently. Take as blocks as before all the singleton sets of the elements of the set $[0,1] \times [0,10]$ with the ends $\{0\} \times [0,1]$ and $\{10\} \times [0,1]$ removed but this time together with all sets of the form $\{(0,x),(10,1-x)\}$, where $x \in [0,1]$. It is here that we introduce the twist. The set of blocks is a Möbius band.

The diagrams below use arrows to show what is glued to what.

is the cylinder, and

is the Möbius band.

Example 3.5.10. The construction of the rational numbers provides a good example of equivalence relations being used to build new sets. Rather than write a rational number like this $\frac{a}{b}$, we shall write it temporarily like this (a,b). We would therefore like to regard rational numbers as elements of $X = \mathbb{Z} \times (\mathbb{Z} \setminus \{0\})$. But this cannot be the whole story, because two different such ordered pairs can represent the same rational number. A little thought shows that (a,b) and (c,d) represent the same rational number if and only if $ad = bc$. We accordingly define the relation $\sim$ on X by

$$(a,b) \sim (c,d) \Leftrightarrow ad = bc.$$

It follows that a rational number is not simply an ordered pair but is in fact an equivalence class $[(a,b)]$, which we simply write as $[a,b]$. Thus when we write $\frac{a}{b}$ we are choosing a representative from the equivalence class $[a,b]$. For example, $(2,4) \in [1,2]$. This name changing business is vital in understanding how addition of rationals works. To add $[a,b]$ and $[c,d]$ we have to choose representatives from each class which have the same denominator: they have to match in some sense. All of this goes a long way to explain why fractions cause children such problems. We shall meet a very similar phenomenon when we define vectors in Chapter 9.

Exercises 3.5

1. Which of the following are partitions of the set $X = \{1,2,\ldots,9\}$?

   (a) $\{\{1,3,5\},\{2,6\},\{4,8,9\}\}$.
   (b) $\{\{1,3,5\},\{2,4,6,8\},\{5,7,9\}\}$.
   (c) $\{\{1,3,5\},\{2,4,6,8\},\{7,9\}\}$.

2. Show that each of the following relations defined on $\mathbb{R}^2$ is an equivalence relation and in each case describe the equivalence classes in geometrical terms.

   (a) Define $(x_1,y_1)\rho(x_2,y_2)$ if $y_1 = y_2$.
   (b) Define $(x_1,y_1)\rho(x_2,y_2)$ if $x_1 = x_2$.
   (c) Define $(x_1,y_1)\rho(x_2,y_2)$ if $y_2 - y_1 = x_2 - x_2$.
   (d) Define $(x_1,y_1)\rho(x_2,y_2)$ if $y_2 - y_1 = 2(x_2 - x_2)$.

3. Find all equivalence relations on the set $X = \{1,2,3,4\}$ both as partitions and as digraphs.

4. Prove that each partition gives rise to an equivalence relation thus completing the proof of Theorem 3.5.4.

5. Complete the missing details in the proof of Theorem 3.5.7.

3.6   ORDER RELATIONS

In Section 3.5, we studied classifying elements of a set by partitioning them. Another way of organizing the elements of set is in a hierarchy. Once again, we shall formalize this notion using relations but they will be slightly different from equivalence relations. A relation ρ defined on a set X is called a *partial order* if it is reflexive, antisymmetric and transitive. A set equipped with a partial order is called a *partially ordered set* or *poset*. The key difference between an equivalence relation and a partial order relation is that equivalence relations are symmetric whereas partial order relations are antisymmetric, the point being that if two elements are such that each is order-related to the other then they should be equal. Partial orders are denoted by symbols such as $\leq, \preceq, \subseteq$ and so forth since this notation contains a visual reminder of the direction of the order: from smaller to larger.

Example 3.6.1. Let X be a set. We show that $\subseteq$ is a partial order on $P(X)$. Let $A \in P(X)$. Then $A \subseteq A$ by definition. Thus $\subseteq$ is reflexive. Let $A, B \in P(X)$ be such that $A \subseteq B$ and $B \subseteq A$. Then it was an earlier exercise (Exercises 3.1) that $A = B$. Thus $\subseteq$ is antisymmetric. Finally, let $A, B, C \in P(X)$ such that $A \subseteq B$ and $B \subseteq C$. Let $a \in A$. Then $a \in B$ by the definition of $A \subseteq B$, and $a \in C$ by the definition of $B \subseteq C$. Thus $a \in C$. It follows that $A \subseteq C$. We have proved transitivity.

Example 3.6.2. There is a concise way of drawing the digraph of partial orders on finite sets called *Hasse diagrams*. We describe them by means of an example. Let $X = \{a, b, c\}$. The digraph of the relation $\subseteq$ on $P(X)$ has a lot of edges, but because partial orders are transitive we do not need to include them all. It is enough to draw an edge from U to V if $U \subseteq V$ and V has exactly one more element than U. In addition, rather than drawing arrows, we can simply draw lines if we use the convention that lower down the page means smaller. It is therefore enough to draw the following graph of inclusions for us to be able to work out all the others.

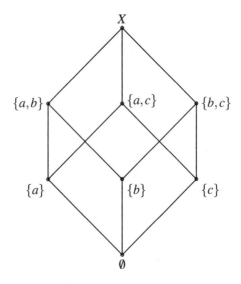

If we are working with an arbitrary partial order relation $\leq$ on a finite set X then we draw a line from x to y if $x \leq y$ and $x \neq y$ and if $x \leq z \leq y$ then either $x = z$ or $y = z$. We say that y *covers* x.

Our definition of a partial order is more general than the notion of order used in everyday life which usually satisfies the following additional property. A partial order $\leq$ on a set X is called a *linear order* if for all $x, y \in X$ we have that either $x \leq y$ or $x = y$ or $y \leq x$. They are called linear orders because we can imagine the elements of X arranged in a line with $x \leq y$ if x is to the left of y, and so y is to the right of x. Linear orders such as this arise when we consider numbers. Thus the number-line pictures the elements of $\mathbb{R}$ arranged along the line according to size.

There is a final piece of notation and terminology that is very common. We make some definitions first. Let ρ be a relation defined on a set X. We say that it is *irreflexive* if $\neg(x\rho x)$ for all $x \in X$. It is *asymmetric* if $x\rho y$ implies that $\neg(y\rho x)$. A relation that is irreflexive, asymmetric and transitive is called a *strict order*. Let $(X, \leq)$ be a poset. Define a new relation $<$ on X by $x < y$ if and only if $x \leq y$ and $x \neq y$. Then $<$ is a strict order.

Exercises 3.6

1. A relation $\preceq$ on a set X is called a *preorder* if it is reflexive and transitive.

   (a) Define the relation $\equiv$ on X by $x \equiv y$ if and only if $x \preceq y$ and $y \preceq x$. Prove that $\equiv$ is an equivalence relation. Denote the $\equiv$-equivalence class containing x by $[x]$.

   (b) On the set $X/\equiv$ define the relation $\leq$ by $[x] \leq [y]$ if and only if $x' \preceq y'$ for some $x' \in [x]$ and some $y' \in [y]$. Prove that $\leq$ is a partial order on the set $X/\equiv$.

2. Let $(A, \leq)$ and $(B, \leq)$ be posets.

   (a) Define a relation on $A \times B$ by $(a', b') \leq (a, b)$ if either $b' < b$ or $b' = b$ and $a' \leq a$. Prove that $\leq$ is a partial order on $A \times B$.

   (b) Prove that if the partial orders on A and B are both linear orders then the partial order defined on $A \times B$ in part (a) is also a linear order.

3. Let X be a non-empty set. Show that there is a bijection between the set of partial order relations on X and the set of strict order relations on X.

3.7 QUANTIFIERS

Mathematics abounds in theorems that state that *some* element of a set exists with certain properties or that *every* element of a set has certain properties. Both of these are examples of quantifiers. Just as with the propositional connectives introduced in Section 3.2, we shall here introduce symbols when explaining their precise mathematical meanings, but use the English words everywhere else.

The phrase *there exists* is represented by the symbol $\exists$, called the *existential quantifier*, and the phrase *for all* by the symbol $\forall$, called the *universal quantifier*. Both of these symbols are then followed by a variable name so we might write $\exists x$ and $\forall x$. If $P(x)$ is some sentence which says something about x then we may write and $(\exists x)P(x)$ and $(\forall x)P(x)$, both of which are now statements.

- The statement $(\exists x)P(x)$ is true when there is *at least one* object a such that $P(a)$ is true.

- The statement $(\forall x)P(x)$ is true when *for every* object a the statement $P(a)$ is true.

The mathematical use of the word *some* can be more parsimonious than the usual meaning. If I said I was going to give you *some dollars*, you would be disappointed if I gave you just *one dollar*, but mathematically I would be within my rights. Usually, we want x to be chosen from some specific set A, say, and so we would write $(\exists x \in A)P(x)$ and $(\forall x \in A)P(x)$. Thus $(\exists x \in A)P(x)$ is true not merely when there is an object a such that $P(a)$ is true but under the additional condition that $a \in A$.

A statement involving a single quantifier or even multiple quantifiers of the same type is not hard to interpret. The difficulty only begins when there are multiple quantifiers of different types. We describe how such statements should be interpreted by means of some examples.

Examples 3.7.1. For each statement below, we translate what it is saying into ordinary English and then determine if what it is saying is true or false.

1. $S_1 = (\forall x \in \mathbb{N})(\exists y \in \mathbb{N})(x < y)$. We read such statements from left-to-right. The first quantifier $(\forall x \in \mathbb{N})$ tells us that we may choose any natural number. Do that and call it n. If we replace x by n we get the statement $(\exists y \in \mathbb{N})(n < y)$. This is now a statement about the specific number n so if there is some choice of y that makes it true that choice will depend on n. This now says that for this choice of n there is a strictly larger natural number than n. This is clearly true and so S_1 is true.

2. $S_2 = (\exists y \in \mathbb{N})(\forall x \in \mathbb{N})(x < y)$. In this example, we have swapped the quantifiers in the example above. For this statement to be true, there must be a specific natural number a such that $(\forall x \in \mathbb{N})(x < a)$ is true. This says that there is a natural number a that is strictly larger than any natural number. This is false.

3. $S_3 = (\exists x \in \mathbb{N})(\forall y \in \mathbb{N})(x \leq y)$. In this example, we have swapped the variables in the example above. For this to be true, there must be a specific natural number a such that $(\forall y \in \mathbb{N})(a \leq y)$. This is true when $a = 0$.

4. $S_4 = (\forall y \in \mathbb{N})(\exists x \in \mathbb{N})(x < y)$. All we have done here is swap the variables in the quantifiers from the first example. This statement now says that given any natural number there is a strictly smaller natural number. This is false since zero is a counterexample.

5. $S_5 = (\forall y \in \mathbb{Z})(\exists x \in \mathbb{Z})(x < y)$. Here we have changed the sets involved from the previous example. This statement now says that for each integer there is a strictly smaller integer. This is true.

The next example is easy and is only included in case you should rush to judgement. In terminology to be introduced in Section 4.1, it says that $(\exists x)$ does not distribute over $\wedge$.

Example 3.7.2. Let $R(x)$ be the phrase x *is red* and let $S(x)$ be the phrase x *is small*. The statement $(\exists x)[R(x) \wedge S(x)]$ is true in some situation if there is something that is both small and red, whereas the statement $(\exists x)R(x) \wedge (\exists x)S(x)$ is true in some situation if there is something that is red and there is also something small, though they need not be one and the same. Hence the statements $(\exists x)[R(x) \wedge S(x)]$ and $(\exists x)R(x) \wedge (\exists x)S(x)$ are not equivalent.

The final example deals with negating quantifiers.

Example 3.7.3. The statement $\neg(\forall x)A(x)$ is true when the statement $(\forall x)A(x)$ is false. But this is false when there is some element, call it a, such that $A(a)$ is false. This means the same thing as saying that $\neg A(a)$ is true. We therefore deduce that $(\exists x)\neg A(x)$ is true. We can turn this argument around, and conclude as a result that $\neg(\forall x)A(x)$ means the same thing as $(\exists x)\neg A(x)$. Similarly we can show that $\neg(\exists x)A(x)$ means the same thing as $(\forall x)\neg A(x)$.

Exercises 3.7

1. The relation ρ is transitive and satisfies the following two conditions: $(\forall x)(\exists y)(x\rho y)$ and $(\forall x)\neg(x\rho x)$. Show that if ρ is defined on a non-empty set then that set must be infinite. Can you give a concrete example of such a relation?

2. Let $S = (\exists x)[A(x) \wedge B(x)]$ and $T = (\exists x)A(x) \wedge (\exists x)B(x)$. Prove that exactly one of $S \Rightarrow T$ and $T \Rightarrow S$ is a theorem.

3.8 PROOF BY INDUCTION

This is a proof technique that has the disadvantage of not delivering much insight into why a statement is true but on occasions is the only option. The basis of this technique is the following property of the natural numbers which should be regarded as an axiom.

Induction principle. Let X be a subset of $\mathbb{N}$ that satisfies the following two conditions: (1) $0 \in X$ and (2) $n \in X \Rightarrow n+1 \in X$. Then $X = \mathbb{N}$.

The induction principle can be used as a proof method in the following situation. Suppose an infinite number of statements $S_0, S_1, S_2, \ldots$ are given which we wish to prove. Denote by I the set of subscripts i such that S_i is true. Then $I \subseteq \mathbb{N}$. By the induction principle, it is enough to do two things to prove that all S_i are true.

1. Prove that S_0 is true. This means show that $0 \in I$.

2. Prove that if S_n is true then S_{n+1} is also true. This means prove that if $n \in I$ then $n+1 \in I$.

It will then follow that S_i is true for all $i \geq 0$. Such a proof is called a *proof by induction*. Proofs by induction always have the following script.

1. *Base step.* Show that the case $n = 0$ holds.

2. *Induction hypothesis (IH).* Assume that the case n holds.

3. *Proof part.* Use (IH) to show that the case $n+1$ holds.

Example 3.8.1. Prove by induction that $k^3 + 2k$ is exactly divisible by 3 for all natural numbers $k \geq 0$.

1. Base step. When $k = 0$, we have that $0^3 + 2 \cdot 0 = 0$ which is exactly divisible by 3.

2. Induction hypothesis. Assume result is true for $k = n$. Thus we assume that $n^3 + 2n$ is exactly divisible by 3.

3. Proof part. Prove that $(n+1)^3 + 2(n+1)$ is exactly divisible by 3 assuming only that $n^3 + 2n$ is exactly divisible by 3. We first expand $(n+1)^3 + 2(n+1)$ to get

$$n^3 + 3n^2 + 3n + 1 + 2n + 2.$$

   This is equal to

$$(n^3 + 2n) + 3(n^2 + n + 1)$$

   which is exactly divisible by 3 using the induction hypothesis.

In practice, a number of simple variants of this principle are used. Rather than the whole set $\mathbb{N}$, we often work with a set of the form

$$\mathbb{N}^{\geq m} = \mathbb{N} \setminus \{0, 1, \ldots, m-1\}$$

where $m \geq 1$. Our induction principle is modified accordingly: a subset X of $\mathbb{N}^{\geq m}$ that contains m, and contains $n+1$ whenever it contains n must be equal to the whole of $\mathbb{N}^{\geq m}$. In our script above, the base step now involves checking the specific case m. Any one of these versions of proof by induction is called *basic induction*.

There are a number of other versions of the induction principle which are sometimes needed and which despite appearances are in fact equivalent to it.

Strong induction principle. Let X be a subset of $\mathbb{N}$ that satisfies the following two conditions: (1) $0 \in X$ and (2) $\{0, 1 \ldots, n\} \subseteq X \Rightarrow \{0, 1 \ldots, n+1\} \subseteq X$. Then $X = \mathbb{N}$.

Well-ordering principle. Every non-empty subset of the natural numbers has a smallest element.

Proposition 3.8.2. *The following are equivalent.*

1. *The induction principle.*

2. *The strong induction principle.*

3. *The well-ordering principle.*

Proof. (1)⇒(2). Let $X \subseteq \mathbb{N}$ be such that $0 \in X$ and if $\{0,1\dots,n\} \subseteq X$ then $\{0,1\dots,n+1\} \subseteq X$. Define $Y \subseteq \mathbb{N}$ to be the set of natural numbers n such that $\{0,1,\dots,n\} \subseteq X$. We have $0 \in Y$ and we have that $n+1 \in Y$ whenever $n \in Y$. By the induction principle $Y = \mathbb{N}$. It follows that $X = \mathbb{N}$.

(2)⇒(3). Let $X \subseteq \mathbb{N}$ be a subset that has no smallest element. We prove that X is empty. Define $Y = \mathbb{N} \setminus X$. We claim that $0 \in Y$. If not, then $0 \in X$ and that would obviously be the smallest element, which is a contradiction. Suppose that $\{0,1,\dots,n\} \subseteq Y$. Then we must have that $n+1 \in Y$ because otherwise $n+1$ would be the smallest element of X. By the strong induction principle $Y = \mathbb{N}$ and so $X = \emptyset$.

(3)⇒(1). Let $X \subseteq \mathbb{N}$ be a subset such that $0 \in X$ and whenever $n \in X$ then $n+1 \in X$. Suppose that $\mathbb{N} \setminus X$ is non-empty. Then it would have a smallest element k say. But then $k-1 \in X$ and so, by assumption, $k \in X$, which is a contradiction. Thus $\mathbb{N} \setminus X$ is empty and so $X = \mathbb{N}$. □

There is a final property that we shall use particularly in showing that an algorithm terminates.

Descending chain condition (DCC). Every strictly descending chain (or sequence) of natural numbers $a_1 > a_2 > a_3 > \dots$ terminates.

Proposition 3.8.3. *The (DCC) is equivalent to the well-ordering principle.*

Proof. Assume that the (DCC) holds. Let $X \subseteq \mathbb{N}$ be a non-empty subset. Suppose that X has no smallest element. By assumption, X is non-empty and so there exists $a_1 \in X$. Since X has no smallest element, there exists $a_2 \in X$ such that $a_1 > a_2$. Continuing in this way, we obtain an infinite descending sequence of natural numbers $a_1 > a_2 > \dots$ which contradicts the (DCC). Conversely, assume that the well-ordering principle holds. Let $a_1 > a_2 > a_3 > \dots$ be a descending sequence of natural numbers. By assumption, the set $\{a_1, a_2, \dots\}$ has a smallest element. This is of the form a_n for some n. This implies that the descending sequence is exactly $a_1 > a_2 > \dots > a_n$ and so terminates. □

Exercises 3.8

1. Prove $n^2 > 2n + 1$ for each natural number $n \geq 3$.

2. Prove $2^n > n^2$ for each natural number $n \geq 5$.

3. Prove that $4^n + 2$ is divisible by 3 for each natural number $n \geq 1$.

4. Prove that
$$1 + 2 + 3 + \dots + n = \frac{n(n+1)}{2}.$$

5. Prove that
$$2 + 4 + 6 + \dots + 2n = n(n+1).$$

6. Prove that
$$1^3 + 2^3 + 3^3 + \dots + n^3 = \left(\frac{n(n+1)}{2}\right)^2.$$

3.9 COUNTING

A basic problem in mathematics is to count the number of elements in a finite set. This is frequently harder than it sounds particularly in applications to probability theory. We therefore need some systematic methods to tackle counting problems. The goal of this section is to introduce just such methods.

Counting is an intuitive process we learn to do as children, but if we want to count complicated sets we need, ironically, to make the process of counting more mathematical. This we do using bijective functions. To say a finite set X has n elements means precisely that there is a bijection from X to the set $\{1,\ldots,n\}$. This is because counting the elements of a set is actually the construction of just such a bijection: choose an element x_1 of X and count 'one', now choose an element x_2 of $X \setminus \{x_1\}$ and count 'two', and so on. We say that finite sets X and Y are *equinumerous*, denoted by $X \cong Y$, if there is a bijection between them.

Lemma 3.9.1. *Finite sets X and Y are equinumerous if and only if they have the same cardinality.*

Proof. Suppose that X and Y are equinumerous. If one is empty so is the other; so we shall assume that both are non-empty. Let $X \cong \{1,\ldots,m\}$ and $Y \cong \{1,\ldots,n\}$. Then $\{1,\ldots,m\} \cong \{1,\ldots,n\}$ by Lemma 3.4.16 and so $m = n$. Conversely, suppose that $X \cong \{1,\ldots,m\}$ and $Y \cong \{1,\ldots,m\}$. Then $X \cong Y$ by Lemma 3.4.16. □

This justifies the first of the counting principles below and the other two will be justified in the theorem that follows.

Counting Principles

1. Sets X and Y contain the same number of elements precisely when there is a bijection between them.

2. *The partition counting principle.* Let $A_1,\ldots,A_n$ be all the blocks of a partition of A. Then
$$|A| = |A_1| + \ldots + |A_n|.$$

3. *The product counting principle.* Let $A_1,\ldots,A_n$ be n sets. Then
$$|A_1 \times \ldots \times A_n| = |A_1|\ldots|A_n|.$$

The first counting principle often arises in constructing a bijection from a set that is hard to count to one that is easy. This often amounts to the 'trick' that enables us to solve the counting problem. We prove principles (2) and (3) below.

Theorem 3.9.2.

 1. *Let $A_1,\ldots,A_n$ be all the blocks of a partition of A. Then $|A| = |A_1| + \ldots + |A_n|$.*

 2. *Let $A_1,\ldots,A_n$ be n sets. Then $|A_1 \times \ldots \times A_n| = |A_1|\ldots|A_n|$.*

Proof. (1) We prove the case $n = 2$ first. Let $X = A \cup B$ and $A \cap B = \emptyset$. To count the elements of X, first count the elements of A and then continue counting the elements of B. There are no repetitions because A and B are disjoint and we count every element of X because the union of A and B is X. Thus $|A \cup B| = |A| + |B|$. The general case is proved by induction which we simply sketch here. Let $A_1, \ldots, A_n, A_{n+1}$ be a partition of X. Then $A_1 \cup \ldots \cup A_n, A_{n+1}$ is a partition of X with two blocks. We use the result proved above to get $|X| = |A_1 \cup \ldots \cup A_n| + |A_{n+1}|$ and then the induction hypothesis to finish off.

(2) We prove the case $n = 2$ first. Specifically, that $|A \times B| = |A| \, |B|$. The set $A \times B$ is the union of the disjoint sets $\{a\} \times B$ where $a \in A$. Now $|\{a\} \times B| = |B|$ because there is a bijection between the set $\{a\} \times B$ and the set B given by $(a, b) \mapsto b$. Thus $A \times B$ is the disjoint union of $|A|$ sets each with $|B|$ elements. Therefore by the partition counting principle from part (1) we have that $|A \times B| = |A| \, |B|$, as required. The general case is proved by induction which we simply sketch here. Let $A_1 \times \ldots \times A_n \times A_{n+1}$ be a product of $n + 1$ sets. Then $(A_1 \times \ldots \times A_n) \times A_{n+1}$ is a product of two sets and so has cardinality $|A_1 \times \ldots \times A_n| \, |A_{n+1}|$ by the result proved above. We now use the induction hypothesis to finish off. □

In the rest of this section, we show how to apply these principles to solve a number of basic counting problems.

Proposition 3.9.3 (Counting n-tuples). *Let A be a set with m elements. Then the number of n-tuples of elements of A is m^n.*

Proof. An element of A^n is an n-tuple of elements of A. Thus the number of such n-tuples is the cardinality of A^n which is m^n by the product counting principle. □

Example 3.9.4. Recall that a binary string of length n is an n-tuple whose elements are taken from the set $\{0, 1\}$. Such n-tuples are usually written without brackets. Thus the set of all binary strings of length 2 is

$$\{00, 01, 10, 11\}.$$

By Proposition 3.9.3, the number of binary strings of length n is 2^n.

Proposition 3.9.5 (Cardinality of power sets). *Let X be a finite set with n elements. Then $|\mathsf{P}(X)| = 2^n$.*

Proof. List the elements of X in some order. A subset of X is determined by saying which elements of X are to be in the subset and which are not. We can indicate these elements by writing a 1 above an element of X in the list if it is in the subset, and a 0 above the element in the list if it is not. Thus a subset determines a binary string of length n where the 1s tell you which elements of X are to appear and the 0s tell you which elements of X are to be omitted. We have therefore defined a bijection between the set of subsets of X and the set of all binary strings of length n. The number of such binary strings is 2^n by Example 3.9.4 which is therefore also the number of subsets of X. □

Example 3.9.6. We illustrate the above proof by considering the case where $X =$ $\{a,b,c\}$. List the elements of X as follows (a,b,c). The empty set corresponds to the binary string 000 and the whole set to the binary string 111. The subset $\{b,c\}$, for example, corresponds to the binary string 011.

Let X be an n-element set. We calculate the number of lists of length k where $k \leq n$ of elements of X where there are *no repetitions*. We call such a list a *k-permutation*. If $k = n$ we usually just speak of *permutations.*[20]

Example 3.9.7. For example, the 2-permutations of the set $\{1,2,3,4\}$ are

$$
\begin{array}{llll}
(1,2) & (1,3) & (1,4) \\
(2,1) & (2,3) & (2,4) \\
(3,1) & (3,2) & (3,4) \\
(4,1) & (4,2) & (4,3).
\end{array}
$$

There are therefore 12 such 2-permutations of a set with 4 elements.

We need the following example of a definition by recursion.[21] Let $n \geq 1$. Define $0! = 1$ and $n! = n \cdot (n-1)!$ when $n \geq 1$. Then we have, in fact, defined $n!$ for all n. The number $n!$ is called *n factorial*.

Proposition 3.9.8 (Counting k-permutations). *Let $0 \leq k \leq n$. Then the number of k-permutations of n elements is $\frac{n!}{(n-k)!}$.*

Proof. We use the partition counting principle. Denote by $P_{(n,k)}$ the set of all k-permutations of an n-element set X. Each such k-permutation of X begins with one of the elements of X. Thus the set of all k-permutations of X is partitioned into n blocks, each block consisting of all k-permutations which begin with one and the same element of X. Take one of these blocks, and remove the first element from each of the k-permutations in that block. The set which results consists of $(k-1)$-permutations of an $(n-1)$-element set. We have proved that

$$
\left| P_{(n,k)} \right| = n \left| P_{(n-1,k-1)} \right|.
$$

Observe that the cardinality of $P_{(m,1)}$ is m. The result now follows from the definition of the factorial. □

The number $\frac{n!}{(n-k)!}$ is sometimes written ^nP_k. This can be read as the number of ways of *ranking* k elements chosen from n elements.

Example 3.9.9. We illustrate the above proof. Let $X = \{a,b,c\}$. The set of permutations which arise is

$$
\{(a,b,c),(a,c,b),(b,a,c),(b,c,a),(c,a,b),(c,b,a)\}.
$$

---

[20]Recall from Section 3.4, that the word permutation is used in two related senses. It is a bijection of a set to itself, and the result of applying that bijection. Here, it is the second use that we need.

[21]This is a powerful way of defining functions particularly important in programming. I do not discuss the theory in this book.

Thus the number of permutations of a 3-element set is 6. Observe that there is a natural partition of this set, where each block contains all permutations that begin with the same letter. Thus there are three blocks

$$\{(a,b,c),(a,c,b)\}, \quad \{(b,a,c),(b,c,a)\}, \quad \{(c,a,b),(c,b,a)\}.$$

If we choose one of the blocks, say the first, and remove the first letter, then we obtain all permutations on the remaining letters.

A subset with k elements of a set A consisting of n elements is called a *k-subset*. It is often called a *combination* of k objects.

Example 3.9.10. Let $A = \{a,b,c,d\}$. The set of 2-subsets of A is

$$\{\{a,b\},\{a,c\},\{a,d\},\{b,c\},\{b,d\},\{c,d\}\}.$$

Thus it has cardinality 6.

Denote the number of k-subsets of an n element set by

$$\binom{n}{k},$$

pronounced 'n choose k'. The number $\binom{n}{k}$ is sometimes written ^nC_k and is read as the number of ways of *choosing* k elements from n elements. Numbers of this form are called *binomial coefficients*.

Example 3.9.11. Let $X = \{a,b,c\}$. There is one 0-subset, namely $\emptyset$, and one 3-subse,t namely X. The 1-subsets are $\{a\},\{b\},\{c\}$ and so there are three of them. The 2-subsets are $\{a,b\}$, $\{a,c\}$, $\{b,c\}$ and so there are three of them. Observe that $1+3+3+1 = 8 = 2^3$.

Proposition 3.9.12 (Counting combinations). *Let $0 \le k \le n$. Then*

$$\binom{n}{k} = \frac{n!}{k!(n-k)!}.$$

Proof. Let P be the set of all k-permutations of a set with n elements. Partition P by putting two permutations into the same block if they both permute the same set of k elements. There is a bijection between the set of blocks of P and the set of k-subsets. Each block contains $k!$ elements by Proposition 3.9.8. It follows by the partition counting principle that the number of blocks is $\frac{|P|}{k!}$. We know that $|P| = \frac{n!}{(n-k)!}$ by Proposition 3.9.8. $\qquad\square$

Example 3.9.13. Algebraic calculations can sometimes be simplified by using the counting principles. Direct calculation shows that

$$\binom{n}{k} = \binom{n}{n-k}$$

but we can explain why this is true in terms of counting subsets of a set. Every time a subset with k elements is chosen from a set with n elements, a subset with $n - k$ elements is not chosen. There is therefore a bijection between the set of subsets with k elements and the set of subsets with $n - k$ elements. It follows that there must be the same number of k-subsets as there are $(n - k)$-subsets.

If we now go back to our proof of Proposition 3.9.5, we may extract the following result using Proposition 3.9.12.

Proposition 3.9.14. *The number of binary strings of length n where the number of 0s is exactly k is $\binom{n}{k}$.*

There is one final basic counting result to prove. Before stating it, we give an example that contains all the ideas needed.

Example 3.9.15. We are interested in products of powers of the algebraic symbols x, y and z, thus expressions like xy^5z^3 and $x^3y^2z^5$. Remember that $x^0 = 1$ corresponds to the case where the variable x is missing, that $x = x^1$ and that $xy^5z^3 = z^3y^5x$ so the order in which the symbols appears does not matter. If $d = a + b + c$ then the expression $x^ay^bz^c$ is said to have *degree d*. We count the number of such expressions of degree 3. Examples are $x^0y^0z^3$ and $x^1y^0z^2$. To count them systematically, we define a bijection between the set of such expressions and a set of binary strings. We map the expression $x^ay^bz^c$ to the binary string

$$\overbrace{1...1}^{a}0\overbrace{1...1}^{b}0\overbrace{1...1}^{c}$$

Thus

$$x^1y^0z^2 \mapsto 10011 \text{ and } x^1y^1z^1 \mapsto 10101.$$

It follows that the number of expressions of degree 3 is equal to the number of such binary strings. Said binary strings consist of all binary strings of length 5 with exactly two 0s. There are therefore $\binom{5}{2} = 10$ of them. Here they all are

$$x^3, y^3, z^3, xy^2, x^2y, y^2z, yz^2, xz^2, x^2z, xyz.$$

We now generalize the above example. Let $x_1, \ldots, x_k$ be k variables. An expression $x_1^{m_1} \ldots x_k^{m_k}$, where $n = m_1 + \ldots + m_k$, and where the order of the variables is ignored, is called a *homogeneous expression of degree n*. The proof of the following is a generalization of the argument used above.

Proposition 3.9.16 (Counting homogeneous expressions). *The number of homogeneous expressions of degree n in k variables is*

$$\binom{n+k-1}{k-1}.$$

We conclude with a famous example.

Example 3.9.17. There is a result of probability theory called the *birthday paradox* that can be derived using the ideas of this chapter. This states that in any group of at least 23 people the probability that two people share the same birthday, by which is meant day and month, is about one half. It is a paradox in the sense that most people would expect that many more people would be needed for the probability of such a coincidence to be so high. Probabilities are not discussed in this book but in reality this is just a counting problem.

Let X be a set of m people, and let $B = \{n : 1 \leq n \leq 365\}$ be the set of all possible birthdays.[22] Assigning a birthday to somebody is therefore a function from X to B. Fix an ordering of the set X. Then a function from X to B is determined by an m-tuple in B^m. Thus by the product counting principle there are 365^m such functions. The injective functions from X to B correspond to groups of people all having different birthdays. The number of injective functions from X to B is $^{365}P_m$. Therefore the number of functions from X to B which are *not* injective is $365^m - {}^{365}P_m$. It follows that the fraction of all functions from X to B which are not injective is

$$\frac{365^m - {}^{365}P_m}{365^m}$$

which can be interpreted as the probability that two people in a group of m people share the same birthday.

The problem now is one of evaluating this fraction for various m but particularly for $m = 23$. This means estimating

$$p_m = 1 - \frac{365!}{365^m(365 - m)!}$$

and here we follow [42]. Define

$$q_m = \left(1 - \frac{1}{365}\right) \cdots \left(1 - \frac{(m-1)}{365}\right).$$

A little manipulation shows that $p_m = 1 - q_m$. To solve our problem we need to find the smallest m such that $q_m \leq \frac{1}{2}$. If x is small then

$$\ln(1 - x) \approx -x.$$

Thus for $1 \leq r \leq m$, we have that $\ln\left(1 - \frac{r}{365}\right) \approx -\frac{r}{365}$. It follows that

$$\ln(q_m) \approx -\frac{m(m-1)}{730}.$$

Now $\ln(0 \cdot 5) \approx -0 \cdot 693147$ and it is routine to check that $m = 23$ works.

---

[22] Ignoring those born on 29th February, I'm afraid.

Exercises 3.9

1. None of these questions is intrinsically interesting. The point is to find the correct translation from the 'real-life' problem to the corresponding mathematical counting problem.

   (a) A menu consists of 2 starters, 3 main courses and 4 drinks. How many possible dinners are there consisting of one starter, one main course and one drink?

   (b) For the purposes of this question, a *date* consists of an ordered triple consisting of the following three components: first component a natural number d in the range $1 \leq d \leq 31$; second component a natural number m in the range $1 \leq m \leq 12$; third component a natural number y in the range $0 < y \leq 3000$. How many possible dates are there?

   (c) In how many ways can 10 books be arranged on a shelf?

   (d) There are 10 contestants in a race. Assuming no ties, how many possible outcomes of the race are there?

   (e) 8 cars are to be ranked first, second and third. In how many ways can this be done?

   (f) In how many ways can a hand of 13 cards be chosen from a pack of 52 cards?

   (g) In how many ways can a committee of 4 people be chosen from 10 candidates?

   (h) A committee of 9 people has to elect a chairman, secretary and treasurer (assumed all different). In how many ways can this be done?

   (i) In a lottery, 6 distinct numbers are chosen from the range 1 to 49 without regard for order. In how many ways can this be done?

   (j) Given the digits 1,2,3,4,5 how many 4-digit numbers can be formed if repetition is allowed?

2. Let X be a set with cardinality n.

   (a) Recall that $T(X)$ is the set of all functions from X to itself. What is the cardinality of $T(X)$?

   (b) Recall that $S(X)$ the set of all bijective functions from X to itself. What is the cardinality of $S(X)$?

3. A novel has 250 pages, each page has 45 lines, and each line consists of 60 symbols. Assume that the number of possible symbols available is 100. How many possible novels are there? We allow avant garde novels that consist of nonsense words or are blank.[23] Express the answer as a power of ten.

---

[23] Or do not contain the letter 'e' such as the novel *Gadsby* by Ernest Vincent Wright.

4. (a) Let A and B be finite sets. Prove that

$$|A \cup B| = |A| + |B| - |A \cap B|.$$

(b) Generalize your result to obtain a similar formula for $|A \cup B \cup C|$.

5. In how many ways can a committee consisting of 3 Montagues and 2 Capulets be chosen from 7 Montagues and 5 Capulets?

6. Prove that

$$\binom{n+1}{r} = \binom{n}{r-1} + \binom{n}{r}$$

for $n \geq r \geq 1$ by using results about sets

3.10 INFINITE NUMBERS

One of Cantor's greatest discoveries was that there are not only infinite numbers but there are infinitely many of them.[24] This can be seen as further vindication for the idea of a set. Despite their apparent distance from everyday life, the ideas of this section have important consequences. We discuss one of them at the end of this section and another in Section 7.10. To define infinite numbers, we take an indirect route and generalize a definition made in Section 3.9. Let X and Y be arbitrary sets, not just finite ones. We say that X is *equinumerous* with Y, denoted by $X \cong Y$, if there is a bijection from X to Y. Clearly, this relation is saying that X and Y contain the same number of elements but obviating the need to define what we mean by 'number of elements' in the infinite case. The proofs of the following are immediate by Lemma 3.4.16.

Lemma 3.10.1.

1. $X \cong X$ for every set X.

2. If $X \cong Y$ then $Y \cong X$.

3. If $X \cong Y$ and $Y \cong Z$ then $X \cong Z$.

If $X \cong \emptyset$ or $X \cong \{1, \ldots, n\}$ we say that X is *finite*; otherwise we say that X is *infinite*. To prove that sets are equinumerous, we need some tools to help in the construction of bijections. Here is one whose proof is left as an exercise.

Lemma 3.10.2. *Let* $X = A \cup B$ *where* $A \cap B = \emptyset$, *and let* $Y = C \cup D$ *where* $C \cap D = \emptyset$. *If* $A \cong C$ *and* $B \cong D$ *then* $X \cong Y$.

We now state our first results about infinite sets. Parts (1) and (2) below are perhaps not that surprising, but part (3) certainly is.

---

[24] And ∞ is not one of them.

Proposition 3.10.3 (Countably infinite sets).

1. $\mathbb{E} \cong \mathbb{O} \cong \mathbb{N}$.

2. $\mathbb{Z} \cong \mathbb{N}$.

3. $\mathbb{Q} \cong \mathbb{Z}$.

Proof. (1). To show that $\mathbb{N} \cong \mathbb{E}$, use the bijective function $n \mapsto 2n$. To show that $\mathbb{N} \cong \mathbb{O}$, use the bijective function $n \mapsto 2n + 1$.

(2) List the elements of $\mathbb{Z}$ as follows

$$0, -1, 1, -2, 2, -3, 3, \ldots.$$

We can now count them off as $0, 1, 2, 3, \ldots$ and so establish a bijection with $\mathbb{N}$.

(3) Split $\mathbb{Q}$ into two disjoint sets: the set $\mathbb{Q}^{\geq 0}$ of non-negative rationals, and the set $\mathbb{Q}^{<0}$ of negative rationals. We split $\mathbb{N}$ into the disjoint sets $\mathbb{E}$, the even numbers, and $\mathbb{O}$, the odd numbers. We show that $\mathbb{Q}^{\geq 0}$ is equinumerous with $\mathbb{E}$, and that $\mathbb{Q}^{\leq 0}$ is equinumerous with $\mathbb{O}$. By Lemma 3.10.2, these two bijections can be glued together to yield a bijection from $\mathbb{Q}$ to $\mathbb{N}$. We prove explicitly that $\mathbb{Q}^{\geq 0}$ is equinumerous with $\mathbb{E}$ since the other case is similar. Set up an array with columns and rows labelled by the non-zero natural numbers. Interpret the entry in row m and column n as the rational number $\frac{n}{m}$.

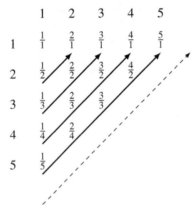

The resulting fractions are counted using the numbers $2, 4, 6, \ldots.$. We start with the fraction $\frac{1}{1}$ and then count along each successive arrow from bottom-left to top-right omitting repetitions. Zero is counted by 0. □

Example 3.10.4. We prove that $\mathbb{R} \cong (0, 1)$. As a first step, we prove that if a and b are any real numbers such that $a < b$ then $(a, b) \cong (0, 1)$. To see why, first apply the function $x \mapsto x - a$ to show that $(a, b) \cong (0, b - a)$. Next apply the function $x \mapsto \frac{x}{b-a}$ to show that $(0, b - a) \cong (0, 1)$. By Lemma 3.10.1, we deduce from $(a, b) \cong (0, b - a)$ and $(0, b - a) \cong (0, 1)$ that $(a, b) \cong (0, 1)$ as claimed. In particular, $(-a, a) \cong (0, 1)$ for any $a > 0$. Now let C be a semicircle of radius a centred on the point $(0, a)$ with end points removed. Then $C \cong (-a, a)$. The following diagram shows how this bijection is constructed.

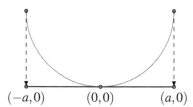

$$(-a,0) \qquad (0,0) \qquad (a,0)$$

Map the point (x,y) on C to the point $(x,0)$. Next we prove that $\mathbb{R} \cong C$. The bijection is described by the following diagram.

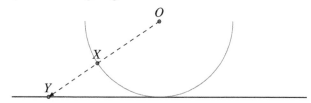

We map the point X to the point Y. But from $\mathbb{R} \cong C$ and $C \cong (-a,a)$ and $(-a,a) \cong (0,1)$ we deduce that $\mathbb{R} \cong (0,1)$.

We now come to a result that is really surprising, and one of Cantor's celebrated theorems.

Theorem 3.10.5 (Cantor diagonalization argument). *The set of reals is not equinumerous with the set of rationals.*

Proof. To make the argument more natural we use the easily proved fact that $\mathbb{N} \cong \mathbb{N}^*$ where the latter set is the set of positive natural numbers. We also use the result of Example 3.10.4 that $\mathbb{R} \cong (0,1)$. Assume that there is a bijection between $\mathbb{N}^*$ and $(0,1)$. This means that we can list *all* the elements of $(0,1)$ exactly once as $r_1, r_2, r_3 \ldots$. Each real number in $(0,1)$ can be expressed as an infinite decimal (if this is not clear see Example 4.6.4), $r_i = 0 \cdot a_{i1} a_{i2} a_{i3} \ldots$. Define a real number R by

$$R = 0 \cdot R_1 R_2 R_3 \ldots$$

where R_i is equal to 0 if a_{ii} is odd, and is equal to 1 otherwise. Observe that $R \neq r_i$ for all i by construction and that $R \in (0,1)$. This contradicts the fact that we have listed *all* the real numbers in $(0,1)$. It follows that no bijection as stated can exist. □

Thus $\mathbb{Q}$ and $\mathbb{R}$ are both infinite sets but there is no bijection between them.

To explore further the equinumerosity relation between infinite sets, it is useful to introduce another relation between sets. If there is an injective function from X to Y then write $X \preceq Y$. The following is another celebrated theorem. Its proof is difficult and so it has been placed in a box.

Theorem 3.10.6 (Cantor-Schröder-Bernstein). *If $X \preceq Y$ and $Y \preceq X$ then $X \cong Y$.*

If X and Y are sets such that $X \preceq Y$ but it is not the case that $X \cong Y$ then we write $X \prec Y$. Intuitively, the set X has strictly fewer elements than the set Y.

Box 3.7: Proof of the Cantor-Schröder-Bernstein Theorem

Let $f\colon X \to Y$ and $g\colon Y \to X$ be the two given injective functions. Let $U_0 = X \setminus \mathrm{im}(g)$. If $U_0 = \emptyset$ then g would already be a bijection and we would be done. So, in what follows $U_0 \neq \emptyset$. The problem is to use f and g to manufacture a bijection from X to Y. Before we go into details, we explain in general terms what we shall do. We construct subsets $U \subseteq X$ and $V \subseteq Y$ and bijections $h\colon U \to V$ and $k\colon Y \setminus V \to X \setminus U$. This enables us to define a bijection $\theta\colon X \to Y$ by

$$\theta(x) = \begin{cases} h(x) & \text{if } x \in U \\ k^{-1}(x) & \text{if } x \in X \setminus U \end{cases}$$

Here are the details of the above construction

- We define the sets U and V. Observe that $gf\colon X \to X$. Define $U_n = (gf)^n(U_0)$ for each n. These are all subsets of X. Define $V_n = f(U_n)$ for each n. These are all subsets of Y. Now define $U = \bigcup_{i=0}^{\infty} U_i$ and $V = \bigcup_{i=0}^{\infty} V_i$. Observe that

$$U_0 \xrightarrow{f} V_0 \xrightarrow{g} U_1 \xrightarrow{f} V_1 \xrightarrow{g} U_2 \xrightarrow{f} \dots.$$

- Define $h\colon U \to V$ by $h(x) = f(x)$. Since f is obtained by restricting the domain of f it must be injective, and it is surjective by construction. Thus h is a bijective function.

- Define $k\colon Y \setminus V \to X \setminus U$ by $k(y) = g(y)$. We need to check first that $g(y) \in X \setminus U$ is true. Suppose that $y \in Y \setminus V$ and $g(y) \in U$. We cannot have $g(y) \in U_0$ since by definition U_0 is the complement of the image of g. It follows that $g(y) \in U_n$ for some $n \geq 1$. Thus $g(y) = (gf)^n(u)$ where $u \in U_0$. Since $n \geq 1$ we may write $g(y) = gf(gf)^{n-1}(u)$. But g is injective and so $y = f(gf)^{n-1}(u)$. It follows that $y \in V_{n-1}$, which is a contradiction. Thus the function k does have domain and codomain as stated. It is injective because g is injective. It remains only to prove that it is surjective and we are done. From $U_0 \subseteq U$ we deduce that $X \setminus U \subseteq X \setminus U_0 = \mathrm{im}(g)$. Let $x \in X \setminus U$. Then there is $y \in Y$ such that $x = g(y)$. Suppose that $y \in V$. Then $y \in V_n$ for some n. Thus $y = f(x')$ for some $x' \in U_n$. This gives $x = g(y) = (gf)(x')$ and so $x \in U_n$ which is a contradiction. Hence $y \in Y \setminus V$ and k is surjective.

The next lemma is another that helps in constructing bijections. The proof is left as an exercise.

Lemma 3.10.7. *If $A \cong B$ and $C \cong D$ then $A \times C \cong B \times D$.*

Example 3.10.8. The real numbers $\mathbb{R}$ can be regarded as the real line and $\mathbb{R}^2$ as the real plane. It is easy to define an injection from $\mathbb{R}$ to $\mathbb{R}^2$. For example, the function $x \mapsto (x, 0)$ clearly works. It follows that $\mathbb{R} \preceq \mathbb{R}^2$. What about defining an injection from $\mathbb{R}^2$ to $\mathbb{R}$? Observe first that $\mathbb{R}^2 \cong (0, 1) \times (0, 1)$ where we use the fact that $\mathbb{R} \cong (0, 1)$, proved in Example 3.10.4, and Lemma 3.10.7. Thus it is enough to define an injective function from $(0, 1) \times (0, 1)$ to $(0, 1)$. This is done as follows. Each real number in $(0, 1)$ can be represented by an infinite decimal $0 \cdot a_1 a_2 \dots$. (See Example 4.6.4 if this is not clear.) Let $(a, b) \in (0, 1) \times (0, 1)$. Then $a = 0 \cdot a_1 a_2 \dots$ and $b = 0 \cdot b_1 b_2 \dots$. Define the number $c \in (0, 1)$ by $c = 0 \cdot a_1 b_1 a_2 b_2 \dots$. In other words, we interleave the two infinite decimals. This defines an injective function. We now apply Cantor-Schröder-Bernstein to deduce that $\mathbb{R} \cong \mathbb{R}^2$. We have proved something that sounds paradoxical: that there are as many points in the real line as there are in the real plane.

> **Box 3.8: The Banach-Tarski Paradox**
>
> This section is about counting, but there are also paradoxes, even more striking, associated with the idea of quantity. These also deal with the infinite but in more subtle ways. For example, a solid the size of an apple can be cut up into a finite number of pieces and then reassembled in such a way as to form another solid the size of the moon. This is known as the Banach-Tarski Paradox (1924). There is no trickery involved here and no sleight of hand. This is clearly pure mathematics: give me a real apple and whatever I do it will remain resolutely apple-sized. The ideas used in understanding this paradox involve such fundamental and seemingly straightforward ones as length, area and volume that have important applications in applied mathematics.

Is that it? Is every infinite set equinumerous to either $\mathbb{N}$ or $\mathbb{R}$? The remarkable answer is no. This follows from another celebrated theorem by Cantor. There is clearly an injective function from the set X to the set $P(X)$ given by $x \mapsto \{x\}$. Thus $X \preceq P(X)$. We now rule out the possibility that $X \cong P(X)$.

Theorem 3.10.9. $X \prec P(X)$ *for every set X.*

Proof. Suppose that there is a bijective function $f \colon X \to P(X)$. We derive a contradiction. Define a subset A of X as follows. It consists of all $x \in X$ such that $x \notin f(x)$. Now $A \in P(X)$, very possibly empty but that does not matter. By assumption the function f is surjective, and so there exists $y \in X$ such that $A = f(y)$. There are two possibilities. Suppose first that $y \in A$. Then by definition $y \notin f(y) = A$, which is a contradiction. Suppose therefore that $y \notin A$. Then by definition $y \in f(y) = A$, which is also a contradiction. We deduce that f cannot exist. $\qquad\square$

Theorem 3.10.9 has an astonishing consequence. Start with $\mathbb{N}$, and define $P^1(\mathbb{N}) = P(\mathbb{N})$ and for $n > 1$ define $P^n(\mathbb{N}) = P(P^{n-1}(\mathbb{N}))$. Then

$$\mathbb{N} \prec P(\mathbb{N}) \prec P^2(\mathbb{N}) \prec \ldots$$

where each of these sets is infinite but no two distinct ones are equinumerous. There are therefore infinitely many infinite sets that, intuitively, all have different sizes.

A set which is equinumerous with either a finite set or the set $\mathbb{N}$ is said to be *countable*, and an infinite countable set is said to be *countably infinite*. A set that is not countable is said to be *uncountable*. Define $|\mathbb{N}| = \aleph_0$. This is our first infinite number and is called *aleph nought*, the symbol $\aleph$ being the first letter of the Hebrew alphabet.[25] Define $|\mathbb{R}| = c$. This is our second infinite number and is called *the cardinality of the continuum*. The infinite sets $\mathbb{N}$ and $\mathbb{R}$ are not equinumerous and from what we have proved $\aleph_0 < c$. Are there any infinite numbers between these two? For the surprising answer, see the box on the Continuum Hypothesis.

The numbers in $\mathbb{N}$ together with $\aleph_0$ and c are examples of what are called *cardinal numbers*. These are what we count with.[26] To conclude this section and this chapter, we sketch the first steps in cardinal arithmetic.

Let X and Y be sets. Define $X \sqcup Y = (X \times \{0\}) \cup (Y \times \{1\})$. This operation is

---

[25] Strictly speaking it is a *betagam* since it doesn't use vowels.

[26] There are other kinds of numbers called *ordinals* used in ranking. In the finite case, they coincide with the cardinals but in the infinite case they are different.

called *disjoint union*. It is a trick for taking two sets which might have a non-empty intersection and replacing them with very similar sets which are disjoint. Clearly, $X \cong X \times \{0\}$ and $Y \cong Y \times \{1\}$. Let m and n be any two cardinals where we also allow infinite ones. We define $m + n$ as follows. Let X be any set such that $|X| = m$, and let Y be any set such that $|Y| = n$. Define $m + n = |X \sqcup Y|$. However, not so fast. We have to be careful, because we chose X and Y to be any sets. For our definition to make sense, it must be independent of the sets chosen. We need the following lemma which is left as an exercise.

Lemma 3.10.10. *If $X \cong X'$ and $Y \cong Y'$ then $X \sqcup Y \cong X' \sqcup Y'$.*

If m and n are finite cardinals then the definition of addition we have made agrees with the usual one. It is therefore the partition counting principle extended to infinite sets.

Example 3.10.11. The *Hilbert Hotel* provides a nice illustration of these ideas. In fact, it tells us how to do addition with $\aleph_0$. The hotel has $\aleph_0$ rooms numbered $1, 2, 3, \ldots$. All the rooms are occupied. It is important to remember that there is no door labelled $\aleph_0$ and certainly no door labelled ∞. A new tourist turns up. How can they be accommodated without anyone having to share a room? We simply tell each person in room n to go into room $n + 1$. This frees up room 1 which is available for the unexpected guest. This can easily be modified to deal with any finite number of new tourists. This proves that if n is any finite cardinal then $n + \aleph_0 = \aleph_0$. Suppose now that $\aleph_0$ tourists from the large Welsh village of Llareggub turn up. How can they be accommodated without anyone having to share a room? This time, we tell each person in room n to go into room $2n$. This frees up all the odd-numbered rooms and there are $\aleph_0$ of those and so the Welsh tourists can be accommodated. This proves that $\aleph_0 + \aleph_0 = \aleph_0$.

We now define multiplication. Let X be any set such that $|X| = m$, and let Y be any set such that $|Y| = n$. Define $mn = |X \times Y|$. This definition makes sense by Lemma 3.10.7. If m and n are finite cardinals then the definition of multiplication we have made agrees with the usual one. It is therefore the product counting principle extended to infinite sets.

By using the results of this section, it is left as an exercise to verify the following two Cayley tables for addition and multiplication of cardinal numbers. In the tables, *finite* stands for any finite non-zero cardinal.

+	0	finite	$\aleph_0$	c
0	0	finite	$\aleph_0$	c
finite	finite	finite	$\aleph_0$	c
$\aleph_0$	$\aleph_0$	$\aleph_0$	$\aleph_0$	c
c	c	c	c	c

×	0	finite	$\aleph_0$	c
0	0	0	0	0
finite	0	finite	$\aleph_0$	c
$\aleph_0$	0	$\aleph_0$	$\aleph_0$	c
c	0	c	c	c

Example 3.10.12. We derive a surprising consequence from counting binary strings. A binary string is a finite sequence of 0s and 1s where the empty binary string is denoted by ε. Thus the set of binary strings is $B = \{\varepsilon, 0, 1, 00, 01, 10, 11, \ldots\}$. We

show that B is countably infinite. To do this we draw a graph whose vertices are labelled by all the binary strings. The vertex ε is placed at the top. For each vertex w we draw two edges to the vertices $w0$ and $w1$ respectively where $w0$ is placed below and to the left of w, and $w1$ is placed below and to the right of w.

The first few steps in drawing this graph are shown below.

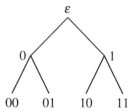

To count the set of binary strings, we linearly order them. Let u and v be binary strings. Define $u < v$ if either u is strictly shorter than v or u and v have the same length and u comes before v in the dictionary order where 0 comes before 1. In terms of the graph, this simply means that within each row strings are ordered from left-to-right, and every string in row i comes before every string in row $i + 1$. This leads to the following infinite sequence of binary strings

$$\varepsilon, 0, 1, 00, 01, 10, 11, 000, 001, 010, 011, 100, 101, 110, 111, \ldots$$

which can then be counted off $0, 1, 2, \ldots.$ This proves that $|\{0,1\}^*| = \aleph_0$. Fix now a programming language such as Python. Every such programme is, as far as a computer is concerned, nothing more than a binary string. Thus the total number of Python programmes is countably infinite. Consider now all those Python programmes that compute a real number. There cannot be more than a countable infinity of such programmes, but there are uncountably many real numbers. We deduce that there are real numbers that cannot be computed by any Python programme.

Exercises 3.10

1. Prove Lemma 3.10.2.

2. Prove Lemma 3.10.7.

3. Prove Lemma 3.10.10.

4. *Verify that the Cayley tables for cardinal arithmetic are correct. If a, b and c are cardinal numbers is it true that $ab = ba$ and that $a(bc) = (ab)c$? If $a^2 = a$ is it true that $a = 0, 1$?

Box 3.9: The Continuum Hypothesis

At this point, we begin to reach the limits of what is known. The *continuum hypothesis (CH)* is the assertion that the only infinite subsets of the reals are either countably infinite or have cardinality of the continuum. The work of Kurt Gödel (1906–1978) and Paul Cohen (1937–2007) clarified the nature of this question. To understand how, you have to know that set theory itself can be axiomatized just as Euclidean geometry can be axiomatized. The resulting system is known as *Zermelo-Fraenkel* or *ZF set theory* after the two mathematicians who worked it out. The result is that (CH) is independent of ZF set theory. This means that you can equally well develop set theory with (CH) or with ¬(CH), but you obtain different set theories as a result.

CHAPTER 4

Algebra redux

"*Arithmétique! Algèbre! Géométrie! Trinité grandiose! Triangle lumineux!*" – Comte de Lautréamont

This chapter is an essential prerequisite for Section II of the book. Algebra deals with the manipulation of symbols. In high school, algebra is based on the properties of the real numbers. This means that when you write x you mean a real number. The properties you use in solving equations involving x are therefore also the properties of real numbers. The problem is that if those properties are not spelt out, then doing algebra becomes instinctive: you simply apply a method without thinking. That is fine if the only algebra you meet is the algebra of the real numbers, but will cause you serious problems as soon as you have to do algebra requiring the application of different properties. For example, in Chapter 8 matrices are introduced where $(A + B)^2 \neq A^2 + 2AB + B^2$, and in Chapter 9 vectors are introduced where $\mathbf{a} \times (\mathbf{b} \times \mathbf{c}) \neq (\mathbf{a} \times \mathbf{b}) \times \mathbf{c}$. The first job, therefore, is to spell out in detail what those properties might be and to make explicit the implicit rules that you know. This is done in Section 4.1. The terminology introduced is essential in appreciating the rest of the book, and although there is a lot to take in and it is abstract, the more examples of algebraic systems you meet the more the terminology will reveal its worth. In Section 4.2, we use this terminology to describe precisely the algebraic properties of the real numbers, the very properties which underlie high-school algebra. Sections 4.3 and 4.4, on the solution of quadratic equations and the binomial theorem respectively, are likely to be familiar, but the material is presented in a possibly unfamiliar way. With that, the main business of this chapter is concluded. *There are then two optional sections included for background but not needed in Section II.* In Section 4.5, Boolean algebras are introduced. They have properties that contrast radically with those of the real numbers. For example, $a + a = a$ always holds as does the unnerving $a + (b \cdot c) = (a + b) \cdot (a + c)$. Historically, Boolean algebras were another milestone in the development of algebra in the nineteenth century. They are also an important tool in circuit design and can be regarded as an algebraic premonition of the computer age. In Section 4.6, we return to the real numbers but this time it is their order-theoretic properties that come under the spotlight. This enables us to connect the material in this book with that in any standard introduction to analysis.

4.1 RULES OF THE GAME

A function from a set A to a set B is a rule that assigns to each element of A a unique element of B. A *binary operation* on a set X is just a special kind of a function: it is a function from $X \times X$ to X. Thus the ordered pair (x, y) is mapped to a single element z. The elements x and y are called the *arguments* of the binary operation. We write binary operations between the arguments using infix notation. The symbol $*$ is sometimes used below to denote an arbitrary binary operation, but on other occasions we use *concatenation notation* and write simply xy rather than $x * y$.

If we want to talk about a set X with respect to a specific binary operation $*$ then we write $(X, *)$.

Example 4.1.1. If we write $(\mathbb{N}, +)$ then this means that we want to think about the natural numbers with respect to the binary operation of addition, whereas if we write $(\mathbb{N}, \times)$ then we want to think about them with respect to multiplication.

Let X be equipped with the binary operation $*$, and suppose that $Y \subseteq X$. There is nothing to say that if $y_1, y_2 \in Y$ then $y_1 * y_2 \in Y$. However, if this always happens whichever elements of Y we choose, then we say that Y is *closed* under the binary operation $*$. Thus Y is a set-theoretic black hole which traps the binary operation.

Example 4.1.2. Consider $(\mathbb{N}, +)$, the natural numbers together with the binary operation of addition. The set of even numbers, $\mathbb{E}$, is closed under addition because the sum of any two even numbers is even. On the other hand, the set of odd numbers, $\mathbb{O}$, is not closed under addition because, for example, $1, 3 \in \mathbb{O}$, but their sum $4 \notin \mathbb{O}$.

The two most important properties a binary relation might have are commutativity and associativity. The binary operation is said to be *commutative* if

$$a * b = b * a$$

for all $a, b \in X$. That is, the order in which you carry out the operation is not important. Addition and multiplication of real, and as we shall see later, complex numbers are commutative. We shall also meet binary operations that are not commutative: both matrix multiplication and the vector product are examples. Commutativity is therefore not automatic. The binary operation is said to be *associative* if

$$(a * b) * c = a * (b * c)$$

for all $a, b, c \in X$. Remember that the brackets tell you how to work out the product. Thus $(a * b) * c$ means first work out $a * b$, call it d, and then work out $d * c$. Almost all the binary operations we shall meet in this book are associative, the one important exception being the vector product. In order to show that a binary operation $*$ is associative, we have to check that all possible products $(a * b) * c$ and $a * (b * c)$ are equal. To show that a binary operation is *not* associative, we simply have to find specific values for a, b and c so that $(a*b)*c \neq (a*b)*c$. Here are examples of both of these possibilities.

Example 4.1.3. Define a new binary operation $\circ$ on the set of real numbers by $a \circ b = a + b + ab$. We prove that it is associative. From the definition, we have to prove that $(a \circ b) \circ c = a \circ (b \circ c)$ for all real numbers a, b and c. To do this, we calculate first the lefthand side and then the righthand side and verify they are equal. Because we are trying to prove a result true for all real numbers, we cannot choose specific values of a, b and c. We first calculate $(a \circ b) \circ c$. We get that

$$(a \circ b) \circ c = (a + b + ab) \circ c = (a + b + ab) + c + (a + b + ab)c$$

which is equal to $a + b + c + ab + ac + bc + abc$. Now we calculate $a \circ (b \circ c)$. We get that

$$a \circ (b \circ c) = a \circ (b + c + bc) = a + (b + c + bc) + a(b + c + bc)$$

which is equal to $a + b + c + ab + ac + bc + abc$. We therefore obtain the same answers however we bracket the product and so we have proved that the binary operation $\circ$ is associative.

Example 4.1.4. Define a new binary operation $\oplus$ on the set of natural numbers by $a \oplus b = a^2 + b^2$. We show that this binary operation is not associative. Calculate first $(1 \oplus 2) \oplus 3$. By definition this is computed as follows

$$(1 \oplus 2) \oplus 3 = (1^2 + 2^2) \oplus 3 = 5 \oplus 3 = 5^2 + 3^2 = 25 + 9 = 34.$$

Now we calculate $1 \oplus (2 \oplus 3)$ as follows

$$1 \oplus (2 \oplus 3) = 1 \oplus (2^2 + 3^2) = 1 \oplus (4 + 9) = 1 \oplus 13 = 1^2 + 13^2 = 1 + 169 = 170.$$

Therefore

$$(1 \oplus 2) \oplus 3 \neq 1 \oplus (2 \oplus 3).$$

It follows that the binary operation $\oplus$ is not associative.

The next result is quite difficult to prove though what it means is easy and important. The proof can be omitted on a first reading.

Proposition 4.1.5 (Generalized associativity). *Let $*$ be any binary operation defined on a set X. If $*$ is associative then however the product*

$$x_1 * \ldots * x_n$$

is bracketed, the same answer will always be obtained.

Box 4.1: Proof of Generalized Associativity

If $x_1, x_2, \cdots, x_n$ are elements of the set X then one particular bracketing will play an important rôle in our proof

$$x_1 * (x_2 * (\cdots (x_{n-1} * x_n) \cdots))$$

which we write as $[x_1 x_2 \ldots x_n]$. The proof is by strong induction on the length n of the product in question. The base case is where $n = 3$ and is just an application of the associative law. Assume that $n \geq 4$ and that for $k < n$, all bracketings of sequences of k elements of X lead to the same answer. This is therefore the induction hypothesis for strong induction. Let X denote any properly bracketed expression obtained by inserting brackets into the sequence $x_1, x_2, \cdots, x_n$. Observe that the computation of such a bracketed product involves computing $n - 1$ products. This is because at each step we can only compute the product of adjacent letters $x_i * x_{i+1}$. Thus at each step of our calculation we reduce the number of letters by one until there is only one letter left. In whatever way the expression is bracketed, the final step in the computation will be of the form $Y * Z$, where Y and Z will each have arisen from properly bracketed expressions. In the case of Y it will involve a bracketing of some sequence $x_1, x_2, \ldots, x_r$, and for Z the sequence $x_{r+1}, x_{r+2}, \ldots x_n$ for some r such that $1 \leq r \leq n - 1$. Since Y involves a product of length $r < n$, we can use the induction hypothesis to deduce that $Y = [x_1 x_2 \ldots x_r]$. Observe that $[x_1 x_2 \ldots x_r] = x_1 * [x_2 \ldots x_r]$. Hence by associativity,

$$X = Y * Z = (x_1 * [x_2 \ldots x_r]) * Z = x_1 * ([x_2 \ldots x_r] * Z).$$

But $[x_2 \ldots x_r] * Z$ is a properly bracketed expression of length $n - 1$ in $x_2, \cdots, x_n$ and so using the induction hypothesis equals $[x_2 x_3 \ldots x_n]$. It follows that $X = [x_1 x_2 \ldots x_n]$. We have therefore shown that all possible bracketings yield the same result in the presence of associativity. The example below illustrates the idea behind the proof.

Example 4.1.6. Let $n = 5$. Then the notation $[x_1 x_2 x_3 x_4 x_5]$ introduced in the above proof means $x_1 * (x_2 * (x_3 * (x_4 * x_5)))$. Consider the product $((x_1 * x_2) * x_3) * (x_4 * x_5)$. Here we have $Y = (x_1 * x_2) * x_3$ and $Z = x_4 * x_5$. By associativity $Y = x_1 * (x_2 * x_3)$. Thus $Y * Z = (x_1 * (x_2 * x_3)) * (x_4 * x_5)$. But this is equal to $x_1 * ((x_2 * x_3) * (x_4 * x_5))$ again by associativity. By the induction hypothesis $(x_2 * x_3) * (x_4 * x_5) = x_2 * (x_3 * (x_4 * x_5))$, and so

$$((x_1 * x_2) * x_3) * (x_4 * x_5) = x_1 * (x_2 * (x_3 * (x_4 * x_5))),$$

as required.

If a binary operation is associative then Proposition 4.1.5 tells us that computing products of elements is straightforward because we never have to worry about how to evaluate them as long as we maintain the order of the elements. The following special case of this result is important. Assume we are working with an associative binary operation denoted by concatenation. Define the nth power a^n of a, where n is a positive natural number, as follows: $a^1 = a$ and $a^n = a^{n-1}a$ for any $n \geq 2$. Generalized associativity tells us that a^n can, in fact, be calculated by bracketing in any way we like.

Lemma 4.1.7 (Laws of exponents). *Assume that the binary operation used is associative. Let $m, n \geq 1$ be natural numbers.*

1. $a^m a^n = a^{m+n}$.

2. $(a^m)^n = a^{mn}$.

Proof. (1) By definition $a^{m+1} = a^m a$ for all m. Suppose that $a^{m+n} = a^m a^n$ for all m and some fixed n, the induction hypothesis. We calculate $a^{m+(n+1)}$. This is equal to $a^{(m+1)+n} = a^{m+1}a^n$ by the induction hypothesis. But $a^{m+1} = a^m a$ by definition, and $aa^n = a^n a$ which is equal to a^{n+1} by definition. We have therefore proved that $a^{m+(n+1)} = a^m a^{n+1}$.

(2) By definition $(a^m)^1 = a^m$ for all m. Suppose that $(a^m)^n = a^{mn}$ for all m and some fixed n, the induction hypothesis. We calculate $(a^m)^{n+1}$. By definition $(a^m)^{n+1} = (a^m)^n a^m$. By the induction hypothesis this is equal to $a^{mn}a^m$, and by part (1) this is also equal to $a^{mn+m} = a^{m(n+1)}$, as required. □

By Lemma 4.1.7, when the binary operation is associative powers of the same element a always commute with one another, that is $a^m a^n = a^n a^m$, since both products equal a^{m+n}.

Let $\{a_1, \ldots, a_n\}$ be a set of n elements. Recall that if we write them all in some order $a_{i_1}, \ldots, a_{i_n}$ without repetitions then this is a permutation of the elements. What the following result says is easy to understand though the proof can be omitted on a first reading.

Proposition 4.1.8 (Generalized commutativity). *Let $*$ be an associative and commutative binary operation on a set X. Let $a_1, \ldots, a_n$ be a list of any n elements of X and let $a_{i_1}, \ldots, a_{i_n}$ be some permutation of these elements. Then*

$$a_{i_1} * \ldots * a_{i_n} = a_1 * \ldots * a_n.$$

Box 4.2: Proof of Generalized Commutativity

The operation is associative and so by generalized associativity we do not need to use brackets. The proof is in two acts. First prove by induction that

$$a_1 * \ldots * a_n * b = b * a_1 * \ldots * a_n.$$

The induction step takes the following form. We have to calculate

$$a_1 * \ldots * a_n * a_{n+1} * b.$$

Bracket as follows

$$a_1 * (a_2 * \ldots * a_n * a_{n+1} * b).$$

By the induction hypothesis we get

$$a_1 * (b * a_2 * \ldots * a_{n+1}).$$

Rebracket and then apply commutativity

$$a_1 * (b * a_2 * \ldots * a_{n+1}) = (a_1 * b) * a_2 * \ldots * a_{n+1} = b * a_1 * a_2 * \ldots * a_{n+1}.$$

We now turn to the second act. The induction step takes the following form. Let $a_1, \ldots, a_n, a_{n+1}$ be $n+1$ elements. Consider the product $a_{i_1} * \ldots * a_{i_n} * a_{i_{n+1}}$. Suppose that $a_{n+1} = a_{i_r}$. Then

$$a_{i_1} * \ldots * \mathbf{a_{i_r}} * \ldots * a_{i_n} * a_{i_{n+1}} = (a_{i_1} * \ldots * a_{i_{n+1}}) * \mathbf{a_{n+1}}$$

where the expression in the brackets is a product of some permutation of the elements $a_1, \ldots, a_n$ and we have used the first act result. By the induction hypothesis, we can write $a_{i_1} * \ldots * a_{i_{n+1}} = a_1 * \ldots * a_{n+1}$ as required.

We now turn to the properties that elements can have in a set X equipped with a binary operation $*$.

- An element $e \in X$ is called an *identity* for the binary operation $*$ if

$$x * e = x = e * x$$

for all $x \in X$.

- An element $z \in X$ is called a *zero* for the binary operation $*$ if

$$x * z = z = z * x$$

for all $x \in X$.

- An element $f \in X$ is called an *idempotent* if $f * f = f$.

Lemma 4.1.9. *Let X be a set equipped with a binary operation written as concatenation.*

1. *If e and f are identities then $e = f$.*

2. *If y and z are zeros then $y = z$.*

3. *If e is an identity then $e^2 = e$.*

4. *If z is a zero then $z^2 = z$.*

Proof. (1) The element e is an identity and so $ef = f$. The element f is also an identity and so $fe = e$. It follows that $e = f$.

(2) Since y is a zero $yz = y$ but z is also a zero and so $yz = z$. Thus $y = z$.

(3) Since e is an identity $ee = e$, as required.

(4) Since z is a zero $zz = z$, as required. □

Of all the properties that elements might have the most important, at least in classical mathematics, is the following. Let X be a set equipped with an associative binary operation, written as concatenation, and with an identity e. An element $x \in X$ is said to be *invertible* if there is an element y such that

$$xy = e = yx.$$

Such an element y is called an *inverse* of x.

Lemma 4.1.10. *Let X be a set equipped with an associative binary operation, written as concatenation, and an identity e. Let y and z be inverses of x. Then $y = z$.*

Proof. By definition $xy = yx = e$ and $xz = zx = e$. Multiply $xy = e$ on the left by z to obtain $z(xy) = ze$. Now e is an identity and so $ze = z$. By associativity, $z(xy) = (zx)y$. By definition $zx = e$. But $ey = y$. Thus $y = z$ and we have therefore proved that if x has an inverse it is unique. □

Lemma 4.1.10 tells us that if an element x has an inverse then that inverse is unique. We often denote the inverse by x^{-1} in the general case and $-x$ in the commutative case. This is primarily for psychological reasons so that equations 'look right'. In the non-commutative case, a multiplicative-type notation tends to be used whereas in the commutative case an additive-type notation tends to be used. The next lemma develops the properties of inverses further.

Lemma 4.1.11. *Let X be a set equipped with an associative binary operation, written as concatenation, and an identity e.*

1. *If x is invertible then x^{-1} is invertible and has inverse x.*

2. *If x and y are invertible then xy is invertible and its inverse is $y^{-1}x^{-1}$.*

3. *If $x_1, \dots, x_n$ are each invertible then $x_1 \dots x_n$ is invertible and its inverse is $x_n^{-1} \dots x_1^{-1}$.*

Proof. (1) By definition $x^{-1}x = e = xx^{-1}$. This also says that x is an inverse of x^{-1}. We have proved that inverses are unique and so $(x^{-1})^{-1} = x$.

(2) This is by direct computation

$$xyy^{-1}x^{-1} = xex^{-1} = xx^{-1} = e \text{ and } y^{-1}x^{-1}xy = y^{-1}ey = y^{-1}y = e.$$

It follows that $y^{-1}x^{-1}$ is an inverse of xy. Inverses are unique and so $(xy)^{-1} = y^{-1}x^{-1}$.

(3) This is a typical application of proof by induction. Here is the induction step. We are given that $x_1, \dots, x_{n+1}$ are invertible. By the induction hypothesis $x_1 \dots x_n$ is invertible with inverse $x_n^{-1} \dots x_1^{-1}$. Thus the product $(x_1 \dots x_n)x_{n+1}$ is invertible by part (2) with inverse $x_{n+1}^{-1}(x_1 \dots x_n)^{-1} = x_{n+1}^{-1}x_n^{-1} \dots x_1^{-1}$, as required. $\square$

Example 4.1.12. Lemma 4.1.11 is significant since it holds for every associative operation with an identity. For example, the set of all functions $T(X)$ of a set X to itself is equipped with the binary operation of composition of functions. This is associative and there is an identity. It is now immediate that if f and g are invertible functions in $T(X)$ then their product fg is invertible and has inverse $g^{-1}f^{-1}$. Of course, we proved this result in Section 3.4 but we can now see that it has nothing to do with functions per se. The result holds by virtue of the fact that the binary operation is associative and the definition of an invertible element. In Chapter 8, we shall see that Lemma 4.1.11 can be applied to invertible matrices. The point is that we should prove general results once rather than prove what amounts to the same result many times over in special cases. We gain generality at the price of abstraction. These are the first steps in *abstract algebra*.

Examples 4.1.13. The properties defined in this section enable the differences between the sets of numbers introduced in Section 3.1 to be described. In passing from $(\mathbb{N}, +)$ to $(\mathbb{Z}, +)$ every element acquires an additive inverse, and in passing from $(\mathbb{Z}, \cdot)$ to $(\mathbb{Q}, \cdot)$ every non-zero element acquires a multiplicative inverse. The algebraic properties introduced so far do not allow us to distinguish between $\mathbb{Q}$ and $\mathbb{R}$. This will turn out to depend on the properties of the usual order relation. See Section 4.6.

We have so far discussed a set equipped with a single binary operation, but more often than not there are two. For example, we would usually regard the set $\mathbb{N}$ with respect to both addition and multiplication. When there are two binary operations, the following axioms become relevant. Let X be a set equipped with two binary operations $*$ and $\circ$. We say that $*$ *distributes over* $\circ$ *on the left* if

$$x * (y \circ z) = (x * y) \circ (x * z)$$

for all $x, y, z \in X$. Similarly, we say that $*$ *distributes over* $\circ$ *on the right* if

$$(y \circ z) * x = (y * x) \circ (z * x)$$

for all $x, y, z \in X$. If both these axioms hold we say simply that $*$ *distributes over* $\circ$.

At this point, we make a definition that is a conceptual leap beyond what has gone before. The rationale for this definition will become clearer as you read the rest of the book. Ultimately it actually unifies what we do. An *algebraic structure* is a set equipped with a selection of unary and binary operations[1] together with some specified elements called constants. Different algebraic structures are distinguished by the different properties their operations satisfy. The study of algebraic structures is called *algebra*.

Box 4.3: Monoids and Groups

A set equipped with an associative binary operation is called a *semigroup*. A semigroup defined on a finite set is called *finite*; otherwise it is called *infinite*. Finite semigroup theory has close connections with theoretical computer science [83]. A semigroup with an identity is called a *monoid*. Monoids are the basic building blocks of much, though not all, of algebra. The set of natural numbers with respect to addition is a monoid, and is the first example of an algebraic structure we learn about as children. There are other monoids that we use as children without being aware that what we are doing is in any way mathematical. When we learn to write, we learn to concatenate symbols in a particular order. Thus we learn to form strings over the alphabet of letters. We also learn to form sentences over the alphabet of all words in our native language, where each word is treated as a single symbol. Mathematically, both of these are examples of free monoids. More generally, a *free monoid* consists of a finite alphabet A of symbols from which we form the set of all finite strings A^* over that alphabet. The set A^* is equipped with a binary operation called concatenation that takes two strings u and v and outputs the string uv in that order. This operation is associative and there is an identity: the empty string ε. If the alphabet A consists of one symbol, say the tally, $|$, then A^* is just $\mathbb{N}$ in disguise. Thus the natural numbers are also examples of free monoids. Though they might look too simple to be interesting, free monoids are the source of complex mathematics. They are useful in the theory of variable-length codes. *Groups* are special kinds of monoids: namely those in which every element is invertible. These algebraic structures have been of great classical importance. They arose from the work of Galois described in Section 7.9.

Example 4.1.14. The natural numbers $\mathbb{N}$ are most naturally regarded as algebraic structures with respect to addition and multiplication and we would probably also want to single out the special rôles played by zero and one. We could package all of this in the 5-tuple $(\mathbb{N}, +, \times, 0, 1)$ which makes clear the set of elements we are interested in, the two binary operations and the two constants.

---

[1] In fact, operations of any arity are allowed but are unusual. In my work, I only once had to deal with a ternary operation.

Box 4.4: Rings

Semigroups, monoids and groups have one binary operation. If there are two binary operations we need additional terminology. Classically, there are two operations usually denoted by + addition, and · multiplication connected by some versions of the distributive law. If the addition operation gives a commutative group with additive identity 0, and the multiplication operation gives a monoid with multiplicative identity 1, then we have a structure called a *ring*. If the multiplication is commutative it is called a *commutative ring*. If each non-zero element has a multiplicative inverse it is called a *division ring*. A commutative division ring is called a *field*. Classically, the study of monoids, groups, rings and fields is called *abstract algebra*. In this book, we will keep things concrete but it will be useful to have the notion of a field defined from scratch in Section 4.2. Do not read too much into words such as 'group', 'ring' and 'field'. They are just convenient labels for bundles of frequently co-occurring properties.

Exercises 4.1

1. Let A be a non-empty set. Define a binary operation on A^2 by

$$(a,b)(c,d) = (a,d).$$

   Prove that this operation is associative.

2. Let A be any non-empty set. Define a binary operation, denoted by concatenation, by $xy = x$ for all $x, y \in A$. Prove that this operation is associative and that every element is idempotent. Under what circumstances can there be an identity element?

3. Let S be a set equipped with an associative and commutative binary operation, denoted by concatenation, in which every element is idempotent. Define the relation $\leq$ on S by $e \leq f$ if $e = ef$. Prove that this relation is a partial order.

4. Let S be a set equipped with an associative binary operation, denoted by concatenation, and an identity 1. Suppose that every element in S is invertible and that $x^2 = 1$ for each $x \in S$. Prove that the binary operation is commutative.

5. *Let S be a set equipped with an associative binary operation denoted by concatenation. Prove that the following are equivalent.

   (a) If $ab = ba$ then $a = b$. Thus the operation is as non-commutative as it can be.

   (b) $aba = a$ for all $a, b \in S$.

   (c) $a^2 = a$ and $abc = ac$ for all $a, b, c \in S$.

6. *The set S is equipped with an associative binary operation that we shall denote by concatenation in which every element is idempotent.

   (a) Suppose that S has at least two distinct elements a and b such that every element can be written as products of as or bs. What is the maximum number of elements S can contain?

(b) Suppose that S has at least three distinct elements a, b, c such that every element can be written as products of as, bs or cs. What is the maximum number of elements S can contain?

Box 4.5: Congruences

Section 3.5 on equivalence relations is needed here. Let S be a set equipped with a binary operation $*$, and let $\equiv$ be an equivalence relation on S. Denote the $\equiv$-class containing the element $a \in S$ by $[a]$. The obvious question is whether we can define a binary operation on the set $S/\equiv$ of equivalence classes. You might think that the answer is obvious. Why do we not simply define $[a] \circ [b] = [a*b]$? The problem is that we have to check that if $[a'] = [a]$ and $[b'] = [b]$ then $[a'*b'] = [a*b]$, and there is no reason why this should be true. As often happens in mathematics, we define our way out of a tight corner. We say that the equivalence relation $\equiv$ is a *congruence* for the binary operation $*$ if $a \equiv a'$ and $b \equiv b'$ implies that $a*b \equiv a'*b'$. If this condition holds then $S/\equiv$ really does have a binary operation $\circ$ when we define $[a] \circ [b] = [a*b]$. Similarly, if the set S is equipped with a unary operation $a \mapsto \bar{a}$ we say that the equivalence relation is a congruence for this unary operation if $a \equiv b$ implies that $\bar{a} \equiv \bar{b}$. We may then define a unary operation on $S/\equiv$ by $[\hat{a}] = [\bar{a}]$. The origins of this idea of a congruence lie in number theory as described in Section 5.4, but congruences are a basic tool in constructing algebraic structures. You will find a further example in Section 7.11.

4.2 ALGEBRAIC AXIOMS FOR REAL NUMBERS

The goal of this section is to describe the algebraic properties of the real numbers using the terminology introduced in Section 4.1. We will only describe the properties of addition and multiplication. Subtraction and division are not viewed as binary operations in their own right. Instead, we *define*

$$a - b = a + (-b).$$

Thus to subtract b means the same thing as adding $-b$. Likewise, we *define*

$$a \backslash b = a \div b = a \times b^{-1} \text{ if } b \neq 0.$$

Thus to divide by b is to multiply by b^{-1}.

Axioms for addition in $\mathbb{R}$

(F1) *Addition is associative.* Let x, y and z be any numbers. Then $(x+y)+z = x+(y+z)$.

(F2) *There is an additive identity.* The number 0 (zero) is the additive identity. This means that $x+0 = x = 0+x$ for any number x.

(F3) *Each number has a unique additive inverse.* For each number x there is a number, denoted $-x$, with the property that $x+(-x) = 0 = (-x)+x$. The number $-x$ is called the *additive inverse* of the number x.

(F4) *Addition is commutative.* Let x and y be any numbers. Then $x+y = y+x$.

Axioms for multiplication in $\mathbb{R}$

(F5) *Multiplication is associative.* Let x, y and z be any numbers. Then $(xy)z = x(yz)$.

(F6) *There is a multiplicative identity.* The number 1 (one) is the multiplicative identity. This means that $1x = x = x1$ for any number x.

(F7) *Each non-zero number has a unique multiplicative inverse.* For each non-zero number x there is a unique number, denoted x^{-1}, with the property that $x^{-1}x = 1 = xx^{-1}$. The number x^{-1} is called the *multiplicative inverse* of x. It is, of course, the number $\frac{1}{x}$, the *reciprocal* of x.

(F8) *Multiplication is commutative.* Let x and y be any numbers. Then $xy = yx$.

Linking axioms for $\mathbb{R}$

(F9) $0 \neq 1$.

(F10) *The additive identity is a multiplicative zero.* This means that $0x = 0 = x0$ for any x.

(F11) *Multiplication distributes over addition on the left and the right.* There are two distributive laws: the *left distributive law* $x(y + z) = xy + xz$ and the *right distributive law* $(y + z)x = yx + zx$.

We have missed out one further ingredient in algebra, and that is the properties of equality. The following are not exhaustive but enough for our purposes.

Axioms for equality

(E1) $a = a$.

(E2) If $a = b$ then $b = a$.

(E3) If $a = b$ and $b = c$ then $a = c$.

(E4) If $a = b$ and $c = d$ then $a + c = b + d$.

(E5) If $a = b$ and $c = d$ then $ac = bd$.

Properties (E1), (E2) and (E3) say that equality is an equivalence relation. See Section 3.5.

Example 4.2.1. When algebra was discussed in the Prolegomena it was mentioned that the usual way of solving a linear equation in one unknown depended on the properties of real numbers. We now show explicitly how to solve $ax + b = 0$ where

$a \neq 0$ using the axioms for the reals. Throughout, we use without comment properties of equality listed above.

$$
\begin{aligned}
ax + b &= 0 \\
(ax + b) + (-b) &= 0 + (-b) \quad \text{by (F3)} \\
ax + (b + (-b)) &= 0 + (-b) \quad \text{by (F1)} \\
ax + 0 &= 0 + (-b) \quad \text{by (F3)} \\
ax &= 0 + (-b) \quad \text{by (F2)} \\
ax &= -b \quad \text{by (F2)} \\
a^{-1}(ax) &= a^{-1}(-b) \quad \text{by (F7) since } a \neq 0 \\
(a^{-1}a)x &= a^{-1}(-b) \quad \text{by (F5)} \\
1x &= a^{-1}(-b) \quad \text{by (F7)} \\
x &= a^{-1}(-b) \quad \text{by (F6).}
\end{aligned}
$$

It is not usually necessary to go into such gory detail when solving equations, but we wanted to show you what actually lay behind the rules that you might have been taught at high school.

Example 4.2.2. We can use our axioms to prove that $-1 \times -1 = 1$, something that is hard to understand in any other way. By definition, -1 is the additive inverse of 1. This means that $1 + (-1) = 0$. We calculate $(-1)(-1) - 1$.

$$
\begin{aligned}
(-1)(-1) - 1 &= (-1)(-1) + (-1) \text{ by definition of subtraction} \\
&= (-1)(-1) + (-1)1 \text{ since 1 is the multiplicative identity} \\
&= (-1)[(-1) + 1] \text{ by the left distributivity law} \\
&= (-1)0 \text{ by properties of additive inverses} \\
&= 0 \text{ by properties of zero.}
\end{aligned}
$$

Hence $(-1)(-1) = 1$. In other words, the result follows from the usual rules of algebra.

The final example explains the reason for the prohibition on dividing by zero.

Example 4.2.3. The following fallacious proof shows that $1 = 2$.

1. Let $a = b$.

2. Then $a^2 = ab$ when we multiply both sides of (1) by a.

3. Now add a^2 to both sides of (2) to get $2a^2 = a^2 + ab$.

4. Subtract $2ab$ from both sides of (3) to get $2a^2 - 2ab = a^2 + ab - 2ab$.

5. Thus $2(a^2 - ab) = a^2 - ab$.

6. We deduce that $2 = 1$ by cancelling the common factor in (5).

The source of the problem is in passing from line (5) to line (6) where we are in fact dividing by zero.

The multiplication of real numbers is associative. Thus the laws of exponents hold as in Lemma 4.1.7. This is an opportune place to review the meaning of algebraic expressions such as $a^{\frac{r}{s}}$ where $\frac{r}{s}$ is any rational number. We shall be guided by the requirement that the laws of exponents for positive natural numbers should continue to hold for rational numbers.

- We define a^0 when $a \neq 0$. Whatever it means we should have that $a^0 a^1 = a^{0+1} = a^1$. That is, $a^0 a = a$ for all non-zero a. This implies that $a^0 = 1$. An oft raised question at this juncture is dealt with in a footnote.[2]

- We define a^{-n} when n is positive. Whatever it means we should have that $a^{-n} a^n = a^0 = 1$. It follows that $a^{-n} = \frac{1}{a^n}$.

- We define $a^{\frac{1}{n}}$ for any non-zero integer n. Whatever it means we should have that $(a^{\frac{1}{n}})^n = a^1 = a$. It follows that $a^{\frac{1}{n}} = \sqrt[n]{a}$.

On the basis of the above analysis, define

$$a^{\frac{r}{s}} = (\sqrt[s]{a})^r .$$

These lucubrations do not tell us, of course, what the number $10^{\sqrt{2}}$ might mean because $\sqrt{2}$ is irrational. It is reasonable to guess that because $\sqrt{2} = 1 \cdot 41421...$, the sequence of numbers $10^{1\cdot4}, 10^{1\cdot41}, 10^{1\cdot414}, ...$ are successive approximations to the value of $10^{\sqrt{2}}$. But there is a more direct route. Let b be any real number and a a positive real number. Whatever we might mean by $r = a^b$, we assume that $\ln r = b \ln a$. This leads us to the following *definition*.

$$a^b = \exp(b \ln a).$$

The laws of exponents continue to hold in this general case.

There is some small change we should also mention here. How do we calculate $(ab)^n$? This is just ab times itself n times. But the order in which we multiply as and bs does not matter and so we can arrange all the as at the front. Thus $(ab)^n = a^n b^n$. More generally, $(a_1 ... a_s)^m = a_1^m ... a_s^m$. We also have similar results for addition. Define $2x = x + x$ and $nx = x + ... + x$ where the x occurs n times. We have $1x = x$ and $0x = 0$.

At this point, it is appropriate to introduce some useful notation. Let $a_1, a_2, ..., a_n$ be n numbers. Their sum is $a_1 + a_2 + ... + a_n$ and because of generalized associativity we do not have to worry about brackets. We now abbreviate this sum as

$$\sum_{i=1}^{n} a_i$$

---

[2]What about 0^0? In this book, it is probably best to define $0^0 = 1$.

where $\sum$ is Greek 'S' and stands for Sum. The letter i is called a *subscript*. The equality $i = 1$ tells us that we start the value of i at 1. The equality $i = n$ tells us that we end the value of i at n. Although we have started the sum at 1, we could, in other circumstances, have started at 0, or any other appropriate number. This notation is useful and can be manipulated using the rules above. If $1 < s < n$, then we can write

$$\sum_{i=1}^{n} a_i = \sum_{i=1}^{s} a_i + \sum_{s+1}^{n} a_i.$$

If b is any number then

$$b\left(\sum_{i=1}^{n} a_i\right) = \sum_{i=1}^{n} ba_i$$

is the *generalized left distributivity law* that can be proved using induction. These uses of sigma-notation should not cause any problems. The most complicated use of $\sum$-notation arises when we have to sum up what is called an *array* of numbers a_{ij} where $1 \leq i \leq m$ and $1 \leq j \leq n$. These arise in matrix theory, for example. For concreteness, we describe the case where $m = 3$ and $n = 4$. We can therefore think of the numbers a_{ij} as being arranged in a 3×4 array as follows.

$$\begin{matrix} a_{11} & a_{12} & a_{13} & a_{14} \\ a_{21} & a_{22} & a_{23} & a_{24} \\ a_{31} & a_{32} & a_{33} & a_{34} \end{matrix}$$

Observe that the first subscript tells you the row and the second subscript tells you the column. Thus a_{23} is the number in the second row and the third column. We now add these numbers up in two different ways. The first way is to add the numbers up along the rows. So we calculate the following sums

$$\sum_{j=1}^{4} a_{1j}, \quad \sum_{j=1}^{4} a_{2j}, \quad \sum_{j=1}^{4} a_{3j}.$$

We then add up these three numbers

$$\sum_{j=1}^{4} a_{1j} + \sum_{j=1}^{4} a_{2j} + \sum_{j=1}^{4} a_{3j} = \sum_{i=1}^{3} \left(\sum_{j=1}^{4} a_{ij}\right).$$

The second way is to add the numbers up along the columns. So, we calculate the following sums

$$\sum_{i=1}^{3} a_{i1}, \quad \sum_{i=1}^{3} a_{i2}, \quad \sum_{i=1}^{3} a_{i3}, \quad \sum_{i=1}^{3} a_{i4}.$$

We then add up these four numbers

$$\sum_{i=1}^{3} a_{i1} + \sum_{i=1}^{3} a_{i2} + \sum_{i=1}^{3} a_{i3} + \sum_{i=1}^{3} a_{i4} = \sum_{j=1}^{4} \left(\sum_{i=1}^{3} a_{ij}\right).$$

The fact that

$$\sum_{i=1}^{3} \left(\sum_{j=1}^{4} a_{ij} \right) = \sum_{j=1}^{4} \left(\sum_{i=1}^{3} a_{ij} \right)$$

is a consequence of the generalized commutativity law. More generally, we have

$$\sum_{i=1}^{m} \left(\sum_{j=1}^{n} a_{ij} \right) = \sum_{j=1}^{n} \left(\sum_{i=1}^{m} a_{ij} \right)$$

or *interchange of summations*.

There is also notation for products. The product $a_1 \ldots a_n$ can be written

$$\prod_{i=1}^{n} a_i,$$

where Π is Greek 'P' and stands for **Product**.

To finish off this section, we make one further important definition. It is a specific example of a class of algebraic structures. Throughout this book, we shall have many occasions to work with sets equipped with two binary operations that satisfy all the axioms (F1)–(F11). It will therefore be convenient to have a term to describe them. That term is *field*. In a field, the usual operations of addition, subtraction, multiplication and division (except by zero) can be carried out and they have the same properties as in high school algebra. In particular, if $ab = 0$ in a field, then at least one of a or b is zero.

Examples 4.2.4.

1. $\mathbb{Q}$ and $\mathbb{R}$ are both fields.

2. The integers modulo a prime $\mathbb{Z}_p$, defined in Section 5.4, form a field.

3. The complex numbers $\mathbb{C}$, introduced in Chapter 6, form a field.

4. The real rational functions $\mathbb{R}(x)$, discussed in Section 7.8, form a field.

Exercises 4.2

1. Prove the following using the axioms for $\mathbb{R}$.

   (a) $(a+b)^2 = a^2 + 2ab + b^2$.

   (b) $(a+b)^3 = a^3 + 3a^2b + 3ab^2 + b^3$.

   (c) $a^2 - b^2 = (a+b)(a-b)$.

   (d) $(a^2 + b^2)(c^2 + d^2) = (ac - bd)^2 + (ad + bc)^2$.

2. Calculate the following.

   (a) 2^3.

    (b) $2^{\frac{1}{3}}$.

    (c) 2^{-4}.

    (d) $2^{-\frac{3}{2}}$.

3. Assume that a_{ij} are assigned the following values

$$
\begin{array}{llll}
a_{11} = 1 & a_{12} = 2 & a_{13} = 3 & a_{14} = 4 \\
a_{21} = 5 & a_{22} = 6 & a_{23} = 7 & a_{24} = 8 \\
a_{31} = 9 & a_{32} = 10 & a_{33} = 11 & a_{34} = 12.
\end{array}
$$

Calculate the following sums.

    (a) $\sum_{i=1}^{3} a_{i2}$.

    (b) $\sum_{j=1}^{4} a_{3j}$.

    (c) $\sum_{i=1}^{3} \left(\sum_{j=1}^{4} a_{ij}^{2} \right)$.

4. Prove the generalized left distributivity law

$$
b \left(\sum_{i=1}^{n} a_i \right) = \sum_{i=1}^{n} b a_i
$$

for real numbers.

5. Let $a, b, c \in \mathbb{R}$. If $ab = ac$ is it true that $b = c$? Explain.

6. *Let $F = \{a + b\sqrt{2} : a, b \in \mathbb{Q}\}$.

    (a) Prove that the set F is closed with respect to the usual operations of addition and multiplication of real numbers.

    (b) Prove that F is a field.

4.3 SOLVING QUADRATIC EQUATIONS

Solving linear equations is routine, as we saw in Example 4.2.1, but solving more complicated equations requires good ideas. The first place where a good idea is needed is in solving quadratic equations. In this section, we explain the idea that leads to the familiar formula for solving a quadratic. We recall some definitions. An expression of the form $ax^2 + bx + c$ where a, b, c are numbers and $a \neq 0$ is called a *quadratic polynomial* or a *polynomial of degree 2*. The numbers a, b, c are called the *coefficients* of the quadratic. A quadratic where $a = 1$ is said to be *monic*. A number r such that $ar^2 + br + c = 0$ is called a *root* of the polynomial. The problem of finding all the roots of a quadratic is called *solving the quadratic*. Usually this problem is stated in the form: 'solve the quadratic equation $ax^2 + bx + c = 0$' (equation because we have set the polynomial equal to zero). We now solve a quadratic equation without having to remember a formula. Observe that if $ax^2 + bx + c = 0$ then $x^2 + \frac{b}{a}x + \frac{c}{a} = 0$.

Thus it is enough to find the roots of monic quadratics. We solve this equation by using the following idea: write $x^2 + \frac{b}{a}x$ as a perfect square plus a number. To see how to do this, represent the expression $x^2 + \frac{b}{a}x$ graphically by means of areas.

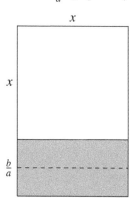

Now cut the shaded rectangle into two equal pieces along the dotted line and rearrange them as shown below.

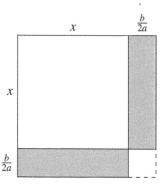

It is now geometrically obvious that if we add the small dotted square, we get a new bigger square. This explains why the procedure is called *completing the square*. We now express algebraically what we have done geometrically

$$x^2 + \frac{b}{a}x = \left(x^2 + \frac{b}{a}x + \frac{\mathbf{b^2}}{\mathbf{4a^2}}\right) - \frac{\mathbf{b^2}}{\mathbf{4a^2}} = \left(x + \frac{b}{2a}\right)^2 - \frac{b^2}{4a^2}.$$

We therefore have that

$$x^2 + \frac{b}{a}x = \left(x + \frac{b}{2a}\right)^2 - \frac{b^2}{4a^2}.$$

Look carefully at what we have done here: we have rewritten the lefthand side as a perfect square plus a number. It follows that

$$x^2 + \frac{b}{a}x + \frac{c}{a} = \left(x + \frac{b}{2a}\right)^2 - \frac{b^2}{4a^2} + \frac{c}{a} = \left(x + \frac{b}{2a}\right)^2 + \frac{4ac - b^2}{4a^2}.$$

Setting the last expression equal to zero and rearranging, we get

$$\left(x + \frac{b}{2a}\right)^2 = \frac{b^2 - 4ac}{4a^2}.$$

Up to this point, we have only used the field axioms (and the fact that 2 is invertible in the field). The next step can only be accomplished if the righthand side of the above equation is non-negative. Take square roots of both sides, remembering that a non-zero non-negative real number has two square roots, to obtain

$$x + \frac{b}{2a} = \pm\sqrt{\frac{b^2 - 4ac}{4a^2}}$$

which simplifies to

$$x + \frac{b}{2a} = \pm\frac{\sqrt{b^2 - 4ac}}{2a}.$$

Thus

$$\boxed{x = \frac{-b \pm \sqrt{b^2 - 4ac}}{2a}}$$

the usual formula for finding the roots of a quadratic. This formula is written down in all cases with the convention that if $b^2 - 4ac < 0$, we say that the quadratic has no real roots.

Example 4.3.1. Solve the quadratic equation

$$2x^2 - 5x + 1 = 0$$

by completing the square. Divide through by 2 to make the quadratic monic giving

$$x^2 - \frac{5}{2}x + \frac{1}{2} = 0.$$

We now want to write

$$x^2 - \frac{5}{2}x$$

as a perfect square plus a number. We get

$$x^2 - \frac{5}{2}x = \left(x - \frac{5}{4}\right)^2 - \frac{25}{16}.$$

Thus our quadratic becomes

$$\left(x - \frac{5}{4}\right)^2 - \frac{25}{16} + \frac{1}{2} = 0.$$

Rearranging and taking roots gives us

$$x = \frac{5 \pm \sqrt{17}}{4}.$$

For the quadratic equation

$$ax^2 + bx + c = 0$$

the number $D = b^2 - 4ac$, called the *discriminant* of the quadratic, plays an important rôle in determining the nature of the roots.

- If $D > 0$ then the quadratic equation has two distinct real solutions.

- If $D = 0$ then the quadratic equation has one repeated real root. In this case, the quadratic is the perfect square $\left(x + \frac{b}{2a}\right)^2$.

- If $D < 0$ then we shall see in Chapters 6 and 7 that the quadratic equation has two complex roots which are complex conjugate to each other. This is called the *irreducible* case.

Put $y = ax^2 + bx + c$ and draw the graph of this equation. The roots of the original quadratic therefore correspond to the points where this graph crosses the x-axis. The diagrams below are illustrative of the three cases that can arise depending on the sign of the discriminant.

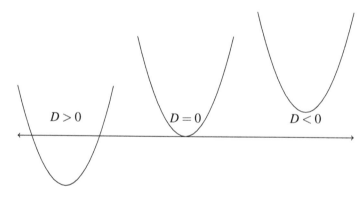

Exercises 4.3

1. Calculate the discriminants of the following quadratics and so determine whether they have distinct roots, repeated roots or no real roots.

   (a) $x^2 + 6x + 5$.
   (b) $x^2 - 4x + 4$.
   (c) $x^2 - 2x + 5$.

2. Solve the following quadratic equations by completing the square.

   (a) $x^2 + 10x + 16 = 0$.
   (b) $x^2 + 4x + 2 = 0$.

(c) $2x^2 - x - 7 = 0$.

3. I am thinking of two numbers x and y. I tell you their sum a and their product b. What are x and y in terms of a and b?

4. Let $p(x) = x^2 + bx + c$ be a monic quadratic with roots x_1 and x_2. Express the discriminant of $p(x)$ in terms of x_1 and x_2.

5. *This question is an interpretation of part of Book X of Euclid. We shall be interested in numbers of the form $a + \sqrt{b}$ where a and b are rational and $b > 0$ where $\sqrt{b}$ is irrational.[3]

   (a) If $\sqrt{a} = b + \sqrt{c}$ where $\sqrt{c}$ is irrational prove that $b = 0$.

   (b) If $a + \sqrt{b} = c + \sqrt{d}$ where a and c are rational and $\sqrt{b}$ and $\sqrt{d}$ are irrational prove that $a = c$ and $\sqrt{b} = \sqrt{d}$.

   (c) Prove that the square roots of $a + \sqrt{b}$ have the form $\pm(\sqrt{x} + \sqrt{y})$.

4.4 BINOMIAL THEOREM

In this section, we prove an important result in algebra using a counting argument from Section 3.9. The goal is to write $(x+y)^n$, where n is any natural number, in terms of powers of x and y. This is called an *expansion*. The expression $x + y$ is called a *binomial* since it consists of two terms. We look at how this expression is expanded for $n = 0, 1, 2, 3, 4$.

$$(x+y)^0 = 1$$
$$(x+y)^1 = 1x + 1y$$
$$(x+y)^2 = 1x^2 + 2xy + 1y^2$$
$$(x+y)^3 = 1x^3 + 3x^2y + 3xy^2 + 1y^3$$
$$(x+y)^4 = 1x^4 + 4x^3y + 6x^2y^2 + 4xy^3 + 1y^4.$$

We have highlighted the numbers that arise: they are called *coefficients* and form what is known as *Pascal's triangle*. Observe that each row of coefficients can be obtained from the preceding one as follows: apart from the 1s at either end, each entry in row $i + 1$ is the sum of two entries in row i, specifically the two numbers above to the left and right. We shall explain why this works later. Look at the last row. The numbers

$$1, \quad 4, \quad 6, \quad 4, \quad 1$$

are precisely the numbers

$$\binom{4}{0}, \quad \binom{4}{1}, \quad \binom{4}{2}, \quad \binom{4}{3}, \quad \binom{4}{4}.$$

---

[3] Remember that *irrational* means *not rational*.

We can therefore write

$$(x+y)^4 = \sum_{i=0}^{4} \binom{4}{i} x^{4-i} y^i.$$

The following theorem says that this result is true for any n not just for $n = 4$.

Theorem 4.4.1 (The Binomial Theorem). *The equality*

$$(x+y)^n = \sum_{i=0}^{n} \binom{n}{i} x^{n-i} y^i$$

holds for any natural number n.

Proof. This is often proved by induction, but we want to give a more conceptual proof. We look at a special case to explain the idea. Calculate

$$(x+y)(x+y)(x+y)$$

in slow motion. Multiplying out the brackets using the distributive law a number of times, but before we carry out any simplifications using the commutative law, we get

$$(x+y)(x+y)(x+y) = xxx + xxy + xyx + xyy + yxx + yxy + yyx + yyy.$$

There are 8 summands[4] here and each summand is a sequence of xs and ys of length 3. When we simplify, all summands containing the same number of ys are collected together. How many summands are there containing i ys? It is here that we can use Proposition 3.9.14 except that we apply it to the set $\{x, y\}$. The answer is

$$\binom{n}{i}.$$

But all summands containing i ys can be simplified using commutativity to look like

$$x^{n-i} y^i.$$

This argument can be generalized to prove the theorem for any n. □

We have also proved that the numbers in Pascal's triangle are the binomial coefficients. The explanation for the rule used in calculating successive rows of Pascal's triangle follows from the next lemma. The proof is left as an exercise in algebraic manipulation.

Lemma 4.4.2. *Let $n \geq r \geq 1$. Then*

$$\binom{n+1}{r} = \binom{n}{r-1} + \binom{n}{r}.$$

Here are some points to bear in mind when using the binomial theorem.

---

[4]A *summand* is something being added in a sum.

- Unless the power you have to calculate is small, the binomial theorem should be used and not Pascal's triangle.

- Always write the theorem down so you have something to work with

$$(x+y)^n = \sum_{i=0}^{n} \binom{n}{i} x^{n-i} y^i.$$

Observe that there is a plus sign between the two terms in the brackets, and that the summation starts at zero.

- What you call x and what you call y does not matter.

- $\binom{n}{i} = \binom{n}{n-i}$.

- It follows that the binomial expansion can equally well be written as

$$(x+y)^n = \sum_{i=0}^{n} \binom{n}{i} x^i y^{n-i}.$$

- The numbers that appear in the terms of the summation involved in the binomial theorem are called *coefficients*.

Exercises 4.4

1. Write out $(1+x)^8$ using Σ-notation.

2. Write out $(1-x)^8$ using Σ-notation.

3. Calculate the coefficient of $a^2 b^8$ in $(a+b)^{10}$.

4. Calculate the coefficient of x^3 in $(3+4x)^6$.

5. Calculate the coefficient of x^3 in $\left(3x^2 - \frac{1}{2x}\right)^9$. What is the value of the constant term?

6. Use the binomial theorem to prove the following.

   (a) $2^n = \sum_{i=0}^{n} \binom{n}{i}$.

   (b) $0 = \sum_{i=0}^{n} (-1)^i \binom{n}{i}$.

   (c) $\left(\frac{3}{2}\right)^n = \sum_{i=0}^{n} \frac{1}{2^i} \binom{n}{i}$.

7. Prove Lemma 4.4.2.

8. Prove the binomial theorem using induction.

9. *This question proves the same result in two different ways.

(a) Use the binomial theorem to prove that $\binom{2n}{n} = \sum_{i=0}^{n} \binom{n}{i}^2$. Hint: Calculate $(x+y)^{2n}$ in two different ways and equate them.

(b) Prove that $\binom{2n}{n} = \sum_{i=0}^{n} \binom{n}{i}^2$ by counting. Hint: It is required to form a committee with n members from a group of n Montagues and n Capulets. Calculate how many such committees there are in two different ways and equate them.

4.5 BOOLEAN ALGEBRAS

One of the themes of this book is that algebra is not just the algebra of real numbers. Despite this, many of the algebraic structures we shall encounter will at least have a familiar feel to them. But there are examples of algebraic structures that look very different. In this section, we describe one such example that originates in the work of George Boole and so are called *Boolean algebras*. They were introduced in the nineteenth century as part of a renaissance in algebra that also included the work of Cayley, Galois, Grassmann, Hamilton and Sylvester, all of which will be touched on in this book. In the twentieth century, Claude Shannon (1916–2001), often known as the father of information theory, described in his master's thesis how Boolean algebras could be used to design certain kinds of circuits.

A *Boolean algebra* is a set B equipped with two binary operations, denoted by $+$ and $\cdot$, together with a unary operation $a \mapsto \bar{a}$ for each $a \in B$, and two constants, denoted by 0 and 1. In addition, the following ten axioms are required to hold.

Axioms for Boolean algebras

(B1) $(x+y)+z = x+(y+z)$.

(B2) $x+y = y+x$.

(B3) $x+0 = x$.

(B4) $(x \cdot y) \cdot z = x \cdot (y \cdot z)$.

(B5) $x \cdot y = y \cdot x$.

(B6) $x \cdot 1 = x$.

(B7) $x \cdot (y+z) = x \cdot y + x \cdot z$.

(B8) $x+(y \cdot z) = (x+y) \cdot (x+z)$.

(B9) $x+\bar{x} = 1$.

(B10) $x \cdot \bar{x} = 0$.

These axioms are organized as follows. The first group of three, (B1), (B2), (B3), describes the properties of $+$ on its own. We can see that the operation is associative, commutative and 0 is the identity. The second group of three, (B4), (B5), (B6), describes the properties of · on its own. We can see that the operation is associative, commutative but this time 1 is the identity. The third group, (B7), (B8), describes how + and · interact. Axiom (B7) looks familiar but axiom (B8) is decidedly odd. The final group, (B9) and (B10), describes the properties of the unary operation $a \mapsto \bar{a}$, called *complementation*.

Example 4.5.1. The axioms for a Boolean algebra should look familiar because they are similar to the properties listed in Theorem 3.2.1. This is not an accident. Let X be any non-empty set. We take as our set B the power set $\mathsf{P}(X)$. Thus the elements of our Boolean algebra are the subsets of X. We interpret $+$ as $\cup$, and $\cdot$ as $\cap$. The unary operation is interpreted as $A \mapsto X \setminus A$. The constant 0 is interpreted as $\emptyset$, and the constant 1 is interpreted as X. The fact that with these definitions $\mathsf{P}(X)$ is a Boolean algebra is now an immediate consequence of Theorem 3.2.1.

Our next result gives some examples of how to prove results from the axioms. It also shows how different algebra is in the Boolean world.

Proposition 4.5.2. *Let B be a Boolean algebra and let $a, b \in B$.*

1. $a^2 = a \cdot a = a$.

2. $a + a = a$.

3. $a \cdot 0 = 0$.

4. $1 + a = 1$.

5. $a + a \cdot b = a$.

6. $a + \bar{a} \cdot b = a + b$.

Proof. We prove (1), (3) and (5) and leave the other three parts as exercises.
(1)

$$
\begin{aligned}
a &= a \cdot 1 \text{ by (B6)} \\
&= a \cdot (a + \bar{a}) \text{ by (B9)} \\
&= a \cdot a + a \cdot \bar{a} \text{ by (B7)} \\
&= a^2 + 0 \text{ by (B10)} \\
&= a^2 \text{ by (B3).}
\end{aligned}
$$

(3)

$$
\begin{aligned}
a \cdot 0 &= a \cdot (a \cdot \bar{a}) \text{ by (B10)} \\
&= (a \cdot a) \cdot \bar{a} \text{ by (B4)} \\
&= a \cdot \bar{a} \text{ by (1) above} \\
&= 0 \text{ by (B10).}
\end{aligned}
$$

(5)

$$a+a\cdot b \;=\; a\cdot 1+a\cdot b \text{ by (B6)}$$
$$=\; a\cdot(1+b) \text{ by (B7)}$$
$$=\; a\cdot 1 \text{ by (4) above}$$
$$=\; a \text{ by (B6).}$$

□

The Boolean algebra in which we shall be most interested is the following.

Example 4.5.3. Put $\mathbb{B} = \{0,1\}$. Let $x,y \in \mathbb{B}$, where these are *Boolean variables* not real ones. Define

$$x+y = \begin{cases} 0 & \text{if } x=y=0 \\ 1 & \text{else} \end{cases}$$

and

$$x\cdot y = \begin{cases} 1 & \text{if } x=y=1 \\ 0 & \text{else.} \end{cases}$$

Define $\bar{x}$ to be the opposite value from x. To prove that $\mathbb{B}$ really does become a Boolean algebra with respect to these operations, observe that it is essentially the same as the Boolean algebra of all subsets of a one-element set as in Example 4.5.1.

To conclude this brief introduction to Boolean algebras, we describe an application to the design of certain kinds of circuits. A computer circuit is a physical, not a mathematical, object but can be modelled by mathematics. Currently, all computers are constructed using binary logic, meaning that their circuits operate using two values of some physical property, such as voltage. Circuits come in two types: *sequential circuits* which have an internal memory and *combinatorial circuits* which do not.[5] A combinatorial circuit with m input wires and n output wires can be modelled by means of a function $f\colon \mathbb{B}^m \to \mathbb{B}^n$ called a *Boolean function*. Our goal is to show that any such Boolean function can be constructed from certain simpler Boolean functions called gates. On a point of notation arbitrary elements of $\mathbb{B}^m$ are denoted by symbols such as **x**. This is an m-tuple whose components are either 0 or 1.

There are three kinds of gates. The *and-gate* is the function $\mathbb{B}^2 \to \mathbb{B}$ defined by $(x,y) \mapsto x\cdot y$. We use the following symbol to represent this function.

The *or-gate* is the function $\mathbb{B}^2 \to \mathbb{B}$ defined by $(x,y) \mapsto x+y$. We use the following symbol to represent this function.

---

[5] I recommend [98] for further details.

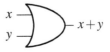

Finally, the *not-gate* is the function $\mathbb{B} \rightarrow \mathbb{B}$ defined by $x \mapsto \bar{x}$. We use the following symbol to represent this function.

The names used for these functions reflect a connection with propositional logic introduced in Section 3.2.[6] Diagrams constructed using gates are called *circuits* and show how Boolean functions can be computed. Such mathematical circuits can be converted into physical circuits with gates being constructed from simpler circuit elements called transistors which operate like electronic switches.

Example 4.5.4. Because of the associativity of $\cdot$, the circuit

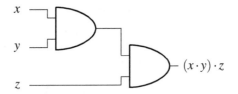

and the circuit

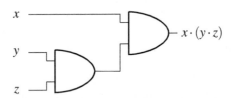

compute the same function. This can obviously be extended using generalized associativity. Similar comments apply to the operation $+$.

Boolean functions can be simplified using the bookkeeping functions described in Examples 3.4.5. A function $f: \mathbb{B}^m \rightarrow \mathbb{B}^n$ can be composed with the n projection functions to yield the n functions $f_i = p_i^n f: \mathbb{B}^m \rightarrow \mathbb{B}$. These are all 1-output functions. On the other hand, given n functions $f_i: \mathbb{B}^m \rightarrow \mathbb{B}$, we can construct the product function $(f_1, \ldots, f_n): \mathbb{B}^m \rightarrow \mathbb{B}^n$, by $\mathbf{x} \mapsto (f_1(\mathbf{x}), \ldots, f_n(\mathbf{x}))$. Using the projection and product functions, the problem of constructing functions $f: \mathbb{B}^m \rightarrow \mathbb{B}^n$ is reduced to the problem of constructing functions $f: \mathbb{B}^m \rightarrow \mathbb{B}$, called *m-input/1-output Boolean functions*. Such a function can be described by means of an *input/output table* that consists of 2^m rows and two columns which record $\mathbf{x}$ and $f(\mathbf{x})$, respectively.

---

[6]A Boolean algebra can be constructed from propositional logic using logical equivalence. This is because logical equivalence is a congruence.

Theorem 4.5.5. *Every Boolean function* $f\colon \mathbb{B}^m \to \mathbb{B}$ *can be constructed from and-gates, or-gates and not-gates.*

Proof. Assume that f is described by means of an input/output table. We deal first with the case where f is the constant function to 0. In this case,

$$f(x_1,\ldots,x_m) = (x_1 \cdot \overline{x_1}) \cdot x_2 \cdot \ldots \cdot x_m.$$

Next we deal with the case where the function f takes the value 1 exactly once. Let $\mathbf{a} = (a_1,\ldots,a_m) \in \mathbb{B}^m$ be such that $f(a_1,\ldots,a_m) = 1$. Define $\mathbf{m} = y_1 \cdot \ldots \cdot y_m$, called the *minterm* associated with $\mathbf{a}$, as follows:

$$y_i = \begin{cases} x_i & \text{if } a_i = 1 \\ \overline{x_i} & \text{if } a_i = 0. \end{cases}$$

Then $f(\mathbf{x}) = y_1 \cdot \ldots \cdot y_m$. Finally, we deal with the case where the function f is none of the above. Let the inputs where f takes the value 1 be $\mathbf{a}_1,\ldots,\mathbf{a}_r$, respectively. Construct the corresponding minterms $\mathbf{m}_1,\ldots,\mathbf{m}_r$, respectively. Then

$$f(\mathbf{x}) = \mathbf{m}_1 + \ldots + \mathbf{m}_r.$$

$\square$

Example 4.5.6. We illustrate the proof of Theorem 4.5.5 by means of the following input/output table.

x	y	z	$f(x,y,z)$
0	0	0	0
0	0	1	1
0	1	0	1
0	1	1	0
1	0	0	1
1	0	1	0
1	1	0	0
1	1	1	0

The three elements of $\mathbb{B}^3$ where f takes the value 1 are $(0,0,1)$, $(0,1,0)$ and $(1,0,0)$. The minterms corresponding to each of these inputs are $\bar{x} \cdot \bar{y} \cdot z$, $\bar{x} \cdot y \cdot \bar{z}$ and $x \cdot \bar{y} \cdot \bar{z}$, respectively. It follows that

$$f(x,y,z) = \bar{x} \cdot \bar{y} \cdot z + \bar{x} \cdot y \cdot \bar{z} + x \cdot \bar{y} \cdot \bar{z}.$$

Example 4.5.7. The input/output table below

x	y	$x \oplus y$
0	0	0
0	1	1
1	0	1
1	1	0

defines *exclusive or* or *xor*. By Theorem 4.5.5

$$x \oplus y = \bar{x} \cdot y + x \cdot \bar{y}.$$

This time we construct the circuit diagram.

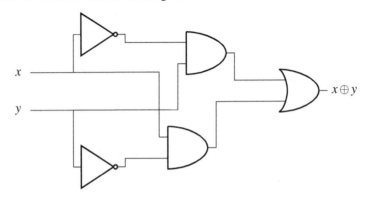

There are two further circuit elements implicit in this diagram. Observe that the line starting at x splits. This is called *fanout*. It has the effect of copying an input as many times as needed in the circuit, twice in this case. It therefore represents the diagonal function from $\mathbb{B}$ to $\mathbb{B}^2$ defined by $x \mapsto (x,x)$. In addition, wires crossing but not otherwise interacting is called *interchange* and forms the second additional circuit element we need. This represents the twist function from $\mathbb{B}^2$ to $\mathbb{B}^2$ defined by $(x,y) \mapsto (y,x)$. For our subsequent examples, it is convenient to abbreviate the circuit for xor by means of a single circuit symbol called an *xor-gate*.

Example 4.5.8. Our next circuit is known as a *half-adder* which has two inputs and two outputs and is defined by the following input/output table.

x	y	c	s
0	0	0	0
0	1	0	1
1	0	0	1
1	1	1	0

This treats the input x and y as numbers in base 2 and then outputs their sum. If you are unfamiliar with this arithmetic, it is discussed in Section 5.1. Observe that $s = x \oplus y$ and $c = x \cdot y$. Thus using the previous example we may construct a circuit that implements this function.

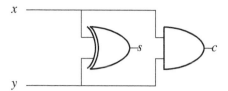

Example 4.5.9. Our final circuit is known as a *full-adder* which has three inputs and two outputs and is defined by the following input/output table.

x	y	z	c	s
0	0	0	0	0
0	0	1	0	1
0	1	0	0	1
0	1	1	1	0
1	0	0	0	1
1	0	1	1	0
1	1	0	1	0
1	1	1	1	1

This treats the three inputs as numbers base 2 and adds them together. The following circuit realizes this behaviour using two half-adders completed with an or-gate.

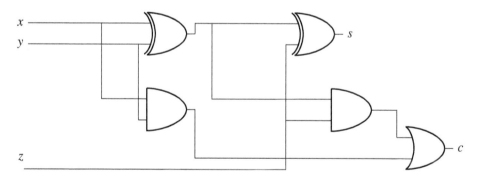

Example 4.5.10. Full-adders are the building blocks from which all arithmetic computer circuits can be built. Specifically, suppose that we want to add two four bit binary numbers together. Denote them by $m = a_3a_2a_1a_0$ and $n = b_3b_2b_1b_0$. The sum $m+n$ in base 2 is computed in a similar way to calculating a sum in base 10. Thus first calculate $a_0 + b_0$, write down the sum bit and pass any carry to be added to $a_1 + b_1$ and so on. Although $a_0 + b_0$ can be computed by a half-adder, subsequent additions may require the addition of three bits because of the presence of a carry bit. For this reason, we actually use four full-adders joined in series with the rightmost full-adder having one of its inputs set to 0.

Exercises 4.5

1. Prove parts (2), (4) and (6) from Proposition 4.5.2.

2. (a) Prove that if $a+b=1$ and $a \cdot b = 0$ then $b = \bar{a}$.

   (b) Prove that $\bar{\bar{a}} = a$.

   (c) Prove that $\overline{(a+b)} = \bar{a} \cdot \bar{b}$ and $\overline{a \cdot b} = \bar{a} + \bar{b}$.

3. Simplify each of the following as much as possible.

   (a) $\bar{x}\bar{y}z + x\bar{y}\bar{z} + x\bar{y}z$.

   (b) $xy + x\bar{y} + \bar{x}y$.

   (c) $x + yz + \bar{x}y + x\bar{y}z$.

4. *Define $a * b = a \cdot \bar{b}$. Show that the other binary operations can be defined in terms of this one. In what sense does it model an electronic switch?

4.6   CHARACTERIZING REAL NUMBERS

In addition to algebraic operations, the real numbers are also ordered. Given two real numbers we can always say whether they are equal or one of them is bigger than the other. We write down first the axioms for order that hold for both rational and real numbers.

Axioms for order in $\mathbb{R}$

(O1) $a \le a$.

(O2) If $a \le b$ and $b \le a$ then $a = b$.

(O3) If $a \le b$ and $b \le c$ then $a \le c$.

(O4) Given a and b then either $a \le b$ or $b \le a$ or $a = b$.

   If $a > 0$ we say a is *positive* and if $a < 0$ we say a is *negative*.

(O5) If $a \le b$ and $c \le d$ then $a + c \le b + d$.

(O6) If $a \le b$ and $0 < c$ then $ac \le bc$.

Axioms (O1)–(O3) simply tell us that we have a partial order and (O4) says that this partial order is linear. What is new are axioms (O5) and (O6) and particular attention needs to be paid to (O6).

   Here is an example of a proof using these axioms.

Example 4.6.1. We prove that $a \leq b$ if and only if $b - a$ is positive. Suppose first that $a \leq b$. By axiom (O5), we may add $-a$ to both sides to get $a + (-a) \leq b + (-a)$. But $a + (-a) = 0$ and $b + (-a) = b - a$, by definition. It follows that $0 \leq b - a$ and so $b - a$ is positive. Now we prove the converse. Suppose that $b - a$ is positive. By definition $0 \leq b - a$. Also by definition, $b - a = b + (-a)$. Thus $0 \leq b + (-a)$. By axiom (O5), we may add a to both sides to get $0 + a \leq (b + (-a)) + a$. But $0 + a = a$ and $(b + (-a)) + a$ quickly simplifies to b. We have therefore proved that $a \leq b$, as required.

The axioms introduced so far apply equally well to both the rationals and the reals. We now introduce some heavy-duty terminology that will enable us to explain the fundamental difference between these two kinds of numbers. Let $(X, \leq)$ be a poset as defined in Section 3.6. An element $0 \in X$ is called a *bottom element* if $0 \leq x$ for all $x \in X$. An element $1 \in X$ is called a *top element* if $x \leq 1$ for all $x \in X$. Top and bottom elements are unique if they exist. Thus a top element is the largest element and a bottom element is the smallest element. It is quite possible for a poset to have no bottom element and no top element.

- Let $Y \subseteq X$. An element $x \in X$ is called an *upper bound* of Y if $y \leq x$ for all $y \in Y$. An element $x \in X$ is called a *least upper bound* or *join* or *supremum* of Y if it is an upper bound of Y, and the bottom element of the set of all upper bounds of Y. This is sometimes denoted by $\bigvee Y$. If $Y = \{a, b\}$, then $\bigvee\{a, b\}$ is written $a \vee b$.

- We say that $x \in X$ is a *lower bound* of Y if $x \leq y$ for all $y \in Y$. An element $x \in X$ is called a *greatest lower bound* or *meet* or *infimum* of Y if it is a lower bound of Y and the top element in the set of all lower bounds of Y. This is sometimes denoted by $\bigwedge Y$. If $Y = \{a, b\}$, then $\bigwedge\{a, b\}$ is written $a \wedge b$.

We use this terminology to reinterpret what $\sqrt{2}$ means.

Example 4.6.2. Let $A = \{a \colon a \in \mathbb{Q} \text{ and } a^2 \leq 2\}$ and let $B = \{a \colon a \in \mathbb{R} \text{ and } a^2 \leq 2\}$. Then $A \subseteq \mathbb{Q}$ and $B \subseteq \mathbb{R}$. Both sets are bounded above: the number $1\frac{1}{2}$, for example, works in both cases. However, we prove that the subset A does not have a least upper bound, whereas the subset B does.

Consider the subset A first. Suppose that r were a least upper bound. We prove that $r^2 = 2$ which is impossible because we have proved that $\sqrt{2}$ is irrational. To do this, we show that neither $r^2 < 2$ nor $2 < r^2$ can hold.

1. Suppose that $r^2 < 2$. We claim there is a rational number r_1 such that $r < r_1$ and $r_1^2 < 2$, contradicting the assumption that r is an upper bound of the set A. To prove the claim we need to find a rational number $h > 0$ such that $(r + h)^2 < 2$ given that r is a rational number such that $r^2 < 2$. A little algebra shows that we need to choose h so that $0 < h(h + 2r) < 2 - r^2$. If we choose $h < 1$ then $h(h + 2r) < h(1 + 2r)$. Thus it is enough to ensure that $h(1 + 2r) < 2 - r^2$ as well. It follows that we choose any rational number h such that $0 < h < 1$ and

$$h < \frac{2 - r^2}{2r + 1}.$$

Put $r_1 = r + h$. By construction $r_1 > r$ and $r_1^2 < 2$ as claimed.

2. Suppose that $2 < r^2$. We claim that we can find a rational number r_1 such that $0 < r_1 < r$ and $2 < r_1^2$ contradicting the assumption that r is a least upper bound of the set A. To prove the claim we need to find a rational number $h > 0$ such that $r - h > 0$ and $2 < (r - h)^2$. A little algebra shows that we need to choose h so that $r^2 - 2 > h(2r - h)$. But $h2r > h(2r - h)$. Thus it is enough to define $h = \frac{r^2 - 2}{2r}$. Put $r_1 = r - h$. By construction $0 < r_1 < r$ and $r_1^2 > 2$ as claimed.

We have therefore proved that if r is a least upper bound of A then $r^2 = 2$ and so $r = \sqrt{2}$. This is impossible because we have proved that $\sqrt{2}$ is irrational. Thus the set A does not have a rational least upper bound. By essentially the same reasoning, the set B does have a real least upper bound: the number $\sqrt{2}$.

The above example motivates the following which is the most sophisticated axiom we shall meet in this book.

The completeness axiom for $\mathbb{R}$

Every non-empty subset of the reals that is bounded above has a least upper bound.

It is this axiom that is needed to develop calculus properly. You might wonder if there are any other important properties of the real numbers still to be described. That there are none is the substance of the following theorem. We shall not prove it here; instead see [114].

Theorem 4.6.3 (Characterization of the real numbers). *The axioms (F1)–(F11) and (O1)–(O6) together with the completeness axiom characterize the real numbers.*

In other words, if you are working with a system that satisfies all of these axioms then you are in fact working with the real numbers.

Example 4.6.4. Finite sums do not pose any difficulties, but infinite sums do. For example, consider the infinite sum

$$S = \sum_{i=1}^{\infty} (-1)^{i+1}.$$

This is just $S = 1 - 1 + 1 - 1 + \dots$. What is S? Your first instinct might be to say 0 because $S = (1 - 1) + (1 - 1) + \dots$. But it could equally well be 1 calculated as follows: $S = 1 + (-1 + 1) + (-1 + 1) + \dots$. In fact, it could even be $\frac{1}{2}$ since $S + S = 1$ and so $S = \frac{1}{2}$. There is clearly something seriously awry here, and it is that infinite sums have to be handled carefully if they are to make sense. Just how is the business of analysis and we will not discuss the theory in this book though it makes essential use of the completeness axiom. We reiterate here that ∞ is not a number. It simply tells us to keep adding on terms for increasing values of i without end so, in particular, we never write

$$\frac{3}{10^{\infty}}.$$

One place where we use infinite sums in everyday life is in the decimal representations of numbers. Thus the fraction $\frac{1}{3}$ can be written as $0 \cdot 3333\ldots$ and this is in fact the infinite sum

$$\sum_{i=1}^{\infty} \frac{3}{10^i}.$$

Even here interesting things occur. Thus the number $1 \cdot 000\ldots$ is equal to the number $0 \cdot 999\ldots$. If we multiply these two numbers by powers of $\frac{1}{10}$ we deduce the following. A finite decimal such as

$$0 \cdot a_1 \ldots a_m,$$

where $a_m \neq 0$, is equal to the infinite decimal

$$0 \cdot a_1 \ldots (a_m - 1)999\ldots.$$

Thus real numbers have, as it were, two different names. If we need to be specific, we can always assume that our decimals are infinite using the above idea. It should be added that there are also *infinite products*

$$\prod_{i=1}^{\infty} a_i$$

which arise in analysis.

Box 4.6: Infinitesimals

There is an old saying that the difference between a physicist and a mathematician is that the former knows that infinitesimals exist whereas the latter knows they do not. Infinitesimals are entities such as $\varepsilon^2 = 0$ but where $\varepsilon \neq 0$. They enable derivatives to be calculated very easily. For example

$$\frac{(x+\varepsilon)^2 - x^2}{\varepsilon} = 2x.$$

They were used in the early development of calculus, much to the disapproval of the Jesuits [2], but by the nineteenth century the game was up and mathematicians in general, not just Jesuits, decided that they did not exist. Clearly no real numbers can have the above properties. This led to the reformulation of infinitesimal calculus into a subject called *analysis* using limits and the technique of *epsilonics* based on the completeness axiom. But physicists persisted in their heresy, and mathematicians secretly pined for them. It is odd that those other mysterious entities, the complex numbers, became an essential tool in mathematics whereas the infinitesimals were banished. Or so it seemed. In the twentieth century, infinitesimals were resurrected from conceptual oblivion. First, in the work of Abraham Robinson (1918–1974), who introduced what is known as *non-standard analysis* [40], and then, from a different perspective, using an advanced branch of category theory known as *topos theory* [10].

Exercises 4.6

1. Prove that if a poset has a top element (respectively, bottom element) then that element is unique.

2. Prove that between any two distinct rational numbers there is another rational number.

3. Prove the following using the order axioms for $\mathbb{R}$.

   (a) If $a \le b$ then $-b \le -a$.

   (b) a^2 is positive for all $a \ne 0$.

   (c) If $0 < a < b$ then $0 < b^{-1} < a^{-1}$.

Box 4.7: The Peano Axioms

Set theory is supposed to be a framework in which all of mathematics can take place. We briefly sketch out how we can construct in particular the real numbers using set theory. The starting point is the Peano axioms studied by Giuseppe Peano (1858–1932). These describe a set P and an operation on this set called the *successor function* which for each $n \in P$ produces a unique element n^+. The following four axioms hold.

(P1) There is a distinguished element of P denoted by 0.

(P2) There is no element $n \in P$ such that $n^+ = 0$.

(P3) If $m, n \in P$ and $m^+ = n^+$ then $m = n$.

(P4) Let $X \subseteq P$ be such that $0 \in X$ and if $n \in X$ then $n^+ \in X$. Then $X = P$.

By using ideas from set theory, one shows that P is essentially the set of natural numbers together with its operations of addition and multiplication. The natural numbers are deficient because it is not always possible to solve equations of the form $a + x = b$ because of the lack of additive inverses. To construct them, we use ordered pairs. The idea is to regard (a, b) as meaning $a - b$. However, there are many names for the same negative number so we should have $(0, 1)$ and $(2, 3)$ and $(3, 4)$ all signifying the same number: namely, -1. To make this work, we use an equivalence relation defined on $\mathbb{N} \times \mathbb{N}$. The set of equivalence classes is denoted by $\mathbb{Z}$. Again using ideas from set theory, the usual operations can be constructed on $\mathbb{Z}$. The integers are deficient because it is not always possible to solve equations of the form $ax + b = 0$ because of the lack of multiplicative inverses. To construct them, we use ordered pairs again. This time (a, b), where $b \ne 0$ and a and b are integers, is interpreted as $\frac{a}{b}$. But again we have the problem of multiple names for what should be the same number. Thus $(1, 2)$ should equal $(-1, -2)$ should equal $(2, 4)$ and so forth. Once again this problem is solved by using an equivalence relation, this time defined on $\mathbb{Z} \times \mathbb{Z}^*$ where $\mathbb{Z}^* = \mathbb{Z} \setminus \{0\}$. The set of equivalence classes is denoted by $\mathbb{Q}$. Again using ideas from set theory, the usual operations can be constructed on $\mathbb{Q}$. The rationals are deficient in not containing numbers like $\sqrt{2}$. The intuitive idea behind the construction of the reals from the rationals is that we want $\mathbb{R}$ to be all the numbers that can be arbitrarily approximated by rational numbers. To do this, we form the set of all subsets X of $\mathbb{Q}$ which have the following characteristics: $X \ne \emptyset$, $X \ne \mathbb{Q}$, if $x \in X$ and $y \le x$ then $y \in X$, and X does not have a biggest element. These subsets are called *Dedekind cuts* and should be regarded as defining the real number r so that X consists of all the rational numbers strictly less than r. In this way, we can construct the real numbers from the Peano axioms using set theory. As we shall show in Section 6.4 the complex numbers can be constructed from ordered pairs of real numbers.

II

THEORIES

CHAPTER 5

Number theory

"*God made the natural numbers, man made everything else.*" – Leopold Kronecker

Number theory is one of the oldest branches of mathematics and deals principally with the properties of the integers, the simplest kinds of numbers. The main result proved in this chapter is that every natural number greater than one can be written as a product of powers of primes, a result known as the fundamental theorem of arithmetic. This shows that the primes are the building blocks, or atoms, from which all natural numbers are constructed. The primes are still the subject of intensive research but their theory goes right back to Euclid's *Elements*.

5.1   REMAINDER THEOREM

We begin with a lemma that can be assumed as an axiom on a first reading. Although simple, it should not be underestimated. Everything in this chapter is unlocked by this result. Recall that $|x|$ is the absolute value of a number.

Lemma 5.1.1 (Remainder theorem). *Let a and b be integers where $b \neq 0$. Then there are unique integers q and r such that $a = bq + r$ and $0 \leq r < |b|$.*

The number q is called the *quotient* and r the *remainder*. For example, if we consider the pair of natural numbers 14 and 3 then $14 = 3 \cdot 4 + 2$ where 4 is the quotient and 2 is the remainder. The case where $r = 0$ is important. Let a and b be integers where $a \neq 0$. We say that a *divides* b or that b *is divisible by* a if there is an integer q such that $b = aq$. We also say that a is a *divisor* or *factor* of b. We write $a \mid b$ to mean the same as 'a divides b'. It is important to remember that $a \mid b$ does not mean the same thing as $\frac{a}{b}$. In fact, $\mid$ is an example of a relation on the set $\mathbb{Z}$, whereas $\frac{a}{b}$ is a number. They are therefore different types of things.

Example 5.1.2. By the remainder theorem, every integer n can be written $n = 2q + r$ where $r = 0$ or $r = 1$ but not both. The numbers where $r = 0$ are called *even* and those where $r = 1$ are called *odd*. We used the properties of odd and even numbers to prove that $\sqrt{2}$ is irrational in Chapter 2. Thus looking at remainders can lead to non-trivial mathematics. We shall generalize this lesson when we study modular arithmetic in Section 5.4.

Box 5.1: Proof of the Remainder Theorem

You might wonder how such a simple result could be proved. We do so using the most basic properties of the natural numbers amongst which we count the well-ordering principle from Section 3.8. We deal first with the case where $b > 0$. Let $X = \{a - nb \colon n \in \mathbb{Z}\}$. If $a > 0$ then X contains positive elements since it contains a. If $a < 0$ then $a - 2a = -a$ is positive and belongs to X. Thus X always contains positive elements. Denote the set of these positive elements by X^+. This is a subset of the natural numbers and so by the well-ordering principle it contains a minimum element r. Thus $r = a - qb \geq 0$ for some $q \in \mathbb{Z}$. Suppose that $r \geq b$. Then $r = b + r'$ where $r' > 0$ and $r' < r$. But $r' = a - (q+1)b \in X^+$, which contradicts the fact that r is the smallest element of X^+. Thus $r < b$. We have therefore proved that $a = bq + r$ where $0 \leq r < b$. It remains to show that q and r are unique with these properties. Suppose that $a = bq' + r'$ where $0 \leq r' < b$. Without loss of generality, assume that $r' - r \geq 0$. Then $r' - r = (q - q')b$. But $r' - r < b$. Thus $q - q' = 0$ and so $q = q'$, and it then follows that $r = r'$. This deals with the case where $b > 0$. The case where $b < 0$ can be dealt with by working with $-b$ and using the result already proved to obtain $a = (-b)q + r$ and then rewriting this as $a = b(-q) + r$.

We describe two applications of the remainder theorem. The following example motivates the first.

Example 5.1.3. By the remainder theorem, every natural number n can be written as $n = 10q + r$ where $0 \leq r \leq 9$. The integer r is nothing other than the units digit in the usual base 10 representation of n. Thus, for example, $42 = 10 \times 4 + 2$.

This leads us to think about the mathematics that lies behind the very way we write numbers. Although numbers have been used for thousands of years, writing them down proved to be a challenge. The simplest way of recording a number is to use a mark like |, called a *tally*, for each thing being counted. So |||||||||| means 10 things. This is a simple system to use but not very efficient, and beyond 6 or 7 tallies it is hard to tell without actually counting whether two sequences of tallies represent the same number or not. An advance on the tally system is the one used by the Romans. This consists of a list of basic symbols

number	1	5	10	50	100	500	1000
symbol	I	V	X	L	C	D	M

where there are more symbols for bigger numbers. Numbers are written down using these symbols according to the *additive principle* where the value of a sequence of symbols is the sum of their values. Thus MMI is 2001.[1] This is an improvement on the tally-system since even large numbers are written compactly, and can easily be compared. One disadvantage of this system is that separate symbols are needed for

---

[1] The custom of also using a *subtractive principle* so that, for example, IX means 9 rather than using VIIII is a more modern innovation.

different powers of 10 and their multiples by 5. This was probably not too inconvenient in the ancient world where the numbers needed on a day-to-day basis were not that big. A common criticism of this system is that it is a hard one in which to do arithmetic, but like us the Romans used mechanical means, such as an abacus, to do their sums. This system has survived down to the present day for purely decorative purposes.

The modern system used to write numbers down is quite different and is called the *positional number system*. It seems to have been in place by the ninth century in India. Its genius is that only 10 symbols, called *digits*,

$$0,1,2,3,4,5,6,7,8,9,$$

are needed. Every natural number can then be written using a string of these digits. The trick, if you will, is to use the *position* of the digit on the page to tell us the value being represented by that digit. In particular, the digit 0 has a two-fold meaning. It represents the number zero and it defines position when needed. Thus whereas the Roman system needed separate symbols for one, ten, a hundred, a thousand and so forth, in the positional system, the zero is used to show the position and therefore the value of the symbol 1 to yield: 1, 10, 100, 1000 and so on. Thus 2001 actually means

$$2 \times 10^3 + 0 \times 10^2 + 0 \times 10^1 + 1 \times 10^0.$$

The positioning function of zero is illustrated below.

10^3	10^2	10^1	10^0
2	0	0	1

This idea can be extended to cope not only with natural numbers but also with integers using the symbol $-$, and with arbitrary real numbers by including negative powers of 10: 10^{-1}, 10^{-2}, 10^{-3}, ... Thus every real number can be encoded by a sign ($\pm$), a finite sequence of digits, a decimal point, and then a potentially infinite sequence of digits. The full decimal system was in place by the end of the sixteenth century.

We investigate now the mathematics behind positional number systems, not just for base 10 as above, but for any number base $d \geq 2$. The main tool is the remainder theorem. If $d \leq 10$ we represent numbers by strings using the symbols

$$\mathbb{Z}_d = \{0,1,2,3,\ldots d-1\}$$

but if $d > 10$ we need new symbols for $10, 11, 12, \ldots$. It is convenient to use A,B,C, ... For example, if we want to write numbers in base 12 we use the symbols

$$\{0,1,\ldots,9,A,B\}$$

whereas if we work in base 16 we use the symbols

$$\{0,1,\ldots,9,A,B,C,D,E,F\}.$$

If x is a string over one of the above sets of symbols then we write x_d to make it clear that we are to interpret it as a number in base d. Thus BAD_{16} is a number in base 16. The symbols in a string x_d, reading from right to left, tell us the contribution each power of d such as $d^0, d^1, d^2, \ldots$ makes to the number the string represents.

Examples 5.1.4. Converting from base d to base 10.

1. BAD_{16} is a number in base 16. This represents the following number in base 10

$$B \times 16^2 + A \times 16^1 + D \times 16^0,$$

which is equal to the number

$$11 \times 16^2 + 10 \times 16 + 13 = 2989.$$

2. 5556_7 is a number in base 7. This represents the following number in base 10

$$5 \times 7^3 + 5 \times 7^2 + 5 \times 7^1 + 6 \times 7^0 = 2001.$$

The examples above demonstrate the general procedure for converting from base d to base 10. To go in the opposite direction and convert from base 10 to base d, we use the remainder theorem. Let

$$n = (a_m \ldots a_1 a_0)_d.$$

Thus $a_m \ldots a_0$ are the base d digits of the number n. This means that

$$n = a_m d^m + a_{m-1} d^{m-1} + \ldots + a_0 d^0$$

and so, in particular,

$$n = qd + a_0$$

where $q = a_m d^{m-1} + \ldots + a_1$. The crucial point to observe is that a_0 is the remainder when n is divided by d and the quotient q has the base d representation $(a_m \ldots a_1)_d$. We can obviously repeat the above procedure. On dividing q by d, the remainder is a_1 and the quotient has base d representation $(a_m \ldots a_2)_d$. Continuing in this way, we can generate the digits of n in base d from *right to left* by repeatedly finding the next quotient and next remainder by dividing the current quotient by d.

Examples 5.1.5. Converting from base 10 to base d.

1. Write 2001 in base 12.

	quotient	remainder
12	2001	
12	166	9
12	13	$10 = A$
12	1	1
	0	1

Thus 2001 in base 12 is 11A9.

2. Write 2001 in base 2.

	quotient	remainder
2	2001	
2	1000	1
2	500	0
2	250	0
2	125	0
2	62	1
2	31	0
2	15	1
2	7	1
2	3	1
2	1	1
	0	1

Thus 2001 in base 2 is 11111010001.

Particular number bases are usually referred to by their classical names. Thus base 2 is *binary*, base 10 is *decimal*, base 16 is *hexadecimal* and base 60 is *sexagesimal*. Binary, of course, is used in computers as we saw in Section 4.5. Sexagesimal was used by the astronomers and mathematicians of ancient Mesopotamia and is still the basis of time and angle measurement today.[2] Arithmetic is performed in base d using the same algorithms as for base 10.

Our second application of the remainder theorem deals with how we can write proper fractions as decimals. To see what is involved, we calculate some examples of decimal fractions.

Examples 5.1.6.

1. $\frac{1}{20} = 0 \cdot 05$. This fraction has a finite decimal representation.

2. $\frac{1}{7} = 0 \cdot 142857142857142857142857142857\ldots$. This fraction has an infinite decimal representation, consisting of the same finite sequence of numbers repeated. We abbreviate this decimal to $0 \cdot \overline{142857}$.

3. $\frac{37}{84} = 0 \cdot 44\overline{047619}$. This fraction has an infinite decimal representation, which consists of a non-repeating part, the digits 44, followed by one which repeats.

These examples illustrate the three kinds of decimal representation of fractions. Those that have a *finite* representation such as (1) above which we shall characterize in Section 5.3. There are then those that have *infinite* representations such as (2) and (3) above, with those in (2) called *purely periodic* representations since they consist of a repeating block of numbers, and those in (3) called *ultimately periodic* representations since they become periodic after a certain point. We regard purely periodic representations as special cases of ultimately periodic ones.

---

[2] We are therefore daily reminded of our cultural links to the earliest civilizations.

Proposition 5.1.7. *An infinite decimal fraction represents a rational number if and only if it is ultimately periodic.*

Proof. Let $\frac{a}{b}$ be a proper fraction representing a rational number. Our goal is to write

$$\frac{a}{b} = 0 \cdot q_1 q_2 \ldots$$

and investigate the nature of the sequence of digits after the decimal point. Mathematically, this means we are trying to write

$$\frac{a}{b} = \sum_{i=1}^{\infty} \frac{q_i}{10^i}.$$

On multiplying both sides of this equality by 10, we get

$$\frac{10a}{b} = q_1 + \sum_{i=2}^{\infty} \frac{q_i}{10^{i-1}} = q_1 + r_1$$

where $r_1 < 1$. It follows that

$$10a = q_1 b + b r_1.$$

Thus q_1 is the quotient when $10a$ is divided by b and $br_1 < b$ is the remainder. We now repeat the above procedure with the fraction $\frac{br_1}{b}$. The crux of the matter is that there are only a finite number of possible remainders when a number is divided by b. There are now two cases. First, we might get the remainder 0 at some point in which case the decimal representation terminates after a finite number of steps. Second, if it does not terminate after a finite number of steps then at some point a remainder will be repeated in which case we obtain an ultimately periodic representation of our fraction. Thus if a rational number has an infinite decimal representation then it is ultimately periodic.

To prove the converse, let

$$r = 0 \cdot a_1 \ldots a_s \overline{b_1 \ldots b_t}.$$

We prove that r is rational. Observe that

$$10^s r = a_1 \ldots a_s \cdot \overline{b_1 \ldots b_t}$$

and that

$$10^{s+t} = a_1 \ldots a_s b_1 \ldots b_t \cdot \overline{b_1 \ldots b_t}.$$

The point of these calculations is that $10^s r$ and $10^{s+t} r$ have exactly the same digits after the decimal point. Thus $10^{s+t} r - 10^s r$ is an integer that we shall call a. It follows that

$$r = \frac{a}{10^{s+t} - 10^s}$$

is a rational number. □

Example 5.1.8. We write the ultimately periodic decimal $0 \cdot 9\overline{4}$. as a proper fraction in its lowest terms. Put $r = 0 \cdot 9\overline{4}$. Then

- $r = 0 \cdot 9\overline{4}$.

- $10r = 9.444\ldots$

- $100r = 94.444\ldots$.

Thus $100r - 10r = 94 - 9 = 85$ and so $r = \frac{85}{90}$. We can simplify this to $r = \frac{17}{18}$ and a check shows that this is correct. Bear in mind that the trick of this method is simply to subtract away the infinite list of digits to the right of the decimal point.

Exercises 5.1

1. Find the quotients and remainders for each of the following pairs of numbers. Divide the smaller into the larger.

   (a) 30 and 6.

   (b) 100 and 24.

   (c) 364 and 12.

2. Write the number 153 in

   (a) Base 5.

   (b) Base 12.

   (c) Base 16.

3. Write the following numbers in base 10.

   (a) DAB_{16}.

   (b) $ABBA_{12}$.

   (c) 443322211_5.

4. Write the following decimals as fractions in their lowest terms.

   (a) $0 \cdot 5\overline{34}$.

   (b) $0 \cdot 2\overline{106}$.

   (c) $0 \cdot \overline{076923}$.

5. Prove the following properties of the division relation on $\mathbb{Z}$.

   (a) If $a \neq 0$ then $a \mid a$.

   (b) If $a \mid b$ and $b \mid a$ then $a = \pm b$.

   (c) If $a \mid b$ and $b \mid c$ then $a \mid c$.

   (d) If $a \mid b$ and $a \mid c$ then $a \mid (b+c)$.

5.2 GREATEST COMMON DIVISORS

It is convenient to represent fractions in their lowest terms. For example, the fraction $\frac{12}{16}$ is equal to the fraction $\frac{3}{4}$ because we can divide both numerator and denominator by the number 4. In this section, we explore this procedure in more detail and as a result prove a lemma, called Euclid's lemma, that is crucial in establishing the fundamental theorem of arithmetic.

Let $a, b \in \mathbb{N}$. A natural number d which divides both a and b is called a *common divisor* of a and b. The largest number which divides both a and b is called the *greatest common divisor* of a and b and is denoted by $\gcd(a, b)$. A pair of natural numbers a and b is said to be *coprime* if $\gcd(a, b) = 1$. For us $\gcd(0, 0)$ is undefined but if $a \neq 0$ then $\gcd(a, 0) = a$.

Example 5.2.1. Consider the numbers 12 and 16. The set of divisors of 12 is $\{1, 2, 3, 4, 6, 12\}$ and the set of divisors of 16 is $\{1, 2, 4, 8, 16\}$. The set of common divisors is $\{1, 2, 4\}$, the intersection of these two sets. The greatest common divisor of 12 and 16 is therefore 4. Thus $\gcd(12, 16) = 4$.

Our next lemma is not surprising. It simply says that if two numbers are divided by their greatest common divisor then the numbers that remain are coprime. This seems intuitively plausible and the proof ensures that our intuition is correct.

Lemma 5.2.2. *Let $d = \gcd(a, b)$. Then $\gcd(\frac{a}{d}, \frac{b}{d}) = 1$.*

Proof. Because d divides both a and b, we can write $a = ud$, for some natural number u, and $b = vd$, for some natural number v. We therefore need to prove that $\gcd(u, v) = 1$. Suppose that $e \mid u$ and $e \mid v$. Then $u = ex$ and $v = ey$ for some natural numbers x and y. Thus $a = exd$ and $b = eyd$. It follows that $ed \mid a$ and $ed \mid b$ and so ed is a common divisor of both a and b. But d is the greatest common divisor and so $e = 1$, as required. $\qquad\square$

Example 5.2.3. Greatest common divisors arise naturally in solving linear equations where the solutions are required to be integers. Consider for example the linear equation $12x + 16y = 5$. If we want our solutions (x, y) to have real number coordinates, then it is of course easy to solve this equation and find infinitely many solutions since the solutions form a line in the plane. Suppose now that we require $(x, y) \in \mathbb{Z}^2$. That is, we want the solutions to be integers. Geometrically this means we want to know whether the line contains any points with integer coordinates. We can see immediately that this is impossible. We have calculated that $\gcd(12, 16) = 4$. Thus if x and y are integers, the number 4 divides the lefthand side of the equation, but clearly 4 does not divide the righthand side. Thus the set

$$\{(x, y) \colon (x, y) \in \mathbb{Z}^2 \text{ and } 12x + 16y = 5\}$$

is empty.

If the numbers a and b are large, then calculating their gcd using the method of Example 5.2.1 is time-consuming and error-prone. We need to find a more efficient method. The following lemma is the basis of just such a method and is another application of the remainder theorem.

Lemma 5.2.4. *Let $a, b \in \mathbb{N}$, where $a \geq b \neq 0$, and let $a = bq + r$ where $0 \leq r < b$. Then $\gcd(a, b) = \gcd(b, r)$.*

Proof. Let d be a common divisor of a and b. Now $r = a - bq$ so that d is also a divisor of r. It follows that any divisor of a and b is also a divisor of b and r. Let d be a common divisor of b and r. Then d divides a since $a = bq + r$. Thus any divisor of b and r is a divisor of a and b. It follows that the set of common divisors of a and b is the same as the set of common divisors of b and r. Hence $\gcd(a, b) = \gcd(b, r)$. $\square$

The point of the above result is that $b \leq a$ and $r < b$. So calculating $\gcd(b, r)$ is easier than calculating $\gcd(a, b)$ because the numbers involved are smaller. Compare

$$\mathbf{a = bq + r} \text{ with } a = bq + r.$$

The above result is the basis of an efficient algorithm for computing greatest common divisors. It was described in Propositions VII.1 and VII.2 of Euclid.

Theorem 5.2.5 (Euclid's algorithm). *Let $a, b \in \mathbb{N}$ be such that $a \geq b \neq 0$. To compute $\gcd(a, b)$, do the following.*

(Step 1). *If $a = b$ then $\gcd(a, b) = a$. Thus in what follows we may assume that $a > b$.*

(Step 2). *Write $a = bq + r$ where $0 \leq r < b$. Then $\gcd(a, b) = \gcd(b, r)$.*

(Step 3). *If $r = 0$ then $b = \gcd(a, b)$ and we are done. If $r \neq 0$ then repeat steps (2) and (3) with b and r until a zero remainder is obtained.*

(Step 4). *The last non-zero remainder is $\gcd(a, b)$.*

Proof. For concreteness, a special case is used to illustrate the general idea. Let

$$
\begin{aligned}
a &= bq_1 + r_1 \\
b &= r_1 q_2 + r_2 \\
r_1 &= r_2 q_3 + r_3 \\
r_2 &= r_3 q_4
\end{aligned}
$$

where $b > r_1 > r_2 > r_3$ and so must eventually become zero by the descending chain condition of Section 3.8. This guarantees that the algorithm always terminates. By Lemma 5.2.4

$$\gcd(a, b) = \gcd(b, r_1) = \gcd(r_1, r_2) = \gcd(r_2, r_3) = \gcd(r_3, 0).$$

Thus $\gcd(a, b) = r_3$, the last non-zero remainder. $\square$

Example 5.2.6. We calculate $\gcd(19, 7)$ using Euclid's algorithm.

$$
\begin{aligned}
19 &= 7 \cdot 2 + 5 \\
7 &= 5 \cdot 1 + 2 \\
5 &= 2 \cdot 2 + 1 * \\
2 &= 1 \cdot 2 + 0.
\end{aligned}
$$

By Lemma 5.2.4 we have that

$$\gcd(19,7) = \gcd(7,5) = \gcd(5,2) = \gcd(2,1) = \gcd(1,0).$$

The last non-zero remainder is 1 and so $\gcd(19,7) = 1$.

There are occasions when we need to extract more information from Euclid's algorithm. The following provides what we need and is due to Etienne Bézout (1730–1783).

Theorem 5.2.7 (Bézout's theorem). *Let a and b be natural numbers. Then there are integers x and y such that $\gcd(a,b) = ax + by$.*

We prove this theorem by describing an algorithm that computes the integers x and y above using the data provided by Euclid's algorithm.

Theorem 5.2.8 (Extended Euclidean algorithm). *Let $a,b \in \mathbb{N}$ where $a \geq b$ and $b \neq 0$. Our goal is to compute numbers $x,y \in \mathbb{Z}$ such that $\gcd(a,b) = ax + by$.*

(Step 1). *Apply Euclid's algorithm to a and b.*

(Step 2). *Using the results from step (1), work from bottom-to-top and rewrite each non-zero remainder in turn beginning with the last non-zero remainder.*

(Step 3). *When this process is completed x and y will have been calculated.*

Proof. Again, a particular case is used to illustrate the general idea. Let

$$
\begin{aligned}
a &= bq_1 + \mathbf{r_1} \\
b &= r_1 q_2 + \mathbf{r_2} \\
r_1 &= r_2 q_3 + \mathbf{r_3} \\
r_2 &= r_3 q_4
\end{aligned}
$$

be an application of Euclid's algorithm, where the non-zero remainders are highlighted. Here $\gcd(a,b) = r_3$. Write the non-zero remainders in reverse order, each one expressed as a difference:

$$
\begin{aligned}
r_3 &= r_1 - r_2 q_3 \\
r_2 &= b - r_1 q_2 \\
r_1 &= a - bq_1.
\end{aligned}
$$

By Euclid's algorithm

$$\gcd(a,b) = r_1 - r_2 q_3.$$

We shall keep the lefthand side fixed but rewrite the righthand side one step at a time. We do this by rewriting each remainder in turn. Thus we rewrite r_2 first. This is equal to $b - r_1 q_2$ so we replace r_2 by this expression to get

$$\gcd(a,b) = r_1 - (b - r_1 q_2)q_3$$

and then simplify

$$\gcd(a,b) = r_1(1+q_2q_3) - bq_3.$$

It remains to rewrite r_1. We get

$$\gcd(a,b) = (a - bq_1)(1+q_2q_3) - bq_3$$

which simplifies to

$$\gcd(a,b) = a(1+q_2q_3) - b(q_1 + q_3 + q_1q_2q_3).$$

Hence $x = 1 + q_2q_3$ and $y = -(q_1 + q_3 + q_1q_2q_3)$. □

Example 5.2.9. The extended Euclidean algorithm is applied to the calculations in Example 5.2.6.

$$
\begin{aligned}
19 &= 7 \cdot 2 + 5 \\
7 &= 5 \cdot 1 + 2 \\
5 &= 2 \cdot 2 + 1.
\end{aligned}
$$

The first step is to rearrange each equation so that the non-zero remainder is alone on the lefthand side.

$$
\begin{aligned}
5 &= 19 - 7 \cdot 2 \\
2 &= 7 - 5 \cdot 1 \\
1 &= 5 - 2 \cdot 2.
\end{aligned}
$$

Next reverse the order of the list

$$
\begin{aligned}
1 &= 5 - 2 \cdot 2 \\
2 &= 7 - 5 \cdot 1 \\
5 &= 19 - 7 \cdot 2.
\end{aligned}
$$

Now begin with the first equation. The lefthand side is the gcd we are interested in. We treat all other remainders as algebraic quantities and systematically substitute them in order. Thus we begin with the first equation

$$1 = 5 - 2 \cdot 2.$$

The next remainder in our list is

$$2 = 7 - 5 \cdot 1$$

so we replace **2** in our first equation by the expression on the right above to get

$$1 = 5 - (7 - 5 \cdot 1) \cdot 2.$$

We now rearrange this equation by collecting up like terms treating the highlighted remainders as algebraic objects to get

$$1 = 3 \cdot 5 - 2 \cdot 7.$$

The next remainder in our list is

$$5 = 19 - 7 \cdot 2$$

so we replace **5** in our new equation by the expression on the right above to get

$$1 = 3 \cdot (19 - 7 \cdot 2) - 2 \cdot 7.$$

Again we rearrange to get

$$1 = 3 \cdot 19 - 8 \cdot 7.$$

The remainders have been used up, the algorithm terminates and so we can write

$$\gcd(19,7) = 3 \cdot 19 + (-8) \cdot 7,$$

as required. Checking is advisable at each stage of the algorithm. A more convenient algorithm will be described in Section 8.7 that simultaneously computes the gcd and finds the integers x and y promised by Bézout's theorem.

Bézout's theorem has a number of applications.

Lemma 5.2.10. *Let a and b be any natural numbers not both zero. Any common divisor of a and b divides $\gcd(a,b)$.*

Proof. By Bézout's theorem there are integers x and y such that $\gcd(a,b) = ax + by$. It is immediate that any number that divides a and b divides $\gcd(a,b)$. □

The next lemma is important and arises whenever coprime numbers are used.

Lemma 5.2.11. *Let a and b be natural numbers. Then a and b are coprime if and only if there are integers x and y such that $1 = ax + by$.*

Proof. Suppose that a and b are coprime. By Bézout's theorem there are integers x and y such that $\gcd(a,b) = 1 = ax + by$. Conversely, suppose that $1 = ax + by$. Then any natural number that divides both a and b divides 1. Hence $\gcd(a,b) = 1$. □

The star lemma of this section is the following.

Lemma 5.2.12 (Gauss's lemma). *Let $a \mid bc$ where a and b are coprime. Then $a \mid c$.*

Proof. By assumption a and b are coprime, and so there are integers x and y such that $1 = ax + by$ by Lemma 5.2.11. Multiply both sides of this equation by c to get $c = axc + byc$. Then a divides the righthand side and so a divides c as required. □

It is worth noting a special case of the above result.

Corollary 5.2.13 (Euclid's lemma). *Let p be a prime. If $p \mid ab$ and p does not divide a then $p \mid b$.*

The results of this section are the key to proving the fundamental theorem of algebra, but they have other applications as well. An important one is in solving linear equations in integers. As intimated in Example 5.2.3, this is more delicate than solving such equations over the rationals, reals or complexes. To conclude this section, we describe how to solve in integers equations of the form

$$ax + by = c,$$

where $a, b, c \in \mathbb{N}$. Incidentally, it is easy to extend the results we prove here to the case where $a, b, c \in \mathbb{Z}$, but this is left as an exercise. By an *integer solution* to the above equation we mean a pair of integers (x_0, y_0) such that $ax_0 + by_0 = c$. This is an example of a *Diophantine equation*. The term 'Diophantine equation' is unusual. The adjective 'Diophantine' refers to the nature of the solutions and not to the nature of the equation. Thus any equation is deemed to be Diophantine if we are only interested in its integer solutions.

Example 5.2.14. The most famous Diophantine equation is $x^n + y^n = z^n$. Fermat's last theorem states that there are no integer solutions where $xyz \neq 0$ for any integer $n \geq 3$.

Lemma 5.2.15 (Existence of a solution). *The equation $ax + by = c$ is such that $a, b, c \in \mathbb{N}$ and non-zero.*

1. *A necessary and sufficient condition for the equation to have an integer solution is that $\gcd(a,b) \mid c$.*

2. *Suppose that $\gcd(a,b) \mid c$. Let $c = q\gcd(a,b)$ for some integer q and let $ax' + by' = \gcd(a,b)$ for some integers x' and y'. Then $(x,y) = (x'q, y'q)$ is a particular solution to the equation.*

Proof. We prove (1) and (2) together. Suppose the equation has the integer solution (x_0, y_0). Then $ax_0 + by_0 = c$. Let $d = \gcd(a,b)$. Then $d \mid ax_0$ and $d \mid by_0$. Thus d divides the lefthand side and so d divides the righthand side. It follows that $d \mid c$. Conversely, suppose that $\gcd(a,b) \mid c$. Then $c = q\gcd(a,b)$ for some integer q. By Theorem 5.2.7, we can find integers x' and y' such that $ax' + by' = \gcd(a,b)$. Multiplying both sides of this equality by q we get that $a(x'q) + b(y'q) = q\gcd(a,b) = c$. Thus $(x'q, y'q)$ is a solution. □

Given that the Diophantine equation does have a solution, it remains to find them all. Let (x,y) denote any integer solution to $ax + by = c$, and let (x_0, y_0) be a particular integer solution. Then $a(x - x_0) + b(y - y_0) = 0$. This motivates the following.

Lemma 5.2.16. *The equation $ax + by = 0$, where $a, b \in \mathbb{N}$, has integer solutions*

$$\left(-\frac{bt}{\gcd(a,b)}, \frac{at}{\gcd(a,b)} \right)$$

where $t \in \mathbb{Z}$.

Proof. Let $d = \gcd(a,b)$. Then $a = a'd$ and $b = b'd$ for integers a' and b', which are coprime by Lemma 5.2.2. Dividing both sides of the equation by d we get

$$a'x + b'y = 0.$$

We therefore reduce to the case where a' and b' are coprime. Clearly $a'x = -b'y$. It follows that a' divides the righthand side. But a' and b' are coprime and so by Gauss's lemma $a' \mid y$. Thus $y = a't$ for some integer t and so also $x = -b't$. It can be checked that $(-b't, a't)$ where $t \in \mathbb{Z}$ really are solutions. ☐

Example 5.2.17. The proof used in Lemma 5.2.16 is really nothing more than the following. Let $\frac{a}{b}$ be a fraction in its lowest terms so that a and b are coprime. Now let $\frac{a}{b} = \frac{c}{d}$. Then $ad = bc$. Since $a \mid bc$ and a and b are coprime, we can use Gauss's lemma to deduce that $c = at$ for some t. It follows immediately that $d = bt$.

We summarize what we have found by combining Lemma 5.2.15 and Lemma 5.2.16.

Theorem 5.2.18 (Diophantine linear equations). *Let $ax + by = c$ be an equation where a, b, c are natural numbers.*

1. *The equation has an integer solution if and only if $\gcd(a,b) \mid c$.*

2. *Let (x_0, y_0) be one particular integer solution to the equation $ax + by = c$. Then all integer solutions have the form*

$$\left(x_0 - \frac{bt}{\gcd(a,b)}, y_0 + \frac{at}{\gcd(a,b)} \right)$$

where $t \in \mathbb{Z}$.

Example 5.2.19. Find all points on the line $2x + 3y = 5$ having integer coordinates. Observe that $\gcd(2,3) = 1$. Thus such points exist. By inspection, $1 = 2 \cdot 2 + (-1)3$ and so $5 = 10 \cdot 2 + (-5)3$. It follows that $(10, -5)$ is one integer solution. Thus the set of all integer solutions is $\{(10 - 3t, -5 + 2t) : t \in \mathbb{Z}\}$.

Exercises 5.2

1. Use Euclid's algorithm to find the gcds of the following pairs of numbers.

    (a) 35, 65.

    (b) 135, 144.

    (c) 17017, 18900.

2. Use the extended Euclidean algorithm to find integers x and y such that $\gcd(a,b) = ax + by$ for each of the following pairs of numbers.

    (a) 112, 267.

(b) 242, 1870.

3. Determine which of the following linear equations have an integer solution and for those that do find all integer solutions.

    (a) $10x + 15y = 7$.

    (b) $5x + 7y = 1$.

    (c) $242x + 1870y = 66$.

4. *This question looks at solving problems not in $\mathbb{Z}$ but in $\mathbb{N}$. New phenomena arise.

    (a) You have an unlimited supply of 3 cent stamps and an unlimited supply of 5 cent stamps. By combining stamps of different values you can make up other values: for example, three 3 cent stamps and two 5 cent stamps make the value 19 cents. What is the largest value you *cannot* make?

    (b) Let $a, b > 0$ be coprime natural numbers. We say that a natural number n can be *represented* by a and b if there are natural numbers, not merely integers, x and y such that $n = ax + by$. Prove that there is a largest number f, called the *Frobenius number* of a and b, such that f *cannot* be represented. Hint: Every non-negative integer n can be written $n = ax + by$ for some integers x and y. Prove that x can be chosen so that $0 \leq x < b$ and that under this condition x and y are unique.

5. *Define $\gcd(a, b, c)$ to be the greatest common divisor of a and b and c jointly. Prove that

$$\gcd(a, b, c) = \gcd(\gcd(a, b), c) = \gcd(a, \gcd(b, c)).$$

Define $\gcd(a, b, c, d)$ to be the greatest common divisor of a and b and c and d jointly. Calculate $\gcd(910, 780, 286, 195)$.

6. *The following question is by Dubisch from *The American Mathematical Monthly* **69** and I learnt about it from [4]. Define $\mathbb{N}^* = \mathbb{N} \setminus \{0\}$. A binary operation $\circ$ defined on $\mathbb{N}^*$ is known to have the following properties.

    (a) $a \circ b = b \circ a$.

    (b) $a \circ a = a$.

    (c) $a \circ (a + b) = a \circ b$.

Prove that $a \circ b = \gcd(a, b)$. Hint: The question is not asking you to prove that $\gcd(a, b)$ has these properties, which it certainly does.

7. *Let a and b be any two natural numbers not both zero. Consider the set

$$I = \{ax + by \colon x, y \in \mathbb{Z}\}.$$

It contains both a and b and so, in particular, contains positive elements. Let d be the smallest positive element of I. Prove that $d = \gcd(a, b)$.

5.3 FUNDAMENTAL THEOREM OF ARITHMETIC

The goal of this section is to state and prove the most basic result about the natural numbers: each natural number, excluding 0 and 1, can be written as a product of powers of primes in essentially one way.

A *proper divisor* of a natural number n is a divisor that is neither 1 nor n. A natural number $n \geq 2$ is said to be *prime* if it has no proper divisors. A natural number $n \geq 2$ which is not prime is said to be *composite*. It is important to remember that the number 1 is not a prime. The only *even prime* is the number 2.

Our first theorem is tangential to the main goal of this section. It generalizes the proof that $\sqrt{2}$ is irrational described in Chapter 2.

Theorem 5.3.1. *The square root of each prime number is irrational.*

Proof. Let p be any prime. Suppose that $\sqrt{p} = \frac{a}{b}$ for some natural numbers a and b. We can assume that $\gcd(a,b) = 1$ by Lemma 5.2.2. Squaring both sides of the equation $\sqrt{p} = \frac{a}{b}$ and multiplying the resulting equation by b^2 we get that

$$pb^2 = a^2.$$

Thus p divides a^2. By Euclid's lemma p divides a. We can therefore write $a = pc$ for some natural number c. Substituting this into our equation above we get that

$$pb^2 = p^2c^2.$$

Dividing both sides of this equation by p gives

$$b^2 = pc^2.$$

Thus p divides b^2 and so p divides b by Euclid's lemma again. We have therefore shown that our assumption that $\sqrt{p}$ is rational leads to both a and b being divisible by p contradicting the fact that $\gcd(a,b) = 1$. It follows that $\sqrt{p}$ is not a rational number. □

The next lemma is the first step in the process of factorizing a number as a product of primes. It was proved as Proposition VII.31 and Proposition VII.32 of Euclid.

Lemma 5.3.2. *Let $n \geq 2$. Either n is prime or its smallest proper divisor is prime.*

Proof. Suppose n is not prime. Let d be the smallest proper divisor of n. If d were not prime then d would have a smallest proper divisor d'. From $d' \mid d$ and $d \mid n$ we obtain $d' \mid n$. But $d' < d$ and this contradicts the choice of d. Thus d must itself be prime. □

The following was proved as Proposition IX.20 of Euclid.

Theorem 5.3.3. *There are infinitely many primes.*

Proof. Let $p_1, \ldots, p_n$ be the first n primes in order. Define $N = (p_1 \ldots p_n) + 1$. If N is a prime, then it is a prime bigger than p_n. If N is composite, then N has a prime divisor p by Lemma 5.3.2. But $p \neq p_i$ for $1 \leq i \leq n$ because N leaves remainder 1 when divided by p_i. It follows that p is a prime bigger than p_n. Thus we can always find a bigger prime and so there must be an infinite number of primes. □

Example 5.3.4. It is interesting to consider some specific cases of the numbers introduced in the proof of Theorem 5.3.3. The first few are already prime.

- $2 + 1 = 3$ prime.

- $2 \cdot 3 + 1 = 7$ prime.

- $2 \cdot 3 \cdot 5 + 1 = 31$ prime.

- $2 \cdot 3 \cdot 5 \cdot 7 + 1 = 211$ prime.

- $2 \cdot 3 \cdot 5 \cdot 7 \cdot 11 + 1 = 2,311$ prime.

- $2 \cdot 3 \cdot 5 \cdot 7 \cdot 11 \cdot 13 + 1 = 30,031 = 59 \cdot 509$.

Box 5.2: The Prime Number Theorem

There are no meaningful formulae to tell us what the nth prime is but there are still some interesting results in this direction. In 1971, Yuri Matiyasevich (b. 1947) found a polynomial in 26 variables of degree 25 with the property that when non-negative integers are substituted for the variables the positive values it takes are all and only the primes [72]. However, this polynomial does not generate the primes in any particular order.

 A different question is to ask how the primes are distributed. For example, are they arranged fairly regularly, or do the gaps between them become ever bigger? We have already noted that there are no formulae which output the nth prime in a usable way, but if we adopt a statistical approach then we can obtain specific results. The idea is that for each natural number n we count the number of primes $\pi(n)$ less than or equal to n. The graph of $\pi(n)$ has a staircase shape and so it is certainly not smooth but as you zoom away it begins to look ever smoother. This raises the question of whether there is a smooth function that is a good approximation to $\pi(n)$. In 1792, the young Gauss observed that $\pi(n)$ appeared to be close to the value of the simple function $\frac{n}{\ln(n)}$. But proving this was always true, and not just an artefact of the comparatively small numbers he looked at, turned out to be difficult. Eventually, in 1896 two mathematicians, Jacques Hadamard (1865–1963) and the spectacularly named Charles Jean Gustave Nicolas Baron de la Vallée Poussin (1866–1962), proved independently of each other that

$$\lim_{x \to \infty} \frac{\pi(x)}{x/\ln(x)} = 1,$$

a result known as the *prime number theorem*. It was proved using complex analysis, that is, calculus using complex numbers. In very rough terms, the prime number theorem really can be interpreted as saying that $\pi(x) \approx \frac{x}{\ln x}$, where $\approx$ means 'approximately equal to', as Gauss surmised. For example, $\pi(1,000,000) = 78,498$ whereas

$$\frac{10^6}{\ln 10^6} = 72,382.$$

Proposition 5.3.5 (Prime test). *To decide whether a number n is prime or composite, check whether any prime $p \leq \sqrt{n}$ divides n. If none of them do, the number n is prime whereas if one of them does, the number is composite.*

Proof. If a divides n then we can write $n = ab$ for some number b. If both $a > \sqrt{n}$ and $b > \sqrt{n}$ then $n = ab > n$, which is nonsense. Thus either $a \leq \sqrt{n}$ or $b \leq \sqrt{n}$. It follows that if n has any proper divisors then at least one of them must be less than or equal to $\sqrt{n}$. Thus if n has no such proper divisors it must be prime. □

Example 5.3.6. We determine whether 97 is prime or not using Proposition 5.3.5. We first calculate the largest whole number less than or equal to $\sqrt{97}$. This is 9. We now carry out trial divisions of 97 by each prime number p where $2 \leq p \leq 9$. If you are not certain which of these numbers is prime: try them all. You will get the right answer although not as efficiently. In this case we carry out trial divisions by $2, 3, 5$ and 7. None of them divides 97, and so 97 is prime.

Box 5.3: Cryptography

Prime numbers play an important rôle in exchanging secret information. In 1976, Whitfield Diffie and Martin Hellman wrote a paper on cryptography [37] that can genuinely be called ground-breaking. They put forward the idea of a *public-key cryptosystem* which would enable

> ...a private conversation ... [to] be held between any two individuals regardless of whether they have ever communicated before.

With considerable farsightedness, Diffie and Hellman realized that such cryptosystems would be essential if communication between computers was to reach its full potential, but they did not describe a practical method for constructing one. It was R. I. Rivest, A. Shamir and L. Adleman (RSA) who found just such a method in 1978 [100]. It is based on the following observation. Given two prime numbers it takes very little time to multiply them together, but given a number that is a product of two primes, it can take a long time to factorize. You might like to think about why in relation to Proposition 5.3.5. After considerable experimentation, RSA showed how to use little more than undergraduate mathematics to put together a public-key cryptosystem that is an essential ingredient in today's e-commerce. Ironically, this secret code had in fact been invented in 1973 at GCHQ, who had kept it secret.

Lemma 5.3.7 (Euclid's lemma: extended play). *Let $p \mid a_1 \ldots a_n$ where p is a prime. Then $p \mid a_i$ for some i.*

Proof. We use induction. The case $i = 2$ is just Euclid's lemma restricted to a prime divisor. Assume the result holds when $n = k > 2$. We prove that it holds for $n = k + 1$. Suppose that $p \mid (a_1 \ldots a_k)a_{k+1}$. From the base case, either $p \mid a_1 \ldots a_k$ or $p \mid a_{k+1}$. We deduce that $p \mid a_i$ for some $1 \leq i \leq k + 1$ using the induction hypothesis. □

We now come to the main theorem of this chapter.

Theorem 5.3.8 (Fundamental theorem of arithmetic). *Every number $n \geq 2$ can be written as a product of primes in one way if we ignore the order in which the primes appear. By* product *we allow the possibility that there is only one prime.*

Proof. Let $n \geq 2$. If n is already a prime then there is nothing to prove, so we suppose that n is composite. By Lemma 5.3.2, the number n has a smallest prime divisor p_1. We can therefore write $n = p_1 n'$ where $n' < n$. Once again, n' is either prime or composite. Continuing in this way leads to n being written as a product of primes in a finite number of steps since the numbers n' arising form a strictly decreasing sequence of natural numbers and so terminates by the descending chain condition. This proves existence. We now prove uniqueness. Suppose that

$$n = p_1 \ldots p_s = q_1 \ldots q_t$$

are two ways of writing n as a product of primes. Now $p_1 \mid n$ and so $p_1 \mid q_1 \ldots q_t$. By Lemma 5.3.7, the prime p_1 must divide one of the q_is and, since they are themselves prime, it must actually equal one of the q_is. By relabelling if necessary, we can assume that $p_1 = q_1$. Cancel p_1 from both sides and repeat with p_2. Continuing in this way, we see that every prime occurring on the lefthand side occurs on the righthand side. Changing sides, we see that every prime occurring on the righthand side occurs on the lefthand side. We deduce that the two prime decompositions are identical. $\square$

When a number is written as a product of primes, it is usual to gather together the same primes into a prime power, and then to write the primes in increasing order giving a unique representation.

Example 5.3.9. Let $n = 999,999$. Write n as a product of primes. There are a number of ways of doing this but in this case there is an obvious place to start. We have that

$$n = 3^2 \cdot 111,111 = 3^3 \cdot 37,037 = 3^3 \cdot 7 \cdot 5,291 = 3^3 \cdot 7 \cdot 11 \cdot 481 = 3^3 \cdot 7 \cdot 11 \cdot 13 \cdot 37.$$

Thus the prime factorization of $999,999$ is $999,999 = 3^3 \cdot 7 \cdot 11 \cdot 13 \cdot 37$ where $3 < 7 < 11 < 13 < 37$.

Box 5.4: Gödel's Theorems

There is an important and subversive application of the unique prime factorization of numbers. We encode words as individual numbers. Begin with a numerical code for letters of the alphabet so $A \leftrightarrow 0$ and $B \leftrightarrow 1$, etc. Thus a word like CAB can be numerically encoded as $2, 0, 1$. We convert this sequence of numbers into a single number. We do this by first writing down the first three primes $2, 3, 5$. Now compute $2^2 3^0 5^1$ to get 20. Thus the number 20 encodes the word CAB. This word is not lost because the uniqueness of the factorization of numbers into powers of primes enables us to find the exponents of the primes and therefore the numerical codes of the letters. For example, we carry out this procedure with the number 40. This can be factorized as $40 = 2^3 \cdot 3^0 \cdot 5^1$ and so we get back the word DAB. This process is called *Gödel numbering*. It seems harmless. Suppose now we encoded the symbols of mathematics in the same way. Then mathematics would be converted into numbers and numbers are themselves part of mathematics, enabling mathematics to talk about mathematics. This idea is a tool in the proof of the most profound results of twentieth century mathematics: *Gödel's theorems* [13]. They can be viewed as either describing the fundamental limitations of axiom systems or as showing the open-ended creative nature of mathematics.

The greatest common divisor of two numbers a and b is the largest number that divides into both a and b. On the other hand, if $a \mid c$ and $b \mid c$ then we say that c is a *common multiple* of a and b. The smallest common multiple of a and b is called the *least common multiple* of a and b and is denoted by $\text{lcm}(a,b)$. You might expect that to calculate the least common multiple we would need a new algorithm, but in fact we can use Euclid's algorithm as the following result shows. Let a and b be natural numbers. Define

$$\min(a,b) = \begin{cases} a & \text{if } a < b \\ b & \text{if } b \le a \end{cases} \quad \text{and} \max(a,b) = \begin{cases} a & \text{if } a > b \\ b & \text{if } b \ge a. \end{cases}$$

Proposition 5.3.10. *Let a and b be natural numbers not both zero. Then*

$$\gcd(a,b) \cdot \text{lcm}(a,b) = ab.$$

Proof. We begin with a special case first. Suppose that $a = p^r$ and $b = p^s$ where p is a prime. Then it is immediate from the properties of exponents that

$$\gcd(a,b) = p^{\min(r,s)} \text{ and } \text{lcm}(a,b) = p^{\max(r,s)}$$

and so, in this special case, we have that $\gcd(a,b) \cdot \text{lcm}(a,b) = ab$. Next suppose that the prime factorizations of a and b are

$$a = p_1^{r_1} \dots p_m^{r_m} \text{ and } b = p_1^{s_1} \dots p_m^{s_m}$$

where the p_i are primes. We can easily determine the prime factorization of $\gcd(a,b)$ when we bear in mind the following points. The primes that occur in the prime factorization of $\gcd(a,b)$ must be from the set $\{p_1, \dots, p_m\}$, and the number $p_i^{\min(r_i,s_i)}$ divides $\gcd(a,b)$ but no higher power does. It follows that

$$\gcd(a,b) = p_1^{\min(r_1,s_1)} \dots p_m^{\min(r_m,s_m)}.$$

A similar argument proves that

$$\text{lcm}(a,b) = p_1^{\max(r_1,s_1)} \dots p_m^{\max(r_m,s_m)}.$$

The proof of the fact that $\gcd(a,b) \cdot \text{lcm}(a,b) = ab$ now follows by multiplying the two prime factorizations together. In the above proof, we assumed that a and b had prime factorizations using the same set of primes. This need not be true in general, but by allowing zero powers of primes we can easily arrange for the same sets of primes to occur and the argument above can then be reused in this case. ☐

The fundamental theorem of arithmetic is just what we need to characterize those proper fractions that can be represented by finite decimals.

Proposition 5.3.11. *A proper rational number $\frac{a}{b}$ in its lowest terms has a finite decimal expansion if and only if $b = 2^m 5^n$ for some natural numbers m and n.*

Proof. Let $\frac{a}{b}$ have the finite decimal representation $0 \cdot a_1 \dots a_n$. This means

$$\frac{a}{b} = \frac{a_1}{10} + \frac{a_2}{10^2} + \dots + \frac{a_n}{10^n}.$$

The righthand side is just the fraction

$$\frac{a_1 10^{n-1} + a_2 10^{n-2} + \ldots + a_n}{10^n}.$$

The denominator contains only the prime factors 2 and 5 and so the reduced fraction will also only contain at most the prime factors 2 and 5.

To prove the converse, consider the proper fraction

$$\frac{a}{2^u 5^v}.$$

If $u = v$ then the denominator is 10^u. If $u < v$ then

$$\frac{a}{2^u 5^v} = \frac{2^{v-u} a}{2^v 5^v} = \frac{2^{v-u} a}{10^v}.$$

If $v < u$ then

$$\frac{a}{2^u 5^v} = \frac{5^{u-v} a}{2^u 5^u} = \frac{5^{u-v} a}{10^u}.$$

In all cases, the resulting fraction has denominator a power of 10. Any fraction with denominator a power of 10 has a finite decimal expansion. □

Box 5.5: Supernatural Numbers

List the primes in order $2,3,5,7,\ldots$. Write each natural number ≥ 2 as a product of *all* primes by including those primes not needed by raising them to the power 0. For example, $10 = 2 \cdot 5 = 2^1 \cdot 3^0 \cdot 5^1 \cdot 7^0 \ldots$. Now encode the number via its exponents. Thus 10 would be written as

$$(1,0,1,0,0,0\ldots).$$

Each natural number ≥ 2 can now be encoded by an infinite sequence of natural numbers that are zero from some point on. Introduce a new symbol ∞ satisfying $a + \infty = \infty = \infty + a$. Define a *supernatural number* to be *any* sequence

$$\mathbf{a} = (a_1, a_2, a_3, \ldots)$$

where the $\mathbf{a}_i = a_i \in \mathbb{N} \cup \{\infty\}$. Denote the set of supernatural numbers by $\mathbb{S}$. Define a *natural number* to be a supernatural number $\mathbf{a}$ where $a_i \in \mathbb{N}$ for all i and $a_i = 0$ for all $i \geq m$ for some m. Let $\mathbf{a}$ and $\mathbf{b}$ be two supernatural numbers. Define their product by $(\mathbf{a} \cdot \mathbf{b})_i = a_i + b_i$. This makes sense because, for example, $10 \cdot 12 = 120$ and

$$(1,0,1,0,0,0\ldots) \cdot (2,1,0,0,0,0\ldots) = (3,1,1,0,0,0\ldots)$$

which encodes $2^3 3^1 5^1 = 120$. Multiplication is associative. Define

$$\mathbf{1} = (0,0,0,\ldots) \text{ and } \mathbf{0} = (\infty, \infty, \infty, \ldots),$$

the identity and zero, respectively. For an application of supernatural numbers see [9].

Example 5.3.12. The fundamental theorem of arithmetic makes the nature of co-prime numbers transparent and explains the term 'coprime'. Let a and b be natural numbers and let $P = \{p_1, \ldots, p_m\}$ be the primes that divide either a or b. Let A be the set of prime factors of a and let B be the set of prime factors of b. Then $P = A \cup B$. The numbers a and b are coprime if and only if $A \cap B = \emptyset$.

Exercises 5.3

1. List the primes less than 100. Hint. Use the *Sieve of Eratosthenes*.[3] This is an algorithm for constructing a table of all primes up to the number N. List all numbers from 2 to N inclusive. Mark 2 as prime and then cross out from the table all numbers which are multiples of 2. The process now iterates. Find the smallest number which is not marked as a prime and which has not been crossed out. Mark it as a prime and cross out all its multiples. If no such number exists then you have determined all primes less than or equal to N.

2. For each of the following numbers use Theorem 5.3.5 to determine whether they are prime or composite. When they are composite find a prime factorization.

   (a) 131.

   (b) 689.

   (c) 5491.

3. Find the lowest common multiples of the following pairs of numbers.

   (a) 22, 121.

   (b) 48, 72.

   (c) 25, 116.

4. Given $2^4 \cdot 3 \cdot 5^5 \cdot 11^2$ and $2^2 \cdot 5^6 \cdot 11^4$, calculate their greatest common divisor and least common multiple.

5. Use the fundamental theorem of arithmetic to show that for each natural number n, the number $\sqrt{n}$ can be written as a product of a natural number and a product of square roots of primes. Calculate the square roots of the following numbers exactly using the above method.

   (a) 10.

   (b) 42.

   (c) 54.

6. *In this exercise, we define some special kinds of prime number.

   (a) Prove that

   $$x^n - y^n = (x-y)(x^{n-1} + x^{n-2}y + \ldots + xy^{n-2} + y^{n-1}).$$

---

[3] Eratosthenes of Cyrene who lived about 250 BCE. He is famous for using geometry and some simple observations to estimate the circumference of the earth.

(b) Prove that

$$x^n + y^n = (x+y)(x^{n-1} - x^{n-2}y + \ldots - xy^{n-2} + y^{n-1})$$

when n is odd.

(c) Prove that

$$x^{mn} - y^{mn} = (x^m - y^m)(x^{m(n-1)} + x^{m(n-2)} + \ldots + y^{m(n-1)}).$$

(d) Prove that a necessary condition for $2^n - 1$ to be prime is that n be prime. Define $M_m = 2^m - 1$, called a *Mersenne number*. If this is prime it is called a *Mersenne prime*. Determining which Mersenne numbers are prime has proved interesting from a computational point of view yielding big prime numbers. Currently (2015) 48 such primes are known. It is not known if there are infinitely many.

(e) Prove that a necessary condition for $2^n + 1$ to be a prime is that n is itself a power of 2. Define $F_m = 2^{2^m} + 1$, called a *Fermat number*. If this is prime it is called a *Fermat prime*. Fermat conjectured that all Fermat numbers were prime. Childs [25] has described this as "one of the least accurate famous conjectures in the history of mathematics". Show that $F_0, \ldots, F_4$ are all prime but that F_5 is composite. No other examples of Fermat primes are known. Despite this, Fermat primes arise naturally in constructing regular polygons using a ruler and compass.

7. *This question describes a method for factorizing numbers that does not involve trial divisions.

   (a) Let n be an odd natural number. Prove that there is a bijection between the set of all pairs of natural numbers (a,b) such that $n = ab$ where $a \geq b > 0$ and the set of all pairs of natural numbers (u,v) such that $n = u^2 - v^2$.

   (b) Factorize
   $$2,027,651,281 = 45,041^2 - 1,020^2$$
   into a non-trivial product.

   (c) Factorize the number $200,819$ by first representing it as a difference of two squares.

8. *This question is about patterns within the primes.

   (a) Prove that every odd prime is either of the form $4n+1$ or $4n+3$ for some n.

   (b) Prove that there are infinitely many primes of the form $4n+3$. [There are also infinitely many primes of the form $4n+1$ but the proof is harder.]

9. *A Pythagorean triple (a,b,c) of positive integers is called *primitive* if $\gcd(a,b,c) = 1$. Prove that exactly one of a and b is even. Without loss of generality assume that b is even. Let p and q be coprime positive integers where $p > q$ exactly one of which is even. Prove that $(p^2 - q^2, 2pq, p^2 + q^2)$ is a primitive Pythagorean triple and that all have this form. Hint: Use Question 9 of Exercises 2.3.

5.4 MODULAR ARITHMETIC

In this section, the equivalence relations of Section 3.5 will be used to construct new kinds of numbers. From an early age, we are taught to think of numbers as being arranged along the *number line*

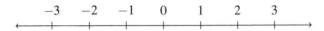

But that is not the only way we count. For example, we count the seasons in a cyclic manner

... autumn, winter, spring, summer ...

likewise the days of the week

... Sunday, Monday, Tuesday, Wednesday, Thursday, Friday, Saturday ...

and the months of the year and the hours in a day. The fact that we use words obscures the fact that we really are counting.[4] In all these cases, counting is not *linear* but *cyclic*. Rather than using a number line to represent this type of counting, we use instead number circles, and rather than using the words above, we use numbers. Here is the number circle for the seasons with numbers replacing words.

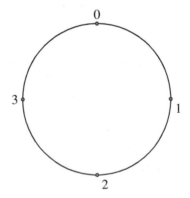

In linear arithmetic, when we add one we move left and when we subtract one we move right. In cyclic arithmetic, when we add one we move clockwise and when we subtract one we move anticlockwise. *Modular arithmetic* is the name given to systems of cyclic counting, and it was Gauss in his 1801 book *Disquisitiones Arithmeticae* who realized that these were mathematically interesting.

The basis of modular arithmetic is the remainder theorem. Let $n \geq 2$ be a fixed natural number which in this context is called the *modulus*. If $a, b \in \mathbb{Z}$ we write $a \equiv b$ (mod n) or just $a \equiv b$ if and only if a and b leave the same remainder when divided by n or, what amounts to the same thing, if and only if $n \mid a - b$.

---

[4]This is clearer in the names for the months since October, November and December were originally the eighth, ninth and tenth months, respectively, until Roman politics led to their being reordered.

Example 5.4.1. If $n = 2$, then $a \equiv b$ if and only if a and b are either both odd or both even. On the other hand, if $n = 10$ then $a \equiv b$ if and only if a and b have the same unit digits.

The symbol $\equiv$ is a modification of the equality symbol $=$ because, as we shall see, it is an equivalence relation. If $a \equiv b$ with respect to n we say that *a is congruent to b modulo n*.

Lemma 5.4.2. *Let $n \geq 2$ be a fixed modulus.*

1. $\equiv$ *is an equivalence relation.*

2. $a \equiv b$ *and* $c \equiv d$ *implies that* $a + c \equiv b + d$.

3. $a \equiv b$ *and* $c \equiv d$ *implies that* $ac \equiv bd$.

Proof. (1) If a is any integer then $n \mid (a - a)$. Thus $\equiv$ is reflexive. If $n \mid (a - b)$ then $n \mid (b - a)$. Thus $\equiv$ is symmetric. We prove that $\equiv$ is transitive. Suppose that $a \equiv b$ and $b \equiv c$. Then $n \mid (a - b)$ and $n \mid (b - c)$. By definition $a - b = nx$ and $b - c = ny$ for some integers x and y. Adding these equations together, we obtain $a - c = nx + ny = n(x + y)$. It follows that $n \mid (a - c)$ and so $a \equiv c$, as required.

(2) Suppose that $a \equiv b$ and $c \equiv d$. Then $n \mid (a - b)$ and $n \mid (c - d)$. By definition $a - b = nx$ and $c - d = ny$ for some integers x and y. Adding these equations together, we obtain $(a + c) - (b + d) = nx + ny = n(x + y)$. It follows that

$$n \mid [(a + c) - (b + d)]$$

and so $a + c \equiv b + d$.

(3) Suppose that $a \equiv b$ and $c \equiv d$. Then $n \mid (a - b)$ and $n \mid (c - d)$. By definition $a - b = nx$ and $c - d = ny$ for some integers x and y. Thus $a = b + nx$ and $c = d + ny$. Multiplying these equations together, we obtain

$$ac = bd + n(yb + xd + nxy)$$

and so $ac \equiv bd$. $\qquad\square$

Here is a simple application of modular arithmetic.

Lemma 5.4.3. *A natural number n is divisible by 9 if and only if the sum of the digits of n is divisible by 9.*

Proof. We work modulo 9. The proof hinges on the fact that $10 \equiv 1$ modulo 9. Thus $10^r \equiv 1$ for all natural numbers $r \geq 1$ by Lemma 5.4.2. Let

$$n = a_n 10^n + a_{n-1} 10^{n-1} + \ldots + a_1 10 + a_0.$$

Then $n \equiv a_n + \ldots + a_0$. Thus n and the sum of the digits of n leave the same remainder when divided by 9, and so n is divisible by 9 if and only if the sum of the digits of n is divisible by 9. $\qquad\square$

We can study the solution of algebraic equations of all kinds within modular arithmetic just as we can within more familiar arithmetic. We focus here on the simplest kinds of linear equation. The results are essentially the same as those from Section 5.2 viewed through the prism of modular arithmetic. For completeness, everything is proved from scratch.

Lemma 5.4.4. *Let a, b and c be integers. Then the following are equivalent.*

1. *The pair (x_1, y_1) is an integer solution to $ax + by = c$ for some y_1.*

2. *The integer x_1 is a solution to the equation $ax \equiv c \pmod{b}$.*

Proof. $(1) \Rightarrow (2)$. Suppose that $ax_1 + by_1 = c$. Then $ax_1 \equiv c \pmod{b}$ by definition.
   $(2) \Rightarrow (1)$. Suppose that $ax_1 \equiv c \pmod{b}$. By definition, $ax_1 - c = bz_1$ for some integer z_1. Thus $ax_1 + b(-z_1) = c$. We may therefore put $y_1 = -z_1$. □

Lemma 5.4.4 shows that solving Diophantine linear equations in two unknowns can be reinterpreted as solving equations in one unknown of the form $ax \equiv b \pmod{n}$.

Lemma 5.4.5. *Consider the linear congruence $ax \equiv b \pmod{n}$. Put $d = \gcd(a, n)$.*

1. *This linear congruence has a solution if and only if $d \mid b$.*

2. *If the condition in part (1) holds and x_0 is any solution, then all solutions have the form*

$$x = x_0 + t\frac{n}{d}$$

where $t \in \mathbb{Z}$.

Proof. (1) Let x_1 be a solution to the linear congruence. By definition $ax_1 - b = nq$ for some integer q. Thus $ax_1 + n(-q) = b$. But $d \mid a$ and $d \mid n$ and so $d \mid b$, as required. We prove the converse. By Bézout's theorem, there are integers u and v such that $au + nv = d$. By assumption, $d \mid b$ and so $b = dw$ for some integer w. It follows that $auw + nvw = dw = b$. Thus $a(uw) \equiv b \pmod{n}$, and we have found a solution.
   (2) Let x_0 be a particular solution to $ax \equiv b \pmod{n}$ and let x be any solution to $ax \equiv b \pmod{n}$. Then $a(x - x_0) \equiv 0 \pmod{n}$. By definition $a(x - x_0) = sn$ for some integer s. Let $a = da'$ and $n = dn'$ for some a' and n' where a' and n' are coprime. Then $a'(x - x_0) = sn'$. By Euclid's lemma, $s = a't$ for some t. It follows that $x = x_0 + n't$, as required. □

There is a special case of the above result that is important. Its proof is immediate.

Corollary 5.4.6. *Let p be a prime. Then the linear congruence $ax \equiv b \pmod{p}$, where a is not congruent to 0 modulo p, always has a solution, and all solutions are congruent modulo p.*

These results lead us to think about modular arithmetic in a different, more holistic, way. We work modulo n and refer to *congruence classes* instead of *equivalence classes*. We denote the set of congruence classes by $\mathbb{Z}/n\mathbb{Z}$ and the congruence class containing a by $[a]$. Define

$$[a] + [b] = [a+b] \text{ and } [a][b] = [ab].$$

At this point, we have to pause and ask ourselves whether these definitions make sense. Take the first. The problem is that if $[a] = [a']$ and $[b] = [b']$ can we be certain that $[a+b] = [a'+b']$? If not, then we have not in fact defined a binary operation at all. However, $[a] = [a']$ means $a \equiv a'$ and $[b] = [b']$ means $b \equiv b'$. By Lemma 5.4.2, this implies that $a+b \equiv a'+b'$ and so we really do have that $[a+b] = [a'+b']$. Similar reasoning applies to the definition of multiplication. Thus the set $\mathbb{Z}/n\mathbb{Z}$ is equipped with two binary operations.

Example 5.4.7. Here are the addition

+	[0]	[1]	[2]	[3]	[4]	[5]
[0]	[0]	[1]	[2]	[3]	[4]	[5]
[1]	[1]	[2]	[3]	[4]	[5]	[0]
[2]	[2]	[3]	[4]	[5]	[0]	[1]
[3]	[3]	[4]	[5]	[0]	[1]	[2]
[4]	[4]	[5]	[0]	[1]	[2]	[3]
[5]	[5]	[0]	[1]	[2]	[3]	[4]

and multiplication tables

·	[0]	[1]	[2]	[3]	[4]	[5]
[0]	[0]	[0]	[0]	[0]	[0]	[0]
[1]	[0]	[1]	[2]	[3]	[4]	[5]
[2]	[0]	[2]	[4]	[0]	[2]	[4]
[3]	[0]	[3]	[0]	[3]	[0]	[3]
[4]	[0]	[4]	[2]	[0]	[4]	[2]
[5]	[0]	[5]	[4]	[3]	[2]	[1]

for $\mathbb{Z}/6\mathbb{Z}$. Addition tables in modular arithmetic are always similar to the one above with the numbers simply being cycled, but the multiplication tables are more interesting.

Define $\mathbb{Z}_n = \{0, 1, \dots, n-1\}$. Every integer is congruent modulo n to exactly one element of this set. Thus $\mathbb{Z}_n$ is an example of a transversal of the congruence classes. The arithmetic of $\mathbb{Z}/n\mathbb{Z}$ is usually carried out in the set $\mathbb{Z}_n$. When calculations lead outside of the set $\mathbb{Z}_n$, we shift them back in by adding or subtracting suitable multiples of n.

Example 5.4.8. Here is the multiplication table of $\mathbb{Z}_5$.

·	0	1	2	3	4
0	0	0	0	0	0
1	0	1	2	3	4
2	0	2	4	1	3
3	0	3	1	4	2
4	0	4	3	2	1

Observe that apart from the zero row, every row is a permutation of the set $\mathbb{Z}_5$.

The following theorem encapsulates a conceptual advance in thinking about modular arithmetic.

Theorem 5.4.9. *Let $n \geq 2$ be any natural number.*

1. *The set $\mathbb{Z}_n$ with the operations of addition and multiplication defined above satisfies all the axioms (F1) – (F11) of Section 4.2 except axiom (F7).*

2. *If n is a prime then (F7) also holds and so $\mathbb{Z}_n$ is a field.*

Proof. The proof of part (2) is a direct consequence of the proof of part (1) and Corollary 5.4.6. We prove part (1) by giving some representative proofs to establish the pattern. To prove the associativity of multiplication, we have to show that $([a][b])[c] = [a]([b][c])$. Consider the lefthand side. By definition $[a][b] = [ab]$ and $[ab][c] = [(ab)c]$. By a similar argument, the righthand side is $[a(bc)]$. We now use the fact that multiplication in $\mathbb{Z}$ is associative and we get the desired result. A similar proof shows that addition is associative. This sets the tone for all the proofs. The properties of $\mathbb{Z}$ seep through to the properties of modular arithmetic except that the resulting system is finite. Observe that $[0]$ is the additive identity and that $[1]$ is the multiplicative identity. The additive inverse of $[a]$ is $[n-a]$. $\qquad\square$

We underline what part (2) of Theorem 5.4.9 is saying. If p is a prime, then $\mathbb{Z}_p$ is a finite arithmetic that enjoys the same properties as the rationals, reals and, as we shall see, complexes. Thus addition, subtraction, multiplication and division (but not by zero!) can be carried out ad lib. This arithmetic uses only the set $\mathbb{Z}_p = \{0, 1, \ldots, p-1\}$.

When doing arithmetic in $\mathbb{Z}_n$ where n is not prime, we have to be more cautious. Addition, subtraction and multiplication can be freely enjoyed, but division is problematic. We shall encounter a similar issue when we study matrix arithmetic in Chapter 8.

These new kinds of arithmetic are not toys.[5] They are important mathematically and also have important applications to information theory, two of which are touched on in this book: the first is an application to error-correcting codes described in Example 8.1.14, and the second to cryptography described in Example 8.5.17.

---

[5]I think this point needs emphasizing. Real mathematics is not only the study of real numbers or, ironically enough, complex numbers. It is also the study of the less familiar sorts of numbers met with at university. Just because they are not familiar does not render them less important.

Example 5.4.10. The mathematical importance of modular arithmetic is illustrated by the following problem. The equation $x^2 - 3y^2 = 19$ represents a hyperbola in the plane. We prove that it has no integer solutions. Suppose that (x_0, y_0) were an integer solution. Then $x_0^2 - 3y_0^2 = 19$. Working modulo 4 we obtain $x_0^2 + y_0^2 \equiv 3$. The squares modulo 4 are 0 and 1. No sum of two of these squares can be congruent to 3. It follows therefore that the equation $x^2 - 3y^2 = 19$ can have no integer solutions.

Multiplication in $\mathbb{Z}_n$ is associative and there is an identity, so it is natural to ask about the invertible elements. We denote them by U_n. By Section 4.1, the set U_n is closed under multiplication. Define $\phi(n) = |U_n|$. The function ϕ is called the *Euler totient function* and plays an important rôle in number theory.

Lemma 5.4.11. $a \in U_n$ *if and only if* $\gcd(a, n) = 1$.

Proof. Suppose that $a \in U_n$. By definition there is $b \in \mathbb{Z}_n$ such that $ab = 1$ which translates into $ab \equiv 1 \pmod{n}$. By definition, $n \mid (ab - 1)$ and so $ab - 1 = nx$ for some integer x. Rearranging, we get $ab - nx = 1$. But then $\gcd(a, n) = 1$ by Lemma 5.2.11. Conversely, suppose that $\gcd(a, n) = 1$. Then $ax + ny = 1$ for some integers x and y by Lemma 5.2.11. Thus $ax \equiv 1 \pmod{n}$. Let $x \equiv b \pmod{n}$ where $b \in \mathbb{Z}_n$. Then $ab = 1$ in $\mathbb{Z}_n$ and so $a \in U_n$, as claimed. □

Example 5.4.12. We determine the invertible elements of $(\mathbb{Z}_{26}, \times)$ and their corresponding inverses. By Lemma 5.4.11, the set U_{26} consists of all natural numbers a such that $1 \leq a \leq 25$ and $\gcd(a, 26) = 1$. The following table lists these numbers and their corresponding inverses.

a	1	3	5	7	9	11	15	17	19	21	23	25
a^{-1}	1	9	21	15	3	19	7	23	11	5	17	25

Thus $U_{26} = \{1, 3, 5, 7, 9, 11, 15, 17, 19, 21, 23, 25\}$. To understand this table, we calculate an example and show that $11^{-1} = 19$. This is equivalent to showing that $11 \times 19 \equiv 1 \pmod{26}$. But $11 \times 19 = 209$ and the remainder when 209 is divided by 26 is 1, as required. The other cases in the table can be checked in a similar way.

We conclude this section with an application of congruences to revealing a hidden pattern in the primes. Let n be a natural number. Clearly $n! \equiv 0 \pmod{n}$, but the value of $(n - 1)!$ modulo n turns out to be interesting.

Theorem 5.4.13 (Wilson's Theorem). *Let n be a natural number. Then n is a prime if and only if*

$$(n - 1)! \equiv n - 1 \pmod{n}.$$

Since $n - 1 \equiv -1 \pmod{n}$ this is usually expressed in the form

$$(n - 1)! \equiv -1 \pmod{n}.$$

Proof. We prove first that if n is prime then $(n-1)! \equiv n-1 \pmod{n}$. When $n = 2$ we have that $(n-1)! = 1$ and $n - 1 = 1$ and so the result holds. We can therefore assume $n \geq 3$. By Corollary 5.4.6, for each $1 \leq a \leq n-1$ there is a unique number $1 \leq b \leq n-1$ such that $ab \equiv 1 \pmod{n}$. If $b \neq a$ then by reordering the terms of the product $(n-1)!$, the numbers a and b effectively cancel each other out. If $a = b$ then $a^2 \equiv 1 \pmod{n}$ which means that $n \mid (a-1)(a+1)$. Since n is a prime either $n \mid (a-1)$ or $n \mid (a+1)$ by Euclid's lemma. This is only possible if $a = 1$ or $a = n-1$. Thus $(n-1)! \equiv n-1 \pmod{n}$, as claimed.

We now prove the converse: if $(n-1)! \equiv n-1 \pmod{n}$ then n is prime. Observe that when $n = 1$ we have that $(n-1)! = 1$ which is not congruent to 0 modulo 1. When $n = 4$, we get that $(4-1)! \equiv 2 \pmod 4$. Suppose that $n > 4$ is not prime. Then $n = ab$ where $1 < a, b < n$ and $a > 2$. If $a \neq b$ then ab occurs as a factor of $(n-1)!$ and so this is congruent to 0 modulo n. If $a = b$ then a occurs in $(n-1)!$ and so does $2a$ (a consequence of our assumption that $n > 4$). Thus n is again a factor of $(n-1)!$. □

Box 5.6: When is a Number Prime?

Wilson's theorem is not just a curiosity. You might think the only way to show a number is prime is to apply the definition and carry out a sequence of trial divisions. Wilson's theorem suggests that you can show a number is prime in a completely different way: in this case, by checking whether n *divides* $(n-1)! + 1$. Although it is not a practical test for deciding whether a number is prime or composite, since $n!$ gets very big very quickly, it opens up the possibility of 'backdoor ways' of showing that a number is prime. Perhaps, also, one of those backdoor ways might be fast. In fact, there is a fast backdoor way of deciding whether a number is prime called the *AKS test*, named for its discoverers Agrawal, Kayal and Saxena [51].

Exercises 5.4

1. Prove that a natural number n is divisible by 3 if and only if the sum of its digits is divisible by 3.

2. Prove that a necessary condition for a natural number n to be the sum of two squares is that $n \equiv 0, 1$ or $2 \pmod 4$.

3. Construct a table of the values of $\phi(n)$ for $1 \leq n \leq 12$.

4. Construct Cayley tables of the following with respect to multiplication.

   (a) U_8.

   (b) U_{10}.

   (c) U_{12}.

5. *This question deals with solving quadratic equations $x^2 + ax + b = 0$ in $\mathbb{Z}_{13}$. Put $\Delta = a^2 + 9b$.

   (a) Determine the elements of $\mathbb{Z}_{13}$ that have square roots.

(b) Prove that $(x+7a)^2 = 10\Delta$.

For each of the following quadratics over $\mathbb{Z}_{13}$ calculate Δ and therefore their roots.

(c) $x^2 + 2x + 1 = 0$.

(d) $x^2 + 3x + 2 = 0$.

(e) $x^2 + 3x + 1 = 0$.

5.5 CONTINUED FRACTIONS

The goal of this section is to show how some of the ideas we have introduced are connected. It is not needed anywhere else in this book. We begin by returning to an earlier calculation. We used Euclid's algorithm to calculate $\gcd(19,7)$ as follows.

$$
\begin{aligned}
19 &= 7 \cdot 2 + 5 \\
7 &= 5 \cdot 1 + 2 \\
5 &= 2 \cdot 2 + 1 \\
2 &= 1 \cdot 2 + 0.
\end{aligned}
$$

We first rewrite each line, except the last, as follows

$$
\begin{aligned}
\frac{19}{7} &= 2 + \frac{5}{7} \\
\frac{7}{5} &= 1 + \frac{2}{5} \\
\frac{5}{2} &= 2 + \frac{1}{2}.
\end{aligned}
$$

Take the first equality

$$
\frac{19}{7} = 2 + \frac{5}{7}.
$$

The fraction $\frac{5}{7}$ is the reciprocal of $\frac{7}{5}$, and from the second equality

$$
\frac{7}{5} = 1 + \frac{2}{5}.
$$

If we combine them, we get

$$
\frac{19}{7} = 2 + \frac{1}{1 + \frac{2}{5}}
$$

however strange this may look. We may repeat the process to get

$$
\frac{19}{7} = 2 + \cfrac{1}{1 + \cfrac{1}{2 + \frac{1}{2}}}
$$

Fractions like this are called *continued fractions*. Given

$$2 + \cfrac{1}{1 + \cfrac{1}{2 + \frac{1}{2}}}$$

you could work out what the usual rational expression was by working from the bottom up. First compute the part in bold below

$$2 + \cfrac{1}{1 + \cfrac{1}{\mathbf{2 + \frac{1}{2}}}}$$

to get

$$2 + \cfrac{1}{1 + \cfrac{1}{\frac{5}{2}}}$$

which simplifies to

$$2 + \cfrac{1}{1 + \frac{2}{5}}.$$

This process can be repeated and we eventually obtain a standard fraction.

The theory of continued fractions will not be developed here, but we shall sketch how they can be used to represent any real number. Let r be a real number. We can write r as $r = m_1 + r_1$ where $0 \le r_1 < 1$. For example, π may be written as $\pi = 3 \cdot 14159265358\ldots$ where here $m = 3$ and $r_1 = 0 \cdot 14159265358\ldots$. If $r_1 \ne 0$ then since $r_1 < 1$, we have that $\frac{1}{r_1} > 1$. We can therefore repeat the above process and write $\frac{1}{r_1} = m_2 + r_2$ where once again $r_2 < 1$. This is an analogue of Euclid's algorithm for real numbers. In fact, we can write

$$r = m_1 + \cfrac{1}{m_2 + r_2},$$

and we can continue the above process with r_2. Thus a continued fraction representation of r can be obtained with the big difference from the first example that it might be infinite. One of the reasons continued fractions are important is that they are a natural, more intrinsic, way of representing real numbers. This should be contrasted with the standard way of representing the decimal part of a real number in base 10 where, mathematically speaking, 10 is arbitrary.

Example 5.5.1. We apply the above process to $\sqrt{3}$. Clearly, $1 < \sqrt{3} < 2$. Thus we can write

$$\sqrt{3} = 1 + (\sqrt{3} - 1)$$

where $\sqrt{3} - 1 < 1$. We focus on

$$\cfrac{1}{\sqrt{3} - 1}.$$

To convert this into a more usable form we *rationalize the denominator* by multiplying numerator and denominator by $\sqrt{3}+1$ to obtain

$$\frac{1}{\sqrt{3}-1} = \frac{1}{2}(\sqrt{3}+1).$$

It is clear that $1 < \frac{1}{2}(\sqrt{3}+1) < 1\frac{1}{2}$. Thus

$$\frac{1}{\sqrt{3}-1} = 1 + \frac{\sqrt{3}-1}{2}.$$

We focus on

$$\frac{2}{\sqrt{3}-1}$$

which simplifies to $\sqrt{3}+1$ by rationalizing the denominator. Clearly

$$2 < \sqrt{3}+1 < 3.$$

Thus $\sqrt{3}+1 = 2+(\sqrt{3}-1)$. However, we have now come full circle. We assemble what we have found. We have that

$$\sqrt{3} = 1 + \cfrac{1}{1 + \cfrac{1}{2+(\sqrt{3}-1)}}.$$

However, we saw above that the pattern repeats at $\sqrt{3}-1$, so what we actually have is

$$\sqrt{3} = 1 + \cfrac{1}{1 + \cfrac{1}{2 + \cfrac{1}{1 + \cfrac{1}{\cdots}}}}.$$

Continued fractions are typographically challenging. The continued fraction representation of $\sqrt{3}$ would usually be written

$$\sqrt{3} = [1; 1, 2, 1, 2, \ldots].$$

We illustrate one way in which algebra and geometry interact with the help of continued fractions. We begin with a problem that looks extremely artificial, but appearances are deceptive. In his book, *Liber Abaci*, Leonardo Fibonacci (c.1170–c.1250) posed the following puzzle.[6]

> "A certain man put a pair of rabbits in a place surrounded on all sides by a wall. How many pairs of rabbits can be produced from that pair in a year if it is supposed that every month each pair begets a new pair which from the second month on becomes productive?"

---

[6]The wording is taken from *MacTutor*.

We spell out the rules explicitly.

1. The problem begins with one pair of immature rabbits.[7]

2. Each immature pair of rabbits takes one month to mature.

3. Each mature pair of rabbits produces a new immature pair at the end of a month.

4. The rabbits are immortal.[8]

The important point is that we must solve the problem using the rules we have been given. To do this, we draw pictures. Represent an immature pair of rabbits by ○ and a mature pair by ●. Rule 2 is represented by

and Rule 3 is represented by

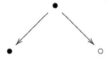

Rule 1 tells us that we start with ○. Applying the rules we obtain the following picture for the first 4 months.

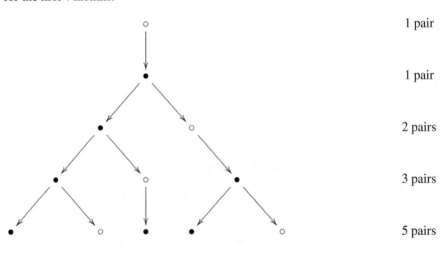

1 pair
1 pair
2 pairs
3 pairs
5 pairs

---

[7]Fibonacci himself seems to have assumed that the starting pair was already mature but we shall not.
[8]These are mathematical idealizations of rabbits.

We start with 1 pair and at the end of the first month we still have 1 pair, at the end of the second month 2 pairs, at the end of the third month 3 pairs, and at the end of the fourth month 5 pairs. Write this as $F_0 = 1$, $F_1 = 1$, $F_2 = 2$, $F_3 = 3$, $F_4 = 5$, and so on. Thus the problem will be solved if we can compute F_{12}. There is an apparent pattern in the sequence of numbers $1, 1, 2, 3, 5, \ldots$ after the first two terms in the sequence, in that each number is the sum of the previous two. We check this pattern really holds. Suppose that the number of immature pairs of rabbits at a given time t is I_t and the number of mature pairs is M_t. Then using our rules at time $t + 1$ we have that $M_{t+1} = M_t + I_t$ and $I_{t+1} = M_t$. Thus

$$F_{t+1} = 2M_t + I_t.$$

Similarly

$$F_{t+2} = 3M_t + 2I_t.$$

It is now easy to check that

$$F_{t+2} = F_{t+1} + F_t.$$

The sequence of numbers such that $F_0 = 1$, $F_1 = 1$ and satisfying the rule $F_{t+2} = F_{t+1} + F_t$ is called the *Fibonacci sequence*. We have that

$$F_0 = 1, F_1 = 1, F_2 = 2, F_3 = 3, F_4 = 5, F_5 = 8, F_6 = 13, F_7 = 21,$$

$$F_8 = 34, F_9 = 55, F_{10} = 89, F_{11} = 144, F_{12} = 233.$$

The solution to the original question is therefore 233 pairs of rabbits.

Fibonacci numbers arise in diverse situations, most famously in *phyllotaxis*, the study of how leaves and petals are arranged on plants.[9] We shall now derive a formula to calculate F_n directly. To figure out how to do this, we apply an idea due to Johannes Kepler (1571–1630), and look at the behaviour of the fractions $\frac{F_{n+1}}{F_n}$ as n gets ever larger. Some calculations are tabulated below.

$\frac{F_1}{F_0}$	$\frac{F_2}{F_1}$	$\frac{F_3}{F_2}$	$\frac{F_4}{F_3}$	$\frac{F_5}{F_4}$	$\frac{F_6}{F_5}$	$\frac{F_7}{F_6}$	$\frac{F_{14}}{F_{13}}$
1	2	$1 \cdot 5$	$1 \cdot 6$	$1 \cdot 625$	$1 \cdot 615$	$1 \cdot 619$	$1 \cdot 6180$

These ratios seem to be going somewhere and the question is where. Observe that

$$\frac{F_{n+1}}{F_n} = \frac{F_n + F_{n-1}}{F_n} = 1 + \frac{F_{n-1}}{F_n} = 1 + \frac{1}{\frac{F_n}{F_{n-1}}}.$$

[9]The Fibonacci numbers also have an unexpected connection with Euclid's algorithm. It is a theorem of Gabriel Lamé (1795–1870) that the number of divisions needed in the application of Euclid's algorithm is less than or equal to five times the number of digits in the smaller of the two numbers whose gcd is being calculated. The proof hinges on the fact that the sequence of non-zero remainders obtained in applying Euclid's algorithm, in reverse order, must be at least as big as the sequence of Fibonacci numbers $F_2, F_3, \ldots$.

For large n we suspect that $\frac{F_{n+1}}{F_n}$ and $\frac{F_n}{F_{n-1}}$ might be almost the same. This suggests, but does not prove, that we need to find the positive solution x to

$$x = 1 + \frac{1}{x}.$$

Thus x is a number with the property that when you take its reciprocal and add 1 you get x back again. This problem is really the quadratic equation $x^2 - x - 1 = 0$ in disguise. This can be solved to give two solutions

$$\phi = \frac{1 + \sqrt{5}}{2} \text{ and } \bar{\phi} = \frac{1 - \sqrt{5}}{2}.$$

The number ϕ is called *the golden ratio*, about which a deal of nonsense has been written. We go back to see if this calculation makes sense. Calculate ϕ and we obtain

$$\phi = 1 \cdot 618033988\ldots$$

and using a calculator

$$\frac{F_{19}}{F_{18}} = \frac{6765}{4181} = 1 \cdot 618033963$$

which is pretty close.

Define

$$f_n = \frac{1}{\sqrt{5}} \left(\phi^{n+1} - \bar{\phi}^{n+1} \right).$$

We prove that $F_n = f_n$. To do this, we use the following equalities

$$\phi - \bar{\phi} = \sqrt{5}, \quad \phi^2 = \phi + 1 \text{ and } \bar{\phi}^2 = \bar{\phi} + 1.$$

- Begin with f_0. We know that

$$\phi - \bar{\phi} = \sqrt{5}$$

and so we really do have that $f_0 = 1$. To calculate f_1 we use the other formulae and again we get $f_1 = 1$.

- We now calculate $f_n + f_{n+1}$.

$$
\begin{aligned}
f_n + f_{n+1} &= \frac{1}{\sqrt{5}} \left(\phi^{n+1} - \bar{\phi}^{n+1} \right) + \frac{1}{\sqrt{5}} \left(\phi^{n+2} - \bar{\phi}^{n+2} \right) \\
&= \frac{1}{\sqrt{5}} \left(\phi^{n+1} + \phi^{n+2} - (\bar{\phi}^{n+1} + \bar{\phi}^{n+2}) \right) \\
&= \frac{1}{\sqrt{5}} \left(\phi^{n+1} (1 + \phi) - \bar{\phi}^{n+1} (1 + \bar{\phi}) \right) \\
&= \frac{1}{\sqrt{5}} \left(\phi^{n+1} \phi^2 - \bar{\phi}^{n+1} \bar{\phi}^2 \right) \\
&= \frac{1}{\sqrt{5}} \left(\phi^{n+3} - \bar{\phi}^{n+3} \right) = f_{n+2}.
\end{aligned}
$$

Because f_n and F_n start in the same place and satisfy the same rules, we have therefore proved that

$$F_n = \tfrac{1}{\sqrt{5}} \left(\phi^{n+1} - \bar{\phi}^{n+1} \right).$$

This is a remarkable formula not least because the numbers on the lefthand side are integers whereas both ϕ and $\bar{\phi}$ are irrational. At this point, we can go back and verify our original idea that the fractions $\frac{F_{n+1}}{F_n}$ seem to approach ϕ as n increases. Define

$$\alpha = \frac{\bar{\phi}}{\phi} \text{ and } \beta = \frac{\phi}{\bar{\phi}}.$$

Then $|\alpha| < 1$ and $|\beta| > 1$. We have that

$$\frac{F_{n+1}}{F_n} = \frac{\phi}{1 - \alpha^{n+1}} - \left(\frac{\bar{\phi}}{\beta^{n+1} - 1} \right).$$

As n increases, the first term approaches ϕ whereas the second term approaches zero. Thus we have proved that $\frac{F_{n+1}}{F_n}$ really is close to ϕ when n is large.

So much for the algebra, now for the geometry. Below is a picture of a regular pentagon with side 1. We claim that the length of a diagonal, such as BE, is equal to ϕ.

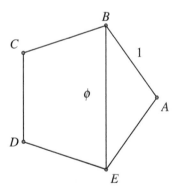

To prove this we use Ptolomy's theorem from Exercises 2.3. Concentrate on the cyclic quadrilateral formed by the vertices $ABDE$.

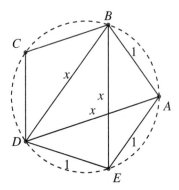

Let the side of a diagonal be x. Then $x^2 = 1 + x$ by Ptolomy's theorem. This is precisely the quadratic equation we solved above. Its positive solution is ϕ and so the length of a diagonal of a regular pentagon with side 1 is ϕ. This raises the question of whether we can see the Fibonacci numbers in the regular pentagon. The answer is almost. Consider the diagram below.

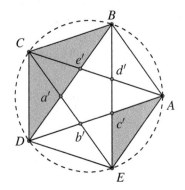

The triangle BCD is similar to the triangle $Ac'E$. This means that they have exactly the same shapes just different sizes. It follows that

$$\frac{Ac'}{AE} = \frac{BC}{BD}.$$

But AE and BC are sides of the pentagon and so have unit lengths, and BD is of length ϕ. Thus

$$Ac' = \frac{1}{\phi}.$$

Now Dc' has the same length as BC which is a side of the pentagon. Thus $Dc' = 1$. We now have

$$\phi = DA = Dc' + Ac' = 1 + \frac{1}{\phi}.$$

Thus, just from the geometry, we get that

$$\phi = 1 + \frac{1}{\phi}.$$

This formula is self-similar, and so may be substituted inside itself to get

$$\phi = 1 + \cfrac{1}{1 + \frac{1}{\phi}}$$

and again

$$\phi = 1 + \cfrac{1}{1 + \cfrac{1}{1 + \frac{1}{\phi}}}$$

and again

$$\phi = 1 + \cfrac{1}{1 + \cfrac{1}{1 + \cfrac{1}{1 + \frac{1}{\phi}}}}$$

ad nauseam. We therefore obtain a continued fraction. For each of these fractions cover up the term $\frac{1}{\phi}$ and then calculate what you see to get

$$1, \quad 1 + \frac{1}{1} = 2, \quad 1 + \cfrac{1}{1 + \frac{1}{1}} = \frac{3}{2}, \quad 1 + \cfrac{1}{1 + \cfrac{1}{1 + \frac{1}{1}}} = \frac{5}{3}, \ldots$$

and the Fibonacci sequence reappears.

CHAPTER 6

Complex numbers

" Entre deux vérités du domaine réel, le chemin le plus facile et le plus court passe bien souvent par le domaine complexe." – Paul Painlevé

Why be one-dimensional when you can be two-dimensional?

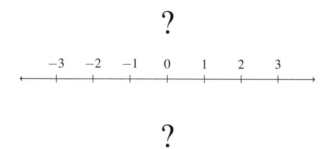

We begin by returning to the familiar number line. The question marks are placed where there appear to be no numbers. We shall rectify this by defining the complex numbers which provide us with a *number plane* rather than just a *number line*. Complex numbers play a fundamental rôle in mathematics. For example, in this chapter they will be used to show that e and π, numbers of radically different origins, are in fact connected.

6.1   COMPLEX NUMBER ARITHMETIC

In the set of real numbers we can add, subtract, multiply and divide at will, but we cannot always extract square roots. For example, the real number 1 has the two real square roots 1 and -1, whereas the real number -1 has no real square roots, the reason being that the square of any real non-zero number is always positive.[1] In this section, we repair this lack of square roots and, as we shall learn, we shall in fact have achieved far more than this. Complex numbers were first studied in the 1500s

---

[1]On a point of notation: if r is a positive real number then $\sqrt{r}$ is usually interpreted to mean the positive square root. If both square roots need to be considered we write $\pm\sqrt{r}$.

but were only fully accepted in the 1800s. We make the following assumption that will be fully justified in Section 6.4.

> *We introduce a new number, denoted by i, whose defining property is that $i^2 = -1$. Observe that i is not merely an algebraic symbol like x and y. We assume that in all other respects it satisfies the usual axioms of high-school algebra. Define a* complex number *to be an expression of the form $z = a + bi$ where $a, b \in \mathbb{R}$. Denote the set of complex numbers by $\mathbb{C}$. Call a the* real part *of z, denoted $Re(z)$, and b the* complex *or* imaginary part *of z, denoted $Im(z)$. Two complex numbers $a + bi$ and $c + di$ are defined to be equal precisely when $a = c$ and $b = d$. In other words, when their real parts are equal and when their complex parts are equal.* **A complex number $a + bi$ is a single entity not two.**

The numbers i and $-i$ are the two 'missing' square roots of -1. Real numbers are complex numbers because if a is real then $a = a + 0i$. Thus $\mathbb{R} \subseteq \mathbb{C}$. Complex numbers of the form bi, where $b \neq 0$, are said to be *purely imaginary*.

Lemma 6.1.1. *The sum, difference and product of complex numbers are complex numbers.*

Proof. Let $a + bi, c + di \in \mathbb{C}$. To add these numbers, we calculate $(a + bi) + (c + di)$. We are allowed to assume that i behaves in all algebraic respects like a real number. Thus we assume both commutativity, associativity and distributivity to obtain

$$(a + bi) + (c + di) = (a + c) + (b + d)i.$$

Similarly

$$(a + bi) - (c + di) = (a - c) + (b - d)i.$$

To multiply these numbers, we calculate $(a + bi)(c + di)$. We use distributivity and associativity to obtain

$$(a + bi)(c + di) = ac + adi + bic + bidi.$$

Commutativity of both addition and multiplication holds and so

$$ac + adi + bic + bidi = ac + bdi^2 + adi + bci.$$

Only at this point do we use the fact that $i^2 = -1$ to get

$$ac + adi + bci + bdi^2 = ac - bd + adi + bci.$$

Finally, using distributivity

$$(a + bi)(c + di) = (ac - bd) + (ad + bc)i.$$

□

Examples 6.1.2.

1. Calculate $(7 - i) + (-6 + 3i)$. We add together the real parts to get 1. Adding together $-i$ and $3i$ we get $2i$. Thus the sum of the two complex numbers is $1 + 2i$.

2. Calculate $(2 + i)(1 + 2i)$. Multiply out the brackets as usual to get $2 + 4i + i + 2i^2$. Now use the fact that $i^2 = -1$ to get $2 + 4i + i - 2$. Simplifying we obtain $0 + 5i = 5i$.

3. Calculate $\left(\frac{1-i}{\sqrt{2}}\right)^2$. Multiply out and simplify to get $-i$.

It is convenient at this point to define a new operation on complex numbers. Let $z = a + bi \in \mathbb{C}$. Define $\bar{z} = a - bi$. The number $\bar{z}$ is called the *complex conjugate* of z.

Lemma 6.1.3.

1. $\overline{z_1 + \ldots + z_n} = \overline{z_1} + \ldots + \overline{z_n}$.

2. $\overline{z_1 \ldots z_n} = \overline{z_1} \ldots \overline{z_n}$.

3. z is real if and only if $\bar{z} = z$.

Proof. (1) We prove the case where $n = 2$. The general case can then be proved by induction. Let $z_1 = a + bi$ and $z_2 = c + di$. Then $z_1 + z_2 = (a + c) + (b + d)i$. Thus

$$\overline{z_1 + z_2} = (a + c) - (b + d)i.$$

But $\overline{z_1} = a - bi$ and $\overline{z_2} = c - di$ and so

$$\overline{z_1} + \overline{z_2} = (a - bi) + (c - di) = (a + c) - (b + d)i.$$

Hence $\overline{z_1 + z_2} = \overline{z_1} + \overline{z_2}$.

(2) We prove the case where $n = 2$. The general case can then be proved by induction. Using the notation from part (1), we have that

$$z_1 z_2 = (ac - bd) + (ad + bc)i.$$

Thus

$$\overline{z_1 z_2} = (ac - bd) - (ad + bc)i.$$

On the other hand,

$$\overline{z_1}\,\overline{z_2} = (ac - bd) - (ad + bd)i.$$

Hence $\overline{z_1 z_2} = \overline{z_1}\,\overline{z_2}$.

(3) If z is real then it is immediate that $\bar{z} = z$. Conversely let $z = a + bi$ and suppose that $\bar{z} = z$. Then $a + bi = a - bi$. Hence $b = -b$ and so $b = 0$. It follows that z is real. $\qquad\square$

To see why this operation is useful, calculate $z\bar{z}$. We have

$$z\bar{z} = (a+bi)(a-bi) = a^2 - abi + abi - b^2 i^2 = a^2 + b^2.$$

Observe that $z\bar{z} = 0$ if and only if $z = 0$. Thus for a non-zero complex number z, the number $z\bar{z}$ is a positive real number.

Theorem 6.1.4. *The complex numbers form a field.*

Proof. In view of Lemma 6.1.1, it only remains to show that each non-zero complex number has a multiplicative inverse. To do so, we use the complex conjugate. We have that

$$\frac{1}{(a+bi)} = \frac{a-bi}{(a+bi)(a-bi)} = \frac{a-bi}{a^2+b^2} = \frac{1}{a^2+b^2}(a-bi)$$

which is a complex number. □

Examples 6.1.5.

1. $\frac{1+i}{i}$. The complex conjugate of i is $-i$. Multiply numerator and denominator of the fraction by this to get $\frac{-i+1}{1} = 1 - i$.

2. $\frac{i}{1-i}$. The complex conjugate of $1 - i$ is $1 + i$. Multiply numerator and denominator of the fraction by this to get $\frac{i(1+i)}{2} = \frac{i-1}{2}$.

3. $\frac{4+3i}{7-i}$. The complex conjugate of $7 - i$ is $7 + i$. Multiply numerator and denominator of the fraction by this to get $\frac{(4+3i)(7+i)}{50} = \frac{1+i}{2}$.

We describe now a way of visualizing complex numbers. A complex number $z = a + bi$ has two components: a and b. It is irresistible to plot these as a *point* in the plane. The plane used in this way is called the *complex plane*: the x-axis is the *real axis* and the y-axis is interpreted as the *complex axis*.

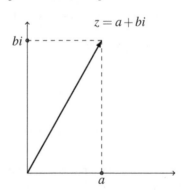

Although a complex number can be regarded as labelling a point in the complex plane, it can also be regarded as labelling the directed line segment from the origin to the point, and this turns out to be more fruitful. By Pythagoras' theorem, the length of this line is $\sqrt{a^2 + b^2}$. Define $|z| = \sqrt{a^2 + b^2}$ where $z = a + bi$. This is called the *modulus*[2] of the complex number z. It is important to observe that this is always a real number. Observe that $|z| = \sqrt{z\bar{z}}$. The modulus has the following important property.

Lemma 6.1.6. $|wz| = |w|\,|z|$.

Proof. Let $w = a + bi$ and $z = c + di$. Then $wz = (ac - bd) + (ad + bc)i$. Now

$$|wz| = \sqrt{(ac - bd)^2 + (ad + bc)^2} \text{ and } |w|\,|z| = \sqrt{(a^2 + b^2)(c^2 + d^2)}.$$

But

$$(ac - bd)^2 + (ad + bc)^2 = (ac)^2 + (bd)^2 + (ad)^2 + (bc)^2 = (a^2 + b^2)(c^2 + d^2).$$

Thus the result follows. $\qquad\square$

There is a corollary to Lemma 6.1.6 that plays an important rôle in number theory.

Corollary 6.1.7 (Products of sums of squares)**.** *If two natural numbers are each a sum of squares then their product is a sum of squares. Specifically,*

$$(a^2 + b^2)(c^2 + d^2) = (ac - bd)^2 + (ad + bc)^2.$$

Box 6.1: Factorizing Primes

Primes cannot be factorized using natural numbers but if certain kinds of complex numbers are used they sometimes can be. For example, $5 = (1 - 2i)(1 + 2i)$. Define the *Gaussian integers*, denoted by $\mathbb{Z}[i]$, to be all complex numbers of the form $m + in$ where m and n are integers. What our example shows is that some primes can be factorized using Gaussian integers. The question is: which ones? Observe that $5 = 1^2 + 2^2$. In other words, the prime can be written as a sum of two squares. Another example of a prime that can be written as a sum of two squares is 13. We have that $13 = 9 + 4 = 3^2 + 2^2$. This prime can also be factorized using Gaussian integers $13 = (3 + 2i)(3 - 2i)$. In fact, any prime p that can be written as a sum of two squares, $p = a^2 + b^2$, can be factorized using Gaussian integers $p = (a + ib)(a - ib)$. This raises the question of exactly which primes can be written as a sum of two squares. It is easy to prove that each odd prime p belongs to one of two sets: those primes $p \equiv 1 \pmod 4$ or those primes $p \equiv 3 \pmod 4$, and it can be proved that each of these two sets contains infinitely many elements (the proof of the second case being elementary). The deep result, first proved by Euler, is that an odd prime p can be written as a sum of two squares if and only if $p \equiv 1 \pmod 4$. A one-sentence proof of this result can be found in [128]. To understand it, you need only know that an *involution* is a function $f\colon X \to X$ such that $f^2 = 1_X$.

The complex numbers are obtained from the reals by simply adjoining one new number, i, a square root of -1. Remarkably, every complex number now has a square root.

---

[2]Plural: *moduli.*

Theorem 6.1.8. *Every nonzero complex number has exactly two square roots.*

Proof. Let $z = a + bi$ be a nonzero complex number. We want to show that there is a complex number w so that $w^2 = z$. Let $w = x + yi$. Then we need to find real numbers x and y such that $(x + yi)^2 = a + bi$. Thus $(x^2 - y^2) + 2xyi = a + bi$, and so equating real and imaginary parts, we have to solve the following two equations

$$x^2 - y^2 = a \text{ and } 2xy = b.$$

Now we actually have enough information to solve our problem, but we can make life easier for ourselves by adding one extra equation. To get it, we use the modulus function. From $(x + yi)^2 = a + bi$ we get that $|x + yi|^2 = |a + bi|$ by Lemma 6.1.6. Now $|x + yi|^2 = x^2 + y^2$ and $|a + bi| = \sqrt{a^2 + b^2}$. We therefore have three equations

$$\boxed{x^2 - y^2 = a \text{ and } 2xy = b \text{ and } x^2 + y^2 = \sqrt{a^2 + b^2}.}$$

If we add the first and third equation together we get

$$x^2 = \frac{a}{2} + \frac{\sqrt{a^2 + b^2}}{2} = \frac{a + \sqrt{a^2 + b^2}}{2}.$$

We can now solve for x and therefore for y using $2xy = b$. The values obtained for x and y can be shown to satisfy all three equations. □

Example 6.1.9. Every negative real number has two square roots. The square roots of $-r$, where $r > 0$, are $\pm i\sqrt{r}$.

Example 6.1.10. We calculate both square roots of $3 + 4i$. Let $x + yi$ be a complex number such that

$$(x + yi)^2 = 3 + 4i.$$

On squaring and comparing real and imaginary parts, we see that the following two equations are satisfied by x and y

$$x^2 - y^2 = 3 \text{ and } 2xy = 4.$$

There is also a third equation

$$x^2 + y^2 = 5$$

obtained from the moduli calculation

$$\left|(x + yi)^2\right| = x^2 + y^2 = |3 + 4i| = \sqrt{3^2 + 4^2} = 5.$$

Adding the first and third equation together we get $x = \pm 2$. Thus $y = 1$ if $x = 2$ and $y = -1$ if $x = -2$. The square roots we want are therefore $2 + i$ and $-2 - i$. Square either square root to check the answer: $(2 + i)^2 = 4 + 4i - 1 = 3 + 4i$, as required.

Observe that the two square roots of a non-zero complex number will have the form w and $-w$. In other words, one root will be -1 times the other.

If we combine our method for solving quadratics from Section 4.3 with our method for determining the square roots of complex numbers, we have a method for finding the roots of quadratics with any coefficients, whether real or complex.

Example 6.1.11. To solve the quadratic equation

$$4z^2 + 4iz + (-13 - 16i) = 0$$

we can use the same methods as in the real case. This is because the complex numbers obey the same algebraic laws as the reals with the bonus that we can take square roots in all cases. Rather than use the formula in this example, we complete the square. First, we convert the equation into a monic one

$$z^2 + iz + \frac{(-13-16i)}{4} = 0.$$

Observe that

$$\left(z + \tfrac{i}{2}\right)^2 = z^2 + iz - \tfrac{1}{4}.$$

Thus

$$z^2 + iz = \left(z + \tfrac{i}{2}\right)^2 + \tfrac{1}{4}.$$

The quadratic equation therefore becomes

$$\left(z + \tfrac{i}{2}\right)^2 + \tfrac{1}{4} + \left(-\tfrac{13}{4} - 4i\right) = 0.$$

We therefore have

$$\left(z + \tfrac{i}{2}\right)^2 = 3 + 4i.$$

Taking square roots of both sides using Example 6.1.10, we have that

$$z + \tfrac{i}{2} = 2 + i \text{ or } -2 - i.$$

It follows that $z = 2 + \tfrac{i}{2}$ or $-2 - \tfrac{3i}{2}$.

One final point on notation. It is usual to write complex numbers in the form $a + bi$ but since b and i commute it is equally correct to write $a + ib$. It is sometimes convenient to use the second convention to avoid confusion.

Exercises 6.1

1. Solve the following problems in complex number arithmetic. In each case, the answer should be in the form $a + bi$ where a and b are real.

   (a) $(2 + 3i) + (4 + i)$.

   (b) $(2 + 3i)(4 + i)$.

   (c) $(8 + 6i)^2$.

   (d) $\frac{2+3i}{4+i}$.

   (e) $\frac{1}{i} + \frac{3}{1+i}$.

   (f) $\frac{3+4i}{3-4i} - \frac{3-4i}{4+4i}$.

2. Find the square roots of each of the following complex numbers.

    (a) $-i$.

    (b) $-1 + \sqrt{24}i$.

    (c) $-13 - 84i$.

3. Solve the following quadratic equations.

    (a) $x^2 + x + 1 = 0$.

    (b) $2x^2 - 3x + 2 = 0$.

    (c) $x^2 - (2 + 3i)x - 1 + 3i = 0$.

4. *Define the *Gaussian integers*, denoted by $\mathbb{Z}[i]$, to be all complex numbers of the form $m + ni$ where m and n are integers. A *perfect square* in $\mathbb{Z}[i]$ is a Gaussian integer that can be written $(a + bi)^2$ for some Gaussian integer $a + bi$. Prove that if $x + yi$ is a perfect square then x and y form part of a Pythagorean triple (see Section 1.1).

6.2 COMPLEX NUMBER GEOMETRY

To make progress in understanding complex numbers, we exploit the natural geometrical way of thinking about them. Recall that angles are most naturally measured in radians rather than degrees.[3] and that positive angles are measured in an anticlockwise direction. As explained, we regard a complex number $z = a + bi$ as describing a directed line segment starting at the origin in the complex plane. Let θ be the angle that this line segment makes with the positive real axis. The length of z is $|z|$, and so by trigonometry $a = |z| \cos \theta$ and $b = |z| \sin \theta$. It follows that

$$z = |z| (\cos \theta + i \sin \theta).$$

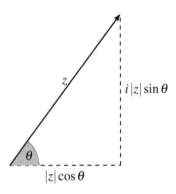

$$\begin{array}{c} i|z| \sin \theta \\ \theta \\ |z| \cos \theta \end{array}$$

---

[3] Since the system of angle measurement based on degrees is an historical accident.

This way of writing complex numbers is called *polar form*. It is at this point that we need to clarify the only feature of complex numbers that ever causes confusion. Suppose that

$$\cos\theta + i\sin\theta = \cos\phi + i\sin\phi.$$

Then $\theta - \phi = 2\pi k$ for some $k \in \mathbb{Z}$. We might say that θ and ϕ are equal *modulo* 2π. For that reason, there is not just one number θ that yields the complex number z but infinitely many: namely, all the numbers $\theta + 2\pi k$ where $k \in \mathbb{Z}$. We therefore define the *argument* of z, denoted by $\arg z$, to be not merely the single angle θ but the *set* of angles

$$\arg z = \{\theta + 2\pi k \colon \text{ where } k \in \mathbb{Z}\}.$$

The angle θ can be chosen in many ways and is not unique. This feature of the argument plays a crucial rôle when we come to calculate nth roots.

Since we are now thinking of complex numbers geometrically, we have to describe what addition and multiplication of complex numbers mean geometrically. We start with addition. To calculate the sum of the complex numbers u and v, complete them to a parallelogram and $u + v$ is the diagonal of that parallelogram as shown.

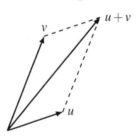

How this result is proved is suggested by the following diagram.

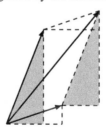

We now derive a geometrical way of interpreting the multiplication of two complex numbers.

Proposition 6.2.1. *Let $w = r(\cos\theta + i\sin\theta)$ and $z = s(\cos\phi + i\sin\phi)$ be two non-zero complex numbers. Then*

$$wz = rs\left(\cos(\theta + \phi) + i\sin(\theta + \phi)\right).$$

Proof. We have that

$$
\begin{aligned}
wz &= rs\left(\cos\theta + i\sin\theta\right)\left(\cos\phi + i\sin\phi\right) \\
&= rs[(\cos\theta\cos\phi - \sin\theta\sin\phi) + i(\sin\theta\cos\phi + \cos\theta\sin\phi)].
\end{aligned}
$$

This reduces to

$$wz = rs\left(\cos(\theta + \phi) + i\sin(\theta + \phi)\right)$$

where we have used the addition formulae

$$\sin(\alpha + \beta) = \sin\alpha\cos\beta + \cos\alpha\sin\beta$$

and

$$\cos(\alpha + \beta) = \cos\alpha\cos\beta - \sin\alpha\sin\beta.$$

□

Proposition 6.2.1 says that to multiply two non-zero complex numbers simply multiply their lengths and add their arguments.

Example 6.2.2. Proposition 6.2.1 helps us to understand the meaning of i. Multiplication by i is the same as a rotation about the origin by a right angle. Multiplication by i^2 is therefore the same as a rotation about the origin by two right angles. But this is exactly the same as multiplication by -1.

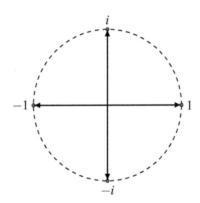

We can apply similar reasoning to explain geometrically why $-1 \times -1 = 1$. Of course, we proved this algebraically in Chapter 3 but the geometric approach delivers more insight. Multiplication by -1 is interpreted as rotation about the origin by $180°$. It follows that doing this twice takes us back to where we started and so is equivalent to multiplication by 1. We usually interpret numbers as representing quantities, but with complex numbers in general, it is more natural to interpret them as representing operations as these two examples show.

The proof of the next theorem follows by induction from Proposition 6.2.1.

Theorem 6.2.3 (De Moivre). *Let n be a positive integer. If $z = \cos\theta + i\sin\theta$ then $z^n = \cos n\theta + i\sin n\theta$.*

Our results above have applications in painlessly obtaining trigonometric identities.

Example 6.2.4. If you remember that when multiplying complex numbers in polar form you add their arguments, then you can easily reconstitute the addition formulae we started with since

$$(\cos\alpha + i\sin\alpha)(\cos\beta + i\sin\beta) = \cos(\alpha + \beta) + i\sin(\alpha + \beta).$$

This is helpful in getting both sines and signs right.

Example 6.2.5. Express $\cos 3\theta$ in terms of $\cos\theta$ and $\sin\theta$. By de Moivre's theorem

$$(\cos\theta + i\sin\theta)^3 = \cos 3\theta + i\sin 3\theta.$$

We can also expand the lefthand side using the binomial theorem to get

$$\cos^3\theta + 3\cos^2\theta\sin\theta i + 3\cos\theta(i\sin\theta)^2 + (i\sin\theta)^3$$

which simplifies to

$$\left(\cos^3\theta - 3\cos\theta\sin^2\theta\right) + i\left(3\cos^2\theta\sin\theta - \sin^3\theta\right)$$

where we use the fact that $i^2 = -1$ and $i^3 = -i$ and $i^4 = 1$. Equating real and imaginary parts we obtain

$$\cos 3\theta = \cos^3\theta - 3\cos\theta\sin^2\theta.$$

The formula

$$\sin 3\theta = 3\cos^2\theta\sin\theta - \sin^3\theta$$

comes for free.

Exercises 6.2

1. Let
   $$S^1 = \{\cos\theta + i\sin\theta : \theta \in \mathbb{R}\}.$$
   Prove that this set is closed under multiplication, and that if $u \in S^1$ then $u^{-1} \in S^1$.

2. Express $\cos 5x$ and $\sin 5x$ in terms of $\cos x$ and $\sin x$.

3. Prove Theorem 6.2.3.

4. By geometry, it follows that $|u + v| \le |u| + |v|$ for all complex numbers u and v. We develop an algebraic proof here.

   (a) Prove that $\mathrm{Re}(z) = \frac{z + \bar{z}}{2}$.
   (b) Prove that $\mathrm{Re}(z) \le |z|$.
   (c) Prove that $|\bar{z}| = |z|$.
   (d) Prove the required inequality starting with $|u + v|^2 = (u + v)\overline{(u + v)}$.

6.3 EULER'S FORMULA

In Chapter 2, mathematical creativity was briefly touched upon. In this section, a concrete example of such creativity is described. Strictly speaking there are no proofs in this section though everything we do can be justified. Instead, we show how playing with mathematics can lead to new results that then challenge mathematicians to prove them.

We have seen that every real number can be written as a whole number plus a possibly infinite decimal part. This suggested to mathematicians, such as Newton, that some functions might be expressed in a similar way. We illustrate how by means of an example. Consider the function e^x. All you need to know about it is that it is equal to its derivative and $e^0 = 1$. We would like to write

$$e^x \;=\; a_0 + a_1 x + a_2 x^2 + a_3 x^3 + \dots \tag{6.1}$$

where the a_i are real numbers that we have yet to determine. This looks like an infinite polynomial and is called a *series expansion*. We can work out the value of a_0 easily by putting $x = 0$. This tells us that $a_0 = 1$. To get the value of a_1, differentiate equation (6.1) to get

$$e^x \;=\; a_1 + 2a_2 x + 3a_3 x^2 + \dots. \tag{6.2}$$

Now put $x = 0$ again and this time we get that $a_1 = 1$. To get the value of a_2, differentiate equation (6.2) to get

$$e^x \;=\; 2a_2 + 2 \cdot 3 \cdot a_3 x + \dots. \tag{6.3}$$

Now put $x = 0$ and we get that $a_2 = \frac{1}{2}$. Continuing in this way we quickly spot the pattern for the values of the coefficients a_n. We find that $a_n = \frac{1}{n!}$. What we have done for e^x we can also do for $\sin x$ and $\cos x$ and we obtain the following series expansions of each of these functions.

- $e^x = 1 + x + \frac{x^2}{2!} + \frac{x^3}{3!} + \frac{x^4}{4!} + \dots.$

- $\sin x = x - \frac{x^3}{3!} + \frac{x^5}{5!} - \frac{x^7}{7!} + \dots.$

- $\cos x = 1 - \frac{x^2}{2!} + \frac{x^4}{4!} - \frac{x^6}{6!} + \dots.$

We have not proved that these equalities really hold, but if you calculate some values of the function on the lefthand side you will find close agreement with the value of the righthand side obtained by calculating a few terms.

What is relevant here is that there are interesting connections between these three series, and we now show that complex numbers help to explain them. Without worrying about the validity of doing so, we calculate the infinite series expansion of $e^{i\theta}$. We have that

$$e^{i\theta} = 1 + (i\theta) + \frac{1}{2!}(i\theta)^2 + \frac{1}{3!}(i\theta)^3 + \dots$$

that is

$$e^{i\theta} = 1 + i\theta - \frac{1}{2!}\theta^2 - \frac{1}{3!}\theta^3 i + \frac{1}{4!}\theta^4 + \dots.$$

By separating out real and complex parts, and using the infinite series we obtained above, we have derived the following remarkable formula of Leonhard Euler (1707–1783).

$$\boxed{e^{i\theta} = \cos\theta + i\sin\theta.}$$

Thus the complex numbers enable us to find the hidden connections between the three most important functions of calculus: the exponential function, the sine function and the cosine function. As a result, every non-zero complex number can be written in the form $re^{i\theta}$ where $r > 0$. If we put $\theta = \pi$ in Euler's formula, we obtain the following theorem, widely regarded as one of the most amazing in mathematics.

Theorem 6.3.1 (Euler's identity). $\boxed{e^{i\pi} = -1.}$

The real numbers π, e and -1 are related to each other, but the complex numbers are needed to show how. This is one of the important rôles of the complex numbers in mathematics in that they enable us to make connections between topics that look different: they form a mathematical hyperspace.[4]

Exercises 6.3

1. Prove the following where x is real.[5]

   (a) $\sin x = \frac{1}{2i}(e^{ix} - e^{-ix})$.

   (b) $\cos x = \frac{1}{2}(e^{ix} + e^{-ix})$.

   Hence show that $\cos^4 x = \frac{1}{8}[\cos 4x + 4\cos 2x + 3]$.

2. *Determine all the values of i^i. What do you notice?

6.4 MAKING SENSE OF COMPLEX NUMBERS

We have assumed that complex numbers exist and that they form a field. How can this be justified? One way that we can understand something new is to relate it to something old we already understand. We certainly understand the real numbers. In 1835, Sir William Rowan Hamilton (1805–1865) used ordered pairs of real numbers to define complex numbers. We now sketch how he did it. It will be convenient to denote ordered pairs of real numbers by bold letters so that $\mathbf{a} = (a_1, a_2)$. Define $\mathbf{0} = (0,0)$, $\mathbf{1} = (1,0)$ and $\mathbf{i} = (0,1)$. Define operations as follows.

- If $\mathbf{a} = (a_1, a_2)$ and $\mathbf{b} = (b_1, b_2)$, define $\mathbf{a} + \mathbf{b} = (a_1 + b_1, a_2 + b_2)$.

---

[4]Some authors are so worried by not being able to prove the above results rigorously at this stage that they simply *define* $e^{i\theta} = \cos\theta + i\sin\theta$. One wonders how such people can sleep at night.

[5]Compare (a) and (b) with $\sinh x = \frac{1}{2}(e^x - e^{-x})$ and $\cosh x = \frac{1}{2}(e^x + e^{-x})$.

- If $\mathbf{a} = (a_1, a_2)$ define $-\mathbf{a} = (-a_1, -a_2)$.

- If $\mathbf{a} = (a_1, a_2)$ and $\mathbf{b} = (b_1, b_2)$, define

$$\mathbf{ab} = (a_1 b_1 - a_2 b_2, a_1 b_2 + a_2 b_1).$$

- If $\mathbf{a} = (a_1, a_2) \neq \mathbf{0}$ define

$$\mathbf{a}^{-1} = \left(\frac{a_1}{a_1^2 + a_2^2}, \frac{-a_2}{a_1^2 + a_2^2} \right).$$

It is now a long exercise to check that the set $\mathbb{R} \times \mathbb{R}$ with respect to these operations is a field. Observe that the element (a_1, a_2) can be written $(a_1, 0)\mathbf{1} + (a_2, 0)\mathbf{i}$ and that $\mathbf{i}^2 = (0, 1)(0, 1) = (-1, 0) = -\mathbf{1}$. You will observe that the set of ordered pairs of real numbers with the above operations behaves like the complex numbers without quite looking like them. This does not matter. Mathematics only deals in behaviour not in looks. There is a bijection between the set of elements of the form $(a, 0)$ with the set of real numbers given by $(a, 0) \mapsto a$, and a bijection between the set of all ordered pairs (a, b) with the set of complex numbers given by $(a, b) \mapsto a + bi$. This proves that the complex numbers exist. We shall return again to work by Hamilton in Section 9.7 because the construction described in this section led him to generalize the complex numbers to higher dimensions.

Box 6.2: Isomorphism

The word *isomorphism* means the 'same shape'. Informally, two algebraic systems are *isomorphic* if they have all the same properties even though they may look different. Using this idea, we can express what we have shown in this section as follows: the set of complex numbers is isomorphic to the set of ordered pairs of real numbers endowed with suitable operations. The concept of isomorphism is vitally important in mathematics. This is because mathematics deals with the properties that an object possesses, that is the rules that it obeys, and not with what it looks like. It was only in the nineteenth century that mathematicians fully grasped this idea. An analogy with a game such as chess may help explain it. It is only aesthetics and tradition that dictate that chess pieces should have a particular appearance. What the pieces actually look like is completely immaterial from the point of view of playing a game of chess as long as it is clear what they are supposed to be. This is because a chess piece is defined by how it moves, the rules that it obeys, rather than by what it looks like. As in chess, so in mathematics.

Exercises 6.4

1. Prove all the claims in this section.

2. On the set $\mathbb{R}^2$ define two binary operations

$$(a, b) + (c, d) = (a + c, b + d) \text{ and } (a, b)(c, d) = (ac, bd).$$

What axioms does this system satisfy? What is the biggest difference between this system and the complex numbers?

CHAPTER 7

Polynomials

"Sauter à pieds joints sur les calculs." – Evariste Galois

The shear volume of mathematics is daunting, but it becomes less so when you recognize that there are patterns. This chapter is mainly about the properties of polynomials but we shall also use it to illustrate one of those patterns. In particular, we show that the ideas introduced in Chapter 5 for the integers may be extended to polynomials. The key is that there is a remainder theorem for polynomials just as there is for integers. Since all our results in number theory were proved using the remainder theorem, this encourages us to look for similar results for polynomials.

7.1 TERMINOLOGY

Let $\mathbb{F}$ be a fixed field. For concreteness, you can think of $\mathbb{F}$ as the complex numbers. An expression

$$p(x) = a_n x^n + a_{n-1} x^{n-1} + \ldots + a_1 x + a_0,$$

where $a_i \in \mathbb{F}$, is called a *polynomial*. The numbers a_i are called the *coefficients*. The set of all polynomials with coefficients from $\mathbb{F}$ is denoted by $\mathbb{F}[x]$. Observe that we enclose the variable in square brackets.

- $\mathbb{Q}[x]$ is the set of *rational polynomials*.

- $\mathbb{R}[x]$ is the set of *real polynomials*.

- $\mathbb{C}[x]$ is the set of *complex polynomials*.

If all the coefficients are zero then the polynomial is identically zero and we call it the *zero polynomial*. We always assume $a_n \neq 0$, and define the *degree* of the polynomial $p(x)$, denoted by $\deg p(x)$, to be n. Observe that the degree of the zero polynomial is left undefined. Two polynomials with coefficients from the same field are *equal* if they have the same degree and the same coefficients.[1] If $a_n = 1$ the polynomial

---

[1] There is one problem. If I write $p(x) = 0$ do I mean that $p(x)$ is the zero polynomial or am I writing an equation to be solved? The notation does not tell you. In situations like this, the context must provide the answer.

is said to be *monic*. The term a_0 is called the *constant term* and $a_n x^n$ is called the *leading term*. Polynomials can be added, subtracted and multiplied in the usual way. Polynomials of degree 1 are said to be *linear*, those of degree 2, *quadratic*, those of degree 3, *cubic*, those of degree 4, *quartic* and those of degree 5, *quintic*.

For most of this book, we shall only deal with polynomials in one variable, but in a few places we shall also need the notion of a polynomial in several variables. Let $x_1, \ldots, x_n$ be n variables. When $n = 1$ we usually take just x, when $n = 2$, we take x, y and when $n = 3$, we take x, y, z. These variables are assumed to commute with each other. It follows that a product of these variables can be written $x_1^{r_1} \ldots x_n^{r_n}$. This is called a *monomial* of *degree* $r_1 + \ldots + r_n$. If $r_i = 0$ then $x_i^{r_i} = 1$ and is omitted. A *polynomial in n variables with coefficients in* $\mathbb{F}$, denoted by $\mathbb{F}[x_1, \ldots, x_n]$, is any sum of a finite number of distinct monomials each multiplied by an element of $\mathbb{F}$. The *degree* of such a non-zero polynomial is the maximum of the degrees of the monomials that appear.

Example 7.1.1. Polynomials in one variable of arbitrary degree are the subject of this chapter whereas the theory of polynomials in two or three variables of degree 2 is described in Chapter 10. The most general polynomial in two variables of degree 2 looks like this

$$ax^2 + bxy + cy^2 + dx + ey + f.$$

It consists of three monomials of degree 2: namely, x^2, xy and y^2; two monomials of degree 1: namely, x and y; and a constant term arising from the monomial of degree 0. Polynomials of degree 1 give rise to linear equations which are discussed in Chapter 8. Polynomials of degree three or more in two or more variables are interesting but outside the scope of this book.

Polynomials form a particularly simple and useful class of functions. For example, complicated functions can often be approximated by polynomial ones. They also have important applications in matrix theory, as we shall see in Chapter 8.

7.2   REMAINDER THEOREM

The addition, subtraction and multiplication of polynomials is easy, so in this section we concentrate on division. Let $f(x), g(x) \in \mathbb{F}[x]$. We say that $g(x)$ *divides* $f(x)$, denoted by

$$g(x) \mid f(x),$$

if there is a polynomial $q(x) \in \mathbb{F}[x]$ such that $f(x) = g(x)q(x)$. We also say that $g(x)$ is a *factor* of $f(x)$. In multiplying and dividing polynomials the following result is key.

Lemma 7.2.1. *Let* $f(x), g(x) \in \mathbb{F}[x]$ *be non-zero polynomials. Then* $f(x)g(x) \neq 0$ *and*

$$\deg f(x)g(x) = \deg f(x) + \deg g(x).$$

Proof. Let $f(x)$ have leading term $a_m x^m$ and let $g(x)$ have leading term $b_n x^n$. Then the leading term of $f(x)g(x)$ is $a_m b_n x^{m+n}$. Now $a_m b_n \neq 0$, because we are working in a field. Thus $f(x)g(x) \neq 0$ and the degree of $f(x)g(x)$ is $m + n$, as required.   □

The following example is a reminder of how to carry out long division of polynomials.

Example 7.2.2. Long division of one polynomial by another is easy once the basic step involved is understood. This step is called *trial division*. Suppose we want to divide $g(x)$ into $f(x)$. Trial division consists in finding a polynomial of the form ax^m such that $f_1(x) = f(x) - ax^m g(x)$ has smaller degree than $f(x)$. For example, let $f(x) = 6x^4 + 5x^3 + 4x^2 + 3x + 2$ and $g(x) = 2x^2 + 4x + 5$. Then here $3x^2 g(x)$ has the same leading term as $f(x)$. This means that $f_1(x) = f(x) - 3x^2 g(x)$ has degree smaller than $f(x)$. In this case it has degree 3. We set out the computation in the following form

$$2x^2 + 4x + 5 \;\big|\; 6x^4 + 5x^3 + 4x^2 + 3x + 2$$

and proceed in stages. To get the term involving $6x^4$ we have to multiply the lefthand side by $3x^2$ as explained above. As a result we write down the following

$$
\begin{array}{r|l}
 & 3x^2 \\
\hline
2x^2 + 4x + 5 & 6x^4 + 5x^3 + 4x^2 + 3x + 2 \\
 & 6x^4 + 12x^3 + 15x^2
\end{array}
$$

We now subtract the lower righthand side from the upper and we get

$$
\begin{array}{r|l}
 & 3x^2 \\
\hline
2x^2 + 4x + 5 & 6x^4 + 5x^3 + 4x^2 + 3x + 2 \\
 & 6x^4 + 12x^3 + 15x^2 \\
\hline
 & -7x^3 - 11x^2 + 3x + 2
\end{array}
$$

We have shown that $f(x) - 3x^2 g(x) = -7x^3 - 11x^2 + 3x + 2$. Trial division is now repeated with the new polynomial.

$$
\begin{array}{r|l}
 & 3x^2 - \frac{7}{2}x \\
\hline
2x^2 + 4x + 5 & 6x^4 + 5x^3 + 4x^2 + 3x + 2 \\
 & 6x^4 + 12x^3 + 15x^2 \\
\hline
 & -7x^3 - 11x^2 + 3x + 2 \\
 & -7x^3 - 14x^2 - \frac{35}{2}x \\
\hline
 & 3x^2 + \frac{41}{2}x + 2
\end{array}
$$

The procedure is repeated once more with the new polynomial

$$
\begin{array}{r|l}
& 3x^2 - \frac{7}{2}x + \frac{3}{2} \text{ quotient} \\
\hline
2x^2 + 4x + 5 & 6x^4 + 5x^3 + 4x^2 + 3x + 2 \\
& 6x^4 + 12x^3 + 15x^2 \\
\hline
& -7x^3 - 11x^2 + 3x + 2 \\
& -7x^3 - 14x^2 - \frac{35}{2}x \\
\hline
& 3x^2 + \frac{41}{2}x + 2 \\
& 3x^2 + \frac{12}{2}x + \frac{15}{2} \\
\hline
& \frac{29}{2}x - \frac{11}{2} \text{ remainder}
\end{array}
$$

This is the end of the line because the new polynomial obtained has degree strictly less than the polynomial we are dividing by. What we have shown is that

$$6x^4 + 5x^3 + 4x^2 + 3x + 2 = \left(2x^2 + 4x + 5\right)\left(3x^2 - \tfrac{7}{2}x + \tfrac{3}{2}\right) + \left(\tfrac{29}{2}x - \tfrac{11}{2}\right).$$

Lemma 7.2.3 (Remainder theorem). *Let $f(x)$ and $g(x)$ be non-zero polynomials in $\mathbb{F}[x]$. Then either $g(x) \mid f(x)$ or $f(x) = g(x)q(x) + r(x)$ where $\deg r(x) < \deg g(x)$ for unique polynomials $q(x), r(x) \in \mathbb{F}[x]$.*

Proof. Example 7.2.2 illustrates the general algorithm for dividing one polynomial by another and is the basis of the proof that if $g(x)$ does not divide $f(x)$ then $f(x) = g(x)q(x) + r(x)$ where $\deg r(x) < \deg g(x)$. To prove uniqueness, suppose that $f(x) = g(x)q'(x) + r'(x)$ where $\deg r'(x) < \deg g(x)$. Then

$$g(x)(q(x) - q'(x)) = r(x) - r'(x).$$

If $q(x) - q'(x) \neq 0$ then the degree of the lefthand side is at least $\deg g(x)$ by Lemma 7.2.1 but by assumption $\deg(r(x) - r'(x)) < \deg g(x)$. Thus we must have that $q(x) = q'(x)$ and so $r(x) = r'(x)$. □

We call $q(x)$ the *quotient* and $r(x)$ the *remainder*. Lemma 7.2.3 is the key to all the theory developed in this chapter. It should be compared with Lemma 5.1.1.

Exercises 7.2

1. Find the quotient and remainder when the first polynomial is divided by the second.

   (a) $2x^3 - 3x^2 + 1$ and x.

   (b) $x^3 - 7x - 1$ and $x - 2$.

   (c) $x^4 - 2x^2 - 1$ and $x^2 + 3x - 1$.

7.3 ROOTS OF POLYNOMIALS

Let $f(x) \in \mathbb{F}[x]$. A number $a \in \mathbb{F}$ is said to be a *root* of $f(x)$ if $f(a) = 0$. The roots of $f(x)$ are the *solutions* of the *equation* $f(x) = 0$. Checking whether a number is a root is easy, but finding a root in the first place is difficult. The next result tells us that when we find a root of a polynomial we are in fact finding a linear factor.

Proposition 7.3.1 (Roots and linear factors). *Let $a \in F$ and $f(x) \in \mathbb{F}[x]$. Then a is a root of $f(x)$ if and only if $(x - a) \mid f(x)$.*

Proof. Suppose that $(x - a) \mid f(x)$. By definition $f(x) = (x - a)q(x)$ for some polynomial $q(x)$. Clearly $f(a) = 0$. To prove the converse, suppose that a is a root of $f(x)$. By the remainder theorem $f(x) = q(x)(x - a) + r(x)$, where either $r(x)$ is the zero polynomial or $r(x) \neq 0$ and $\deg r(x) < \deg(x - a) = 1$. Suppose that $r(x)$ is not the zero polynomial. Then $r(x)$ is a non-zero constant (that is, just a number). Call this number $b \neq 0$. If we calculate $f(a)$ we must get 0 since a is a root, but in fact $f(a) = b \neq 0$, a contradiction. Thus $(x - a) \mid f(x)$ as required. □

A root a of a polynomial $f(x)$ is said to have *multiplicity m* if $(x - a)^m$ divides $f(x)$ but $(x - a)^{m+1}$ does not divide $f(x)$. A root is always counted according to its multiplicity.[2]

Example 7.3.2. The polynomial $x^2 + 2x + 1$ has -1 as a root and no other roots. However $(x + 1)^2 = x^2 + 2x + 1$ and so the root -1 occurs with multiplicity 2. Thus the polynomial has two roots counting multiplicities. This is the sense in which we can say that a quadratic equation always has two roots.

The following result is important because it provides an upper bound to the number of roots a polynomial may have.

Theorem 7.3.3. *A non-constant polynomial of degree n with coefficients from any field has at most n roots.*

Proof. This is proved by induction. We show the induction step. Let $f(x)$ be a non-zero polynomial of degree $n > 0$. Suppose that $f(x)$ has a root a. Then by Proposition 7.3.1 $f(x) = (x - a)f_1(x)$ and the degree of $f_1(x)$ is $n - 1$ by Lemma 7.2.1. By assumption $f_1(x)$ has at most $n - 1$ roots and so $f(x)$ has at most n roots. □

7.4 FUNDAMENTAL THEOREM OF ALGEBRA

In this section, polynomials will be defined only over the complex (or real) numbers. The big question we have so far not answered is whether such a polynomial need have any roots at all. The answer is given by the following theorem. Its name reflects its importance when first discovered, though not its significance in modern algebra, since algebra is no longer just the study of polynomial equations. It was first proved by Gauss.

---

[2]One of the reasons for doing this is so that a complex polynomial of degree n will then always have exactly n roots.

Theorem 7.4.1 (Fundamental theorem of algebra). *Every non-constant polynomial of degree n with complex coefficients has at least one root.*

Box 7.1: Liouville's Theorem

We cannot give a complete proof of the fundamental theorem of algebra here, but we can at least describe the ingredients of the most popular one. Calculus can be generalized from real variables to complex variables. The theory that results, called complex analysis, is profoundly important in mathematics. A complex function that can be differentiated everywhere is said to be *entire*. An example of a differentiable function that is not entire is $z \mapsto \frac{1}{z}$. The problem is that this function is not defined at $z = 0$. Suppose that $z \mapsto f(z)$ is an entire function. We are interested in the behaviour of the real-valued function $z \mapsto |f(z)|$. If there is some real number $B > 0$ such that $|f(z)| \leq B$ for all $z \in \mathbb{C}$, we say that the $f(z)$ function is *bounded*. Clearly, constant functions $z \mapsto a$, where $a \in \mathbb{C}$ is fixed, are bounded entire functions. We can now state *Liouville's theorem*: every bounded entire function is constant. This can be used to prove the fundamental theorem of algebra as follows. Let $z \mapsto p(z)$ be a polynomial function, not constant, and assume that $p(z)$ has no roots. Then $z \mapsto \frac{1}{p(z)}$ is an entire function. We now proceed to get a contradiction using Liouville's theorem. If you look at how polynomials are defined, it is not hard to believe, and it can be proved, that there are real numbers $r > 0$ and $B > 0$ such that $|p(z)| \geq B$ for all $|z| > r$. This means that $\left|\frac{1}{p(z)}\right| \leq B$ for all $|z| > r$. We now have to look at the behaviour of the function $z \mapsto \frac{1}{p(z)}$ when $|z| \leq r$. It is a theorem that just by virtue of this function being continuous it is bounded. It follows that the function $z \mapsto \frac{1}{p(z)}$ is a bounded entire function and so by Liouville's theorem must be constant, which is nonsense. We conclude that the polynomial $p(z)$ must have a root.

Example 7.4.2. In this book, we do not always need the full weight of the fundamental theorem of algebra. Quadratic equations were dealt with in Section 4.3 and in Chapter 6. A real cubic must have at least one real root simply by thinking about its values when x is large and positive and x is large and negative and then using the fact that polynomials are continuous. The equation $z^n - a = 0$ can be handled just by using the geometric properties of complex numbers as we show in the next section.

The fundamental theorem of algebra has the following important consequence using Theorem 7.3.3.

Corollary 7.4.3. *Every non-constant polynomial with complex coefficients of degree n has exactly n complex roots (counting multiplicities). Thus every such polynomial can be written as a product of linear polynomials.*

Proof. This is proved by induction. We describe the induction step. Let $f(x)$ be a non-constant polynomial of degree n. By the fundamental theorem of algebra, this polynomial has a root r_1. Thus $f(x) = (x - r_1)f_1(x)$ where $f_1(x)$ is a polynomial of degree $n - 1$. By assumption, $f_1(x)$ has exactly $n - 1$ roots and so $f(x)$ has exactly n roots. In addition, $f_1(x)$ can be written as a product of linear polynomials and therefore so can $f(x)$. □

Example 7.4.4. The quartic $x^4 - 5x^2 - 10x - 6$ has roots $-1, 3, i - 1$ and $-1 - i$. We can therefore write

$$x^4 - 5x^2 - 10x - 6 = (x + 1)(x - 3)(x + 1 + i)(x + 1 - i).$$

In many practical examples, polynomials have real coefficients and we therefore want any factors of the polynomial to likewise be real. Corollary 7.4.3 does not do this because it could produce complex factors. However, we can rectify this situation at a small price. We use the complex conjugate of a complex number introduced in Chapter 6.

Lemma 7.4.5. *Let $f(x)$ be a polynomial with* real coefficients. *If the complex number z is a root then so too is $\bar{z}$.*

Proof. Let

$$f(x) = a_n x^n + a_{n-1} x^{n-1} + \ldots + a_1 x + a_0$$

where the a_i are real numbers. Let z be a complex root. Then

$$0 = a_n z^n + a_{n-1} z^{n-1} + \ldots + a_1 z + a_0.$$

Take the complex conjugate of both sides and use the properties listed in Lemma 6.1.3 together with the fact that the coefficients are real to get

$$0 = a_n \bar{z}^n + a_{n-1} \bar{z}^{n-1} + \ldots + a_1 \bar{z} + a_0$$

and so $\bar{z}$ is also a root. ☐

Example 7.4.6. In Example 7.4.4, we obtained the factorization

$$x^4 - 5x^2 - 10x - 6 = (x+1)(x-3)(x+1+i)(x+1-i)$$

of the real polynomial. Observe that the complex roots $-1-i$ and $-1+i$ are complex conjugates of each other.

In the next result, we refer to our discussion of quadratic equations in Section 4.4.

Lemma 7.4.7.

1. *Let z be a complex number which is not real. Then $(x-z)(x-\bar{z})$ is an irreducible quadratic with real coefficients.*

2. *If $x^2 + bx + c$ is an irreducible quadratic with real coefficients then its roots are non-real and complex conjugates of each other.*

Proof. (1) Multiply out to get

$$(x-z)(x-\bar{z}) = x^2 - (z+\bar{z})x + z\bar{z}.$$

Observe that $z + \bar{z}$ and $z\bar{z}$ are both real numbers. The discriminant of this polynomial is $(z - \bar{z})^2$. You can check that if z is complex and non-real then $z - \bar{z}$ is purely imaginary. It follows that its square is negative. We have therefore shown that our quadratic is irreducible.

(2) This follows from the formula for the roots of a quadratic combined with the fact that the square roots of a negative real number have the form $\pm ai$ where a is real. ☐

Example 7.4.8. We continue with Example 7.4.6. We have that

$$x^4 - 5x^2 - 10x - 6 = (x+1)(x-3)(x+1+i)(x+1-i).$$

Multiply out $(x+1+i)(x+1-i)$ to get $x^2 + 2x + 2$. Thus

$$x^4 - 5x^2 - 10x - 6 = (x+1)(x-3)(x^2 + 2x + 2)$$

with all the polynomials involved being real.

The following theorem can be used to help solve problems involving real polynomials.

Theorem 7.4.9 (Fundamental theorem of algebra for real polynomials). *Every non-constant polynomial with real coefficients can be written as a product of polynomials with real coefficients which are either linear or irreducible quadratic.*

Proof. By Corollary 7.4.3, we can write the polynomial as a product of linear polynomials. Move the real linear factors to the front. The remaining linear polynomials will have complex coefficients. They correspond to roots that come in complex conjugate pairs by Lemma 7.4.5. Multiplying together those complex linear factors corresponding to complex conjugate roots we get real irreducible quadratics by Lemma 7.4.7. ☐

Any real polynomial is equal to a real number times a product of monic real linear and monic irreducible real quadratic factors. This result is the basis of the method of partial fractions used in integrating rational functions in calculus. This is discussed in Section 7.8.

Finding the exact roots of a polynomial is difficult in general, but the following leads to an algorithm for finding the rational roots of polynomials with integer coefficients. It is a nice, and perhaps unexpected, application of the number theory we developed in Chapter 5.

Theorem 7.4.10 (Rational root theorem). *Let*

$$f(x) = a_n x^n + a_{n-1} x^{n-1} + \ldots + a_1 x + a_0$$

be a polynomial with integer *coefficients. If $\frac{r}{s}$ is a root with r and s coprime then $r \mid a_0$ and $s \mid a_n$. In particular, if the polynomial is monic then any rational roots must be integers and divide the constant term.*

Proof. Substituting $\frac{r}{s}$ into $f(x)$ we have, by assumption, that

$$0 = a_n \left(\tfrac{r}{s}\right)^n + a_{n-1} \left(\tfrac{r}{s}\right)^{n-1} + \ldots + a_1 \left(\tfrac{r}{s}\right) + a_0.$$

Multiplying through by s^n we get

$$0 = a_n r^n + a_{n-1} s r^{n-1} + \ldots + a_1 s^{n-1} r + a_0 s^n.$$

This equation can be used to show that (1) $r \mid a_0 s^n$ and that (2) $s \mid a_n r^n$. We deal first with (1). Observe that r and s^n are coprime because if p is a prime that divides r and s^n then by Euclid's lemma, p divides r and s which is a contradiction since r and s are coprime. It follows, again by Euclid's lemma, that $r \mid a_0$. A similar argument applied to (2) yields $s \mid a_n$. □

Example 7.4.11. Find all the roots of the following polynomial

$$x^4 - 8x^3 + 23x^2 - 28x + 12.$$

The polynomial is monic with integer coefficients and so the only possible rational roots are integers and must divide 12. Thus the only possible rational roots are

$$\pm 1, \pm 2, \pm 3, \pm 4, \pm 6, \pm 12.$$

We find immediately that 1 is a root and so $(x - 1)$ must be a factor. Dividing out by this factor we get the quotient

$$x^3 - 7x^2 + 16x - 12.$$

We check this polynomial for rational roots and find 2 works. Dividing out by $(x - 2)$ we get the quotient

$$x^2 - 5x + 6.$$

Once we get down to a quadratic we can solve it directly. In this case it factorizes as $(x - 2)(x - 3)$. We therefore have that

$$x^4 - 8x^3 + 23x^2 - 28x + 12 = (x - 1)(x - 2)^2(x - 3).$$

Exercises 7.4

1. Find all roots using the information given.

   (a) 4 is a root of $3x^3 - 20x^2 + 36x - 16$.
   (b) $-1, -2$ are both roots of $x^4 + 2x^3 + x + 2$.

2. Find a cubic having roots $2, -3, 4$.

3. Find a quartic having roots i, $-i$, $1 + i$ and $1 - i$.

4. The cubic $x^3 + ax^2 + bx + c$ has roots x_1, x_2 and x_3. Show that a, b, c can each be written in terms of the roots.

5. $3 + i\sqrt{2}$ is a root of $x^4 + x^3 - 25x^2 + 41x + 66$. Find the remaining roots.

6. $1 - i\sqrt{5}$ is a root of $x^4 - 2x^3 + 4x^2 + 4x - 12$. Find the remaining roots.

7. Find all the roots of the following polynomials.

   (a) $x^3 + x^2 + x + 1$.

(b) $x^3 - x^2 - 3x + 6$.

(c) $x^4 - x^3 + 5x^2 + x - 6$.

8. Write each of the following polynomials as a product of real linear or real irreducible quadratic factors.

(a) $x^3 - 1$.

(b) $x^4 - 1$.

(c) $x^4 + 1$.

9. *Let $\lambda_1, \ldots, \lambda_r$ be r real numbers. Construct the *Lagrange polynomials* $p_i(x)$, where $1 \leq i \leq r$, that have the following property

$$p_i(\lambda_j) = \begin{cases} 1 & \text{if } j = i \\ 0 & \text{if } j \neq i. \end{cases}$$

7.5 ARBITRARY ROOTS OF COMPLEX NUMBERS

The number one is called *unity*.[3] A complex number z such that $z^n = 1$ is called an *nth root of unity*. These are therefore exactly the roots of the equation $z^n - 1 = 0$. We denote the set of nth roots of unity by C_n. We have proved that every non-zero complex number has two square roots. More generally, by the fundamental theorem of algebra, we know that the equation $z^n - 1 = 0$ has exactly n roots as does the equation $z^n - a = 0$ where a is any complex number. The main goal of this section is to prove these results *without assuming the fundamental theorem of algebra*. To do this, we shall need to think about complex numbers in a geometric, rather than an algebraic, way. We only need Theorem 7.3.3: every polynomial of degree n has at most n roots.

There is also an important definition that we need to give at this point which gets to the heart of the classical theory of polynomial equations. The word *radical* simply means a square root, or a cube root, or a fourth root and so on. The four basic operations of algebra, that is addition, subtraction, multiplication and division, together with the extraction of nth roots are regarded as purely algebraic operations. Although failing as a precise definition, we say informally that a *radical expression* is an algebraic expression involving nth roots. For example, the formula for the roots of a quadratic describes those roots as radical expressions in terms of the coefficients of the quadratic. A radical expression is an explicit description of a complex number in terms of algebraic operations applied to the rationals. The following table gives

---

[3]There is a *Through the Looking-Glass* feel to this nomenclature

some easy to find radical expressions for the sines and cosines of a number of angles.

θ	$\sin\theta$	$\cos\theta$
$0°$	0	1
$30°$	$\frac{1}{2}$	$\frac{\sqrt{3}}{2}$
$45°$	$\frac{1}{\sqrt{2}}$	$\frac{1}{\sqrt{2}}$
$60°$	$\frac{\sqrt{3}}{2}$	$\frac{1}{2}$
$90°$	1	0

We begin by finding all the nth roots of unity. The next example contains the ideas we need.

Example 7.5.1. We find the three cube roots of unity. An equilateral triangle with 1 as one of its vertices can be inscribed in the unit circle about the origin in the complex plane. The vertices of this triangle are the cube roots of unity. The first is the number 1 and the other two are $\omega_1 = \cos\frac{2\pi}{3} + i\sin\frac{2\pi}{3}$, and $\omega_2 = \cos\frac{4\pi}{3} + i\sin\frac{4\pi}{3}$. You can see that they really are cube roots of unity by cubing them using de Moivre's theorem, remembering what we said about the argument of a complex number. If we put $\omega = \omega_1$ then $\omega_2 = \omega^2$.

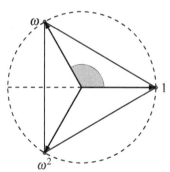

The root $\omega = \cos\frac{2\pi}{3} + i\sin\frac{2\pi}{3}$, for example, is in *trigonometric form*. If we need a numerical value for ω, it can easily be calculated from this form to obtain $\omega \approx -0 \cdot 5 + 0 \cdot 866i$. In this case, it is also easy to write down radical expressions for the roots since we already have radical expressions for $\sin 60°$ and $\cos 60°$. We therefore have that

$$\omega = \tfrac{1}{2}\left(-1 + i\sqrt{3}\right) \text{ and } \omega^2 = -\tfrac{1}{2}\left(1 + i\sqrt{3}\right).$$

The general case is solved in a similar way to our example above using regular n-gons[4] inscribed in the unit circle about the origin in the complex plane where one of the vertices is 1.

---

[4]A polygon with n sides.

Theorem 7.5.2 (Roots of unity). *The n roots of unity are given by the formula*

$$\cos \tfrac{2k\pi}{n} + i \sin \tfrac{2k\pi}{n}$$

for $k = 1, 2, \ldots, n$. These complex numbers are arranged uniformly on the unit circle and include 1 and form a regular polygon with n sides: the cube roots of unity form an equilateral triangle, the fourth roots form a square, the fifth roots form a pentagon, and so on.

- The root $\omega = \cos \tfrac{2\pi}{n} + i \sin \tfrac{2\pi}{n}$ is called the *principal nth root*.

- The *n*th roots of unity are $\omega, \omega^2, \ldots, \omega^{n-1}, \omega^n = 1$.

One feature of Theorem 7.5.2 needs emphasizing here. It is always possible, and easy, to write down trigonometric expressions for the *n*th roots of unity. Using such an expression, we can then write down numerical values of the *n*th roots to any desired degree of accuracy. Thus, from a purely practical point of view, we can find the *n*th roots of unity. It is also always possible to write down radical expressions for the *n*th roots of unity but this is far from easy in general. In fact, it forms part of the advanced subject known as *Galois theory* and is outside the scope of this book.

Example 7.5.3. Gauss proved the following highly non-trivial result. You can verify that it is possibly true using a calculator. It is a nice example of a radical expression where, on this occasion, the only radicals that occur are square roots. The theory Gauss developed showed that the form taken by this radical expression implied that the 17-gon could be constructed using only a ruler and compass.

$$16 \cos \tfrac{2\pi}{17} = -1 + \sqrt{17} + \sqrt{34 - 2\sqrt{17}}$$
$$+ \sqrt{\left(68 + 12\sqrt{17} - 16\sqrt{34 + 2\sqrt{17}} - 2(1 - \sqrt{17})\sqrt{34 - 2\sqrt{17}} \right)}.$$

The *n*th roots of unity play a pivotal rôle in finding arbitrary *n*th roots. The next example illustrates the idea.

Example 7.5.4. We find the three cube roots of 2. If you use your calculator you will simply find $\sqrt[3]{2}$, a real number. There should be two others: where are they? The explanation is that the other two cube roots are complex. Let ω be the principal cube root of 1 described in Example 7.5.1. Then the three cube roots of 2 are the following

$$\sqrt[3]{2}, \ \omega \sqrt[3]{2}, \ \omega^2 \sqrt[3]{2}$$

as you can easily check.

Theorem 7.5.5 (nth roots). *Let $z = r(\cos \theta + i \sin \theta)$ be a non-zero complex number. Put*

$$u = \sqrt[n]{r} \left(\cos \tfrac{\theta}{n} + i \sin \tfrac{\theta}{n} \right),$$

the principal *n*th root *of z, and put*

$$\omega = \cos \frac{2\pi}{n} + i\sin \frac{2\pi}{n},$$

the principal *n*th root of unity. *Then the nth roots of z are*

$$u, u\omega, \ldots, u\omega^{n-1}.$$

The nth roots of z $= r(\cos\theta + i\sin\theta)$ *can therefore be written in the form*

$$\sqrt[n]{r}\left(\cos\left(\frac{\theta}{n} + \frac{2k\pi}{n}\right) + i\sin\left(\frac{\theta}{n} + \frac{2k\pi}{n}\right)\right)$$

for $k = 0, 1, 2, \ldots, n-1.$

Proof. By de Moivre's theorem $u^n = z$. Because $\omega^n = 1$, it is easy to check that each of the complex numbers below

$$u, u\omega, \ldots, u\omega^{n-1}$$

raised to the power n equals z. These numbers are distinct since they are obtained by rotating u through multiples of the angle $\frac{2\pi}{n}$. They are therefore n solutions to the polynomial equation $x^n - z = 0$. But this equation has at most n solutions and so we have found them all. □

Example 7.5.6. Theorem 7.5.5 explains why every non-zero number has two square roots that differ by a multiple of -1. This is because the square roots of unity are 1 and -1.

Example 7.5.7. Consider the polynomial $z^n + 1$ where n is odd. The roots of this polynomial are the nth roots of -1. Because n is odd, an obvious root is -1. Let ω be a principal nth root of unity. Then the n roots of $z^n + 1$ are

$$-1, -\omega, \ldots, -\omega^{n-1}.$$

Thus

$$z^n + 1 = (z+1)(z+\omega)\ldots(z+\omega^{n-1}).$$

Put $z = \frac{x}{y}$ and multiply both sides of the equation obtained by y^n. This gives us

$$x^n + y^n = (x+y)(x+y\omega)\ldots(x+y\omega^{n-1}).$$

Fermat's last theorem states that there are no non-trivial integer solutions to the equation $x^n + y^n = z^n$ when $n \geq 3$. The factorization we have obtained for $x^n + y^n$ leads to one approach to trying to prove Fermat's last theorem for odd primes. Although it does not work for all such primes, it turns out to be quite fruitful and led to new developments in algebra.

We return to the nth roots of unity to view them from a more theoretical perspective. The next example provides the motivation.

Example 7.5.8. Let $C_4 = \{1, -1, i, -i\}$ be the 4th roots of unity. By definition, they all have the property that if $z \in C_4$ then $z^4 = 1$ but they are differentiated by some other properties. In particular, the roots 1 and -1 are also square roots of unity. In fact $C_2 = \{1, -1\}$. Whereas if $z = i$ or $z = -i$ then $C_4 = \{z, z^2, z^3, z^4\}$. We knew that i had this property, because it is the principal root, but so too does $-i$.

Any nth root of unity z such that $C_n = \{z^i : 1 \leq i \leq n\}$ is called a *primitive root*. Thus in C_4 the primitive roots are $\pm i$. Our next result characterizes the primitive nth roots and provides us with another connection with Chapter 5.

Theorem 7.5.9 (Primitive roots of unity). *Let ω be the principal nth root of unity. Let $1 \leq k \leq n$. Then ω^k is a primitive root if and only if $\gcd(n, k) = 1$.*

Proof. Suppose that $\gcd(n, k) = 1$. By Lemma 5.2.11 there are integers a and b such that $1 = na + kb$. Then $p = pna + pkb$ for any natural number p. Thus

$$\omega^p = \omega^{anp + bkp} = \omega^{anp} \omega^{bkp} = \left(\omega^k\right)^{bp}$$

since $\omega^n = 1$. We have proved that ω^k is a primitive root.

Conversely, suppose that ω^k is a primitive root. Then $(\omega^k)^p = \omega$ for some natural number p. Using the fact that $\omega^s = 1$ if and only if $s \equiv 0 \pmod{n}$, we deduce that $kp \equiv 1 \pmod{n}$. It follows that $kp - 1 = nq$ for some q and so $kp - nq = 1$. Thus $\gcd(n, k) = 1$ by Lemma 5.2.11. □

One important point that emerges from this section is that $z \mapsto \sqrt[n]{z}$ is not a function but a relation. How to treat this and similar complex relations is dealt with in complex analysis.

Exercises 7.5

1. Find the 4th roots of unity as radical expressions.

2. Find the 6th roots of unity as radical expressions.

3. Find the 8th roots of unity as radical expressions.

4. Find radical expresssions for the roots of $x^5 - 1$, and so show that

$$\cos 72° = \frac{\sqrt{5} - 1}{4} \quad \text{and} \quad \sin 72° = \frac{\sqrt{10 + 2\sqrt{5}}}{4}.$$

Hint: Consider the equation

$$x^4 + x^3 + x^2 + x + 1 = 0.$$

Divide through by x^2 to get

$$x^2 + \frac{1}{x^2} + x + \frac{1}{x} + 1 = 0.$$

Put $y = x + \frac{1}{x}$. Show that y satisfies the quadratic

$$y^2 + y - 1 = 0.$$

You can now find all four values of x.

5. In the following, express your answers in trigonometric form, and in radical form if possible.

  (a) Find the cube roots of $-8i$.

  (b) Find the fourth roots of 2.

  (c) Find the sixth roots of $1 + i$.

7.6   GREATEST COMMON DIVISORS OF POLYNOMIALS

Let $f(x), g(x) \in \mathbb{F}[x]$. A *common divisor* of $f(x)$ and $g(x)$ is any polynomial $p(x) \in \mathbb{F}[x]$ such that $p(x) \mid f(x)$ and $p(x) \mid g(x)$. The set of common divisors of $f(x)$ and $g(x)$ has elements of maximum degree, any one of which is called *a greatest common divisor*. We have to use the indefinite article because if $d(x)$ is a greatest common divisor of $f(x)$ and $g(x)$ so too is $ad(x)$ where $a \in \mathbb{F}$ is any non-zero element. There is, however, a unique monic common divisor of $f(x)$ and $g(x)$ of maximum degree. This is termed *the greatest common divisor* and is denoted by $\gcd(f(x), g(x))$. If a greatest common divisor of $f(x)$ and $g(x)$ is a constant, then 1 is the greatest common divisor and we say that $f(x)$ and $g(x)$ are *coprime*.

Proposition 7.6.1. *Let $f(x), g(x) \in \mathbb{F}[x]$. Define*

$$I = \{a(x)f(x) + b(x)g(x) : a(x), b(x) \in \mathbb{F}[x]\}.$$

Let $p(x) \in I$ be a monic polynomial of smallest degree.

  1. $p(x)$ divides every element of I and so in particular $f(x)$ and $g(x)$.

  2. Any common divisor of $f(x)$ and $g(x)$ divides $p(x)$.

  3. $p(x) = \gcd(f(x), g(x))$.

Proof. (1) Let $h(x) \in I$. By the remainder theorem, $h(x) = p(x)d(x) + r(x)$. If $r(x)$ were not zero, then $h(x) - p(x)d(x)$ would be an element of I of strictly smaller degree than $p(x)$, which is a contradiction.

(2) This follows from the fact that $p(x) = a(x)f(x) + b(x)g(x)$ for some choice of $a(x)$ and $b(x)$.

(3) This is immediate by (2).   □

The remainder theorem for the integers leads to Euclid's algorithm for computing greatest common divisors of integers. In exactly the same way, the remainder theorem for polynomials leads to Euclid's algorithm for computing greatest common divisors of polynomials. Here is an example. No new ideas are needed.

Examples 7.6.2.

1. Compute
$$\gcd(x^2 + 3x + 2, x^3 + 5x^2 + 7x + 3)$$
using Euclid's algorithm for polynomials.

   (a) $x^3 + 5x^2 + 7x + 3 = (x+2)(x^2 + 3x + 2) - x - 1$.
   (b) $x^2 + 3x + 2 = (-x-1)(-x-2)$.

   The last non-zero remainder is $-x - 1$ and so the monic polynomial required is $x + 1$ which is the gcd.

2. Compute
$$\gcd(x^5 - 2x^4 - 2x^3 + 8x^2 - 7x + 2, x^4 - 4x + 3)$$
using Euclid's algorithm for polynomials.[5]

   (a) $x^5 - 2x^4 - 2x^3 + 8x^2 - 7x + 2 = (x-2)(x^4 - 4x + 3) + (-2x^3 + 12x^2 - 18x + 8)$.
   (b) $x^4 - 4x + 3 = (-\frac{1}{2}x - 3)(-2x^3 + 12x^2 - 18x + 8) + (27x^2 - 54x + 27)$.
   (c) $-2x^3 + 12x^2 - 18x + 8 = -\frac{2}{27}(x-4)27(x^2 - 2x + 1)$.

   The required gcd is therefore $x^2 - 2x + 1$.

Exercises 7.6

1. Calculate the greatest common divisors of the following pairs of polynomials.[6]

   (a) $x^4 + 4x^3 + 8x^2 + 8x + 3$ and $x^3 + x^2 + x - 3$.
   (b) $x^3 + 2x^2 + 3x + 4$ and $x^2 + x + 1$.
   (c) $4x^4 + 26x^3 + 41x^2 - 2x - 24$ and $3x^4 + 20x^3 + 32x^2 - 8x - 32$.

7.7   IRREDUCIBLE POLYNOMIALS

In this chapter, we have proved two factorization results.

1. Every non-constant complex polynomial can be written as a product of complex linear polynomials.

2. Every non-constant real polynomial can be written as a product of real linear and real irreducible quadratic polynomials.

---

[5] Taken from [26].
[6] All taken from [26].

These results are reminiscent of the fundamental theorem of arithmetic in Chapter 5. In case (1), the 'primes' are the linear polynomials, and in case (2), the 'primes' are either real linear or real irreducible quadratic polynomials. We outline how to make this similarity precise. Rather than the word 'prime' it is usual to employ a different term. A polynomial $f(x) \in \mathbb{F}[x]$ of degree at least one is said to be *irreducible* if $f(x) = g(x)h(x)$, where $g(x), h(x) \in \mathbb{F}[x]$, implies that either $g(x)$ or $h(x)$ is a constant.

Example 7.7.1. It is important to pay attention to the field $\mathbb{F}$ in the definition of irreducibility. One and the same (looking) polynomial can be irreducible when the coefficients are regarded as belonging to one field and reducible when a different field is chosen. Thus to ask whether the polynomial $x^2 + 1$ is irreducible or not is meaningless unless the field of coefficients is specified. In this case, the polynomial is irreducible in $\mathbb{R}[x]$ and reducible in $\mathbb{C}[x]$. It is also reducible in $\mathbb{Z}_2[x]$ since here $(x+1)^2 = x^2 + 1$, a special case of the *freshman lemma*. High school mathematics leads us to believe that there is just one arena in which mathematics takes place: either the reals or possibly the complexes. In fact, there are multiple arenas.

The definition of irreducibility enables us to restate the two factorization results.

Example 7.7.2.

1. The irreducible polynomials in $\mathbb{C}[x]$ are the linear ones.

2. The irreducible polynomials in $\mathbb{R}[x]$ are the linear and irreducible quadratic ones.

The proofs of parts (1) and (2) below are immediate by Proposition 7.6.1.

Lemma 7.7.3.

1. *If $p(x) = \gcd(f(x), g(x))$, then there are polynomials $a(x), b(x) \in \mathbb{F}[x]$ such that*

$$p(x) = a(x)f(x) + b(x)g(x).$$

2. *Let $f(x), g(x) \in \mathbb{F}[x]$. Then $f(x)$ and $g(x)$ are coprime if and only if there are polynomials $a(x), b(x) \in \mathbb{F}[x]$ such that*

$$1 = a(x)f(x) + b(x)g(x).$$

The proof of the following is immediate by Lemma 7.7.3, the definitions, and induction.

Lemma 7.7.4 (Euclid's lemma for polynomials). *Let $p(x)$ be an irreducible polynomial in $\mathbb{F}[x]$. Suppose that $p(x) \mid a_1(x)\ldots a_m(x)$ where $a_i(x) \in \mathbb{F}[x]$. Then $p(x) \mid a_i(x)$ for some i.*

By Lemma 7.7.4, the following analogue of the fundamental theorem of arithmetic for polynomials can be proved using similar ideas to the natural number case. As mentioned above, pay attention to the field $\mathbb{F}$.

Theorem 7.7.5 (Primary factorization of polynomials). *Every non-constant polynomial $f(x)$ in $\mathbb{F}[x]$ can be factorized*

$$f(x) = a p_1(x) \ldots p_m(x)$$

where $a \in \mathbb{F}$ and the $p_i(x)$ are monic irreducible polynomials in $\mathbb{F}[x]$. Furthermore, such a factorization is unique if the order of the irreducibles that occur is ignored.

We have so far not discussed the irreducible rational polynomials. Here things become more complicated. In fact, this is another example of a feature of algebra that was discussed in Chapter 5. The sequence — complex numbers, real numbers, rational numbers, integers and natural numbers — represents a progressive simplification in our notion of number and a progressive complication in the algebra that results. In the case of rational polynomials, there is no simple characterization of the irreducibles as there is in the case of real and complex polynomials. For example, it can be proved that the rational polynomial

$$x^{p-1} + x^{p-2} + \ldots + x + 1,$$

where p is a prime, is always an irreducible rational polynomial.

Example 7.7.6. We prove directly that the rational polynomial

$$x^4 + x^3 + x^2 + x + 1$$

is irreducible in $\mathbb{Q}[x]$. By the rational root theorem, the only possible rational roots are ± 1. Since neither of them works, the polynomial has no rational linear factors. We therefore attempt to write the polynomial as a product of two rational quadratic polynomials. Therefore the only possible factorization would be

$$(x^2 + ax + b)(x^2 + cx + d)$$

where a, b, c, d are all rational. However, the only such factorization that results is

$$\left(x^2 - x \left(\tfrac{\sqrt{5}-1}{2} \right) + 1 \right) \left(x^2 + x \left(\tfrac{\sqrt{5}+1}{2} \right) + 1 \right)$$

which does not have rational coefficients.

7.8 PARTIAL FRACTIONS

The practical method for finding partial fractions may well be familiar to you (though it does not matter if it is not). What might be surprising is that the method needs to be justified and that gcds are just the ticket for doing so.[7]

A *rational function over* $\mathbb{F}$ is simply a quotient $\frac{f(x)}{g(x)}$ where $f(x)$ and $g(x)$ are any polynomials in $\mathbb{F}[x]$, the polynomial $g(x)$ of course not being equal to the zero polynomial. If $\deg f(x) < \deg g(x)$, the rational function is *proper*. The set of all

---

[7]I learnt about this theory from [4].

rational functions $\mathbb{F}(x)$ can be added, subtracted, multiplied and divided. In fact, they satisfy all the algebraic laws of high school algebra and so form a field[8]. The goal of this section is to prove that every real rational function can be written as a sum of simpler rational functions in a specific way. We begin with a lemma.

Lemma 7.8.1. *Let $f(x)$ and $g(x)$ be coprime polynomials in $\mathbb{F}[x]$ both at least linear. Let $h(x)$ be any non-zero polynomial in $\mathbb{F}[x]$ such that*

$$\deg h(x) < \deg f(x) + \deg g(x).$$

Then we can write

$$h(x) = a(x)f(x) + b(x)g(x)$$

where if $a(x)$ is non-zero then $\deg a(x) < \deg g(x)$ and if $b(x)$ is non-zero then $\deg b(x) < \deg f(x)$. The polynomials $a(x)$ and $b(x)$ are unique satisfying these two conditions.

Proof. We prove explicitly the case where neither $a(x)$ nor $b(x)$ is zero. The polynomials $f(x)$ and $g(x)$ are coprime, so that

$$h(x) = f(x)p(x) + g(x)q(x)$$

for some polynomials $p(x)$ and $q(x)$ by Lemma 7.7.3. The question therefore is whether we can manipulate things to get the right degrees as stated. For any polynomial $c(x)$, we have that

$$[p(x) - c(x)g(x)]f(x) + [q(x) + c(x)f(x)]g(x) = h(x).$$

Choose $c(x)$ to be the quotient when $p(x)$ is divided by $g(x)$ so that by the remainder theorem $p(x) - c(x)g(x)$ has degree strictly less than that of $g(x)$. Put $a(x) = p(x) - c(x)g(x)$, with this choice for $c(x)$, and put $b(x) = q(x) + c(x)f(x)$. Then $h(x) = a(x)f(x) + b(x)g(x)$. Now

$$\deg a(x)f(x) = \deg a(x) + \deg f(x) < \deg g(x) + \deg f(x).$$

Since by assumption $\deg h(x) < \deg f(x) + \deg g(x)$, it follows that $\deg b(x)g(x) < \deg f(x) + \deg g(x)$. Thus

$$\deg b(x) + \deg g(x) < \deg f(x) + \deg g(x)$$

and so $\deg b(x) < \deg f(x)$, as required.

We now turn to the proof of uniqueness. Let

$$a(x)f(x) + b(x)g(x) = h(x) = a'(x)f(x) + b'(x)g(x)$$

where $\deg a(x) < \deg g(x)$ and $\deg b(x) < \deg f(x)$, and $\deg a'(x) < \deg g(x)$ and $\deg b'(x) < \deg f(x)$. We therefore have that

$$[a(x) - a'(x)]f(x) = [b'(x) - b(x)]g(x).$$

---

[8]Heed the brackets which are round here unlike the square brackets for the set of polynomials.

If $a(x) - a'(x) \neq 0$ then $b'(x) - b(x) \neq 0$. Since $f(x)$ and $g(x)$ are coprime, we have $f(x) \mid (b'(x) - b(x))$ by the polynomial version of Euclid's lemma. But the degree of $b'(x) - b(x)$ is strictly less than the degree of $f(x)$ and so this is impossible. It follows that $a(x) = a'(x)$ and $b(x) = b'(x)$. □

Our goal is to write an arbitrary rational function $\frac{f(x)}{g(x)}$ as a sum of simpler rational functions. We proceed first in general and then restrict to the real case at the end. If $\deg f(x) > \deg g(x)$ then we can apply the remainder theorem and write

$$\frac{f(x)}{g(x)} = q(x) + \frac{r(x)}{g(x)}$$

where $\deg r(x) < \deg g(x)$. Thus without loss of generality, we can assume that $\deg f(x) < \deg g(x)$ in what follows and deal only with proper rational functions. The proof of (1) below follows immediately by Lemma 7.8.1, and the proof of part (2) follows from part (1) by induction.

Proposition 7.8.2 (First partial fraction result). *We work with polynomials in* $\mathbb{F}[x]$.

1. *Let* $\frac{a(x)}{b(x)c(x)}$ *be a proper rational function where* $b(x)$ *and* $c(x)$ *are coprime and at least linear. Then there are unique polynomials* $p(x)$ *and* $q(x)$ *such that*

$$\frac{a(x)}{b(x)c(x)} = \frac{p(x)}{b(x)} + \frac{q(x)}{c(x)}$$

where the rational functions on the righthand side are proper.

2. *Let* $\frac{f(x)}{g(x)}$ *be a proper rational function. Let* $g(x) = p_1(x)\ldots p_m(x)$ *be a product of pairwise coprime polynomials each of which is at least linear. Then there are unique polynomials* $a_i(x)$ *such that*

$$\frac{f(x)}{g(x)} = \sum_{i=1}^{m} \frac{a_i(x)}{p_i(x)},$$

where the rational functions on the righthand side are proper.

Our next result is analogous to writing a number in a number base. Compare the following two examples, where the second is proved in the proposition that follows.

1. Let d be a number base and $n < d^k$. Then we can write

$$n = q_1 d^{k-1} + \ldots + q_{k-1}d + q_k$$

where $0 \leq q_i < d$.

2. Let $f(x)$ be a non-constant polynomial, and let $h(x)$ be a non-zero polynomial such that $\deg h(x) < \deg f(x)^k$. Then we can write

$$h(x) = q_1(x)f(x)^{k-1} + q_2(x)f(x)^{k-2} + \ldots + q_{k-1}(x)f(x) + q_k(x)$$

where $\deg q_i(x) < \deg f(x)$. Thus the polynomial $f(x)$ is the 'base' and the set of polynomials of degree strictly less than $\deg f(x)$ is the set of 'digits'.

Proposition 7.8.3 (Second partial fraction result). *Let $\frac{h(x)}{f(x)^k}$ be a proper rational function where $f(x)$ has degree m and $k > 0$. Then*

$$\frac{h(x)}{f(x)^k} = \frac{q_1(x)}{f(x)} + \ldots + \frac{q_k(x)}{f(x)^k}$$

for unique polynomials $q_i(x)$ which if non-zero are such that $\deg q_i(x) < m$ for $1 \leq i \leq k$.

Proof. By the remainder theorem, we can write

$$h(x) = f(x)^{k-1} q_1(x) + r_1(x)$$

where if $r_1(x)$ is non-zero then $\deg r_1(x) < \deg f(x)^{k-1}$ and $\deg q_1(x) < m$. If $r_1(x)$ is zero we are done; otherwise we now divide $f(x)^{k-2}$ into $r_1(x)$. Continuing in this way, we conclude that

$$h(x) = q_1(x) f(x)^{k-1} + q_2(x) f(x)^{k-2} + \ldots + q_{k-1}(x) f(x) + q_k(x).$$

Now divide through by $f(x)^k$ to get the first part of the result.

To show uniqueness, suppose that

$$\frac{h(x)}{f(x)^k} = \frac{s_1(x)}{f(x)} + \ldots + \frac{s_k(x)}{f(x)^k}$$

where if non-zero $\deg s_i(x) < m$ for $1 \leq i \leq k$. Subtracting the two expressions for $\frac{h(x)}{f(x)^k}$ and then multiplying by $f(x)^k$ we get

$$(q_k(x) - s_k(x)) + \ldots + f(x)^{k-2}(q_2(x) - s_2(x)) + f(x)^{k-1}(q_1(x) - s_1(x)) = 0.$$

We can write this as

$$(q_k(x) - s_k(x)) + f(x)t(x) = 0$$

for some polynomial $t(x)$. But $\deg(q_k(x) - s_k(x)) < m$, and $f(x)$ has degree m. Thus $q_k(x) = s_k(x)$ and $t(x)$ is the zero polynomial. Examining what $t(x)$ is, the argument may be repeated to show successively that the coefficients are equal. □

We now focus on the specific case of real rational functions. By the fundamental theorem for real polynomials, we can write $g(x)$ as a product of distinct factors of the form $(x - a)^r$ or $(x^2 + ax + b)^s$. Using this decomposition of $g(x)$, the rational function $\frac{f(x)}{g(x)}$ can be written as a sum of simpler rational functions which have the following forms.

- For each factor of $g(x)$ of the form $(x - a)^r$, there is a sum of the form

$$\frac{a_1}{x - a} + \ldots + \frac{a_{r-1}}{(x - a)^{r-1}} + \frac{a_r}{(x - a)^r}$$

where the a_i are real numbers.

- For each factor of $g(x)$ of the form $(x^2 + ax + b)^s$, there is a sum of the form

$$\frac{a_1 x + b_1}{x^2 + ax + b} + \cdots + \frac{a_{s-1} x + b_{s-1}}{(x^2 + ax + b)^{s-1}} + \frac{a_s x + b_s}{(x^2 + ax + b)^s}.$$

This leads to what is called the *partial fraction* representation of $\frac{f(x)}{g(x)}$ which we illustrate by means of a number of examples. The practical method used to find particular partial fraction decompositions is to begin by writing down the expected form the decomposition should take in terms of unknowns and then proceed from there by using any legitimate methods to determine them. The point of Proposition 7.8.2 and Proposition 7.8.3 is that if the forms are correct any equations (and they will be the linear equations of Section 8.3) that arise can always be solved.

Examples 7.8.4.

1. Write $\frac{5}{(x+3)(x-2)}$ in partial fractions. Clearly $x+3$ and $x-2$ are coprime. Thus by Proposition 7.8.2, there are unique real numbers a and b such that

$$\frac{5}{(x+3)(x-2)} = \frac{a}{x+3} + \frac{b}{x-2}.$$

The righthand side is equal to

$$\frac{a(x-2) + b(x+3)}{(x+3)(x-2)}.$$

We can equate the lefthand numerator with the righthand numerator to get the following equality of polynomials.

$$5 = a(x-2) + b(x+3).$$

It is important to observe that this equality holds for all values of x. This enables us to write down a set of linear equations to determine the unknowns. Putting $x = 2$ we get $b = 1$, and putting $x = -3$ we get $a = -1$. Thus

$$\frac{5}{(x+3)(x-2)} = \frac{-1}{x+3} + \frac{1}{x-2}.$$

2. Write $\frac{16x}{(x-2)(x+2)(x^2+4)}$ in partial fractions. Clearly $(x-2)$, $(x+2)$ and (x^2+4) are coprime and $x^2 + 4$ is an irreducible quadratic. Thus by Proposition 7.8.2, there are unique real numbers a, b, c, d such that

$$\frac{16x}{x^4 - 16} = \frac{a}{x-2} + \frac{b}{x+2} + \frac{cx+d}{x^2+4}.$$

As above, this leads to the following equality of polynomials

$$16x = a(x+2)(x^2+4) + b(x-2)(x^2+4) + (cx+d)(x-2)(x+2).$$

Using appropriate values of x, we get that $a = 1$, $b = 1$, $c = -2$ and $d = 0$. Thus

$$\frac{16x}{x^4 - 16} = \frac{1}{x-2} + \frac{1}{x+2} - \frac{2x}{x^2+4}.$$

3. Write $\frac{x^2+x+1}{(x+1)^3}$ in partial fractions. By Proposition 7.8.3, we can write

$$\frac{x^2+x+1}{(x+1)^3} = \frac{a}{x+1} + \frac{b}{(x+1)^2} + \frac{c}{(x+1)^3}$$

for unique real numbers a,b,c. A simple calculation shows that

$$\frac{x^2+x+1}{(x+1)^3} = \frac{1}{x+1} - \frac{1}{(x+1)^2} + \frac{1}{(x+1)^3}.$$

4. Write $\frac{9}{(x-1)(x+2)^2}$ in partial fractions. The polynomials $(x-1)$ and $(x+2)^2$ are coprime. The polynomial $(x+2)^2$ is a square of a linear polynomial. The theory therefore tells us to expect a partial fraction decomposition of the form

$$\frac{9}{(x-1)(x+2)^2} = \frac{a}{x-1} + \frac{b}{x+2} + \frac{c}{(x+2)^2}.$$

As above, we get the following equality of polynomials

$$9 = a(x+2)^2 + b(x-1)(x+2) + c(x-1).$$

Putting $x=1$, we get $a=1$, putting $x=-2$, we get $c=-3$ and putting $x=-1$ and using the values we have for a and c we get $b=-1$. Thus

$$\frac{9}{(x-1)(x+2)^2} = \frac{1}{x-1} - \frac{1}{x+2} - \frac{3}{(x+2)^2}.$$

5. Write $\frac{3x^2+2x+1}{(x+2)(x^2+x+1)^2}$ in partial fractions. The polynomial x^2+x+1 is an irreducible quadratic as can be checked by calculating its discriminant. The polynomials $(x+2)$ and $(x^2+x+1)^2$ are clearly coprime. We therefore expect a solution of the form

$$\frac{3x^2+2x+1}{(x+2)(x^2+x+1)^2} = \frac{a}{x+2} + \frac{bx+c}{x^2+x+1} + \frac{dx+e}{(x^2+x+1)^2}.$$

This leads to the following equality of polynomials

$$3x^2+2x+1 = a(x^2+x+1)^2 + (bx+c)(x+2)(x^2+x+1) + (dx+e)(x+2).$$

Somewhat lengthy computations show that

$$\frac{3x^2+2x+1}{(x+2)(x^2+x+1)^2} = \frac{1}{x+2} + \frac{1-x}{x^2+x+1} - \frac{1}{(x^2+x+1)^2}.$$

Partial fractions can be used to calculate

$$\int \frac{f(x)}{g(x)} dx$$

exactly, when $f(x)$ and $g(x)$ are real polynomials. Recall that if a_i are real numbers and $f_i(x)$ are integrable functions then

$$\int \sum_{i=1}^{n} a_i f_i(x)dx = \sum_{i=1}^{n} a_i \int f_i(x)dx$$

ignoring the usual constant of integration. This property is known as *linearity*. We refer the reader to standard calculus books [30, 56] for the explanations of the non-algebraic parts of the proof of the following.

Theorem 7.8.5 (Integrating rational functions). *The integration of an arbitrary real rational function can be reduced to integrals of the following five types.*

(I) $\int x^n dx = \frac{1}{n+1}x^{n+1}$ *where $n \geq 0$.*

(II) $\int \frac{dx}{x^n} = \frac{1}{(1-n)x^{n-1}}$ *where $n > 1$.*

(III) $\int \frac{dx}{x} = \ln|x|$.

(IV) *If $n > 1$ define $I_n = \int \frac{dx}{(1+x^2)^n}$. Then $I_n = \frac{x}{2(n-1)(1+x^2)^{n-1}} + \frac{2n-3}{2(n-1)}I_{n-1}$.*

(V) $I_1 = \int \frac{dx}{1+x^2} = \arctan x$.

Proof. The proof is broken down into a number of steps.

(1) Suppose that $\frac{f(x)}{g(x)}$ is a rational function where $\deg f(x) > \deg g(x)$. By the remainder theorem for polynomials we can write

$$\frac{f(x)}{g(x)} = q(x) + \frac{r(x)}{g(x)}$$

where $\deg r(x) < \deg g(x)$. By linearity of integration

$$\int \frac{f(x)}{g(x)}dx = \int q(x)dx + \int \frac{r(x)}{g(x)}dx.$$

Thus to integrate an arbitrary rational function it is enough to know how to integrate polynomials and proper rational functions.

(2) By linearity of integration, integrating arbitrary polynomials can be reduced to integrating the following

$$\int x^n dx$$

where $n \geq 0$. This is equal to $\frac{1}{n+1}x^{n+1}$. This deals with case (I). From now on we focus on proper rational functions.

(3) Let $\frac{f(x)}{g(x)}$ be a proper rational function. We factorize $g(x)$ into a product of real linear polynomials and real irreducible quadratic polynomials and then write $\frac{f(x)}{g(x)}$ in partial fractions and so as a sum of rational functions of one of the following two forms

$$\frac{a}{(x-b)^n} \quad \text{and} \quad \frac{px+q}{(x^2+bx+c)^n}$$

where a, b, c, p, q are real, $n \geq 1$, and where the quadratic has a pair of complex conjugate roots so that $b^2 < 4ac$. By the linearity of integration, this reduces calculating

$$\int \frac{f(x)}{g(x)} dx$$

to calculating integrals of the following two types

$$\int \frac{a}{(x-b)^n} dx \text{ and } \int \frac{px+q}{(x^2+bx+c)^n} dx.$$

By linearity of integration again, this reduces to being able to calculate the following three integrals

$$\int \frac{1}{(x-b)^n} dx, \quad \int \frac{x}{(x^2+bx+c)^n} dx \text{ and } \int \frac{1}{(x^2+bx+c)^n} dx.$$

(4) We concentrate first on the two integrals involving quadratics. By completing the square, we can write

$$x^2 + bx + c = \left(x + \frac{b}{2}\right)^2 + \left(c - \frac{b^2}{4}\right).$$

By assumption $b^2 - 4c < 0$. Put $e^2 = \frac{4c-b^2}{4} > 0$, the notation being simply a way of signalling that this number is positive. Thus

$$x^2 + bx + c = \left(x + \frac{b}{2}\right)^2 + e^2.$$

We now use a technique of calculus known as substitution and put $y = x + \frac{b}{2}$. Doing this, and returning to x as the variable, we need to be able to evaluate the following three integrals

$$\int \frac{1}{(x-b)^n} dx, \quad \int \frac{x}{(x^2+e^2)^n} dx \text{ and } \int \frac{1}{(x^2+e^2)^n} dx.$$

(5) The second integral above can be converted into the first by means of the substitution $x^2 = u$ and then returning to x as the variable. The first integral can be simplified by using the substitution $y = x - b$ and then returning to the variable x. The third integral can be simplified by using the substitution $y = \frac{x}{e}$ and then returning to the variable x. Thus we need only evaluate the integrals

$$\int \frac{dx}{x^n} \text{ and } \int \frac{dx}{(x^2+1)^n}.$$

(6) For the first integral, there are two cases. If $n > 1$ then

$$\int \frac{dx}{x^n} = \frac{1}{(1-n)x^{n-1}}.$$

If $n = 1$ then

$$\int \frac{dx}{x} = \ln|x|.$$

This deals with cases (II) and (III).

(7) The second integral can be evaluated using a recurrence technique and integration by parts. Define

$$I_n = \int \frac{dx}{(x^2+1)^n},$$

where $n > 1$. Then

$$I_n = \frac{x}{2(n-1)(x^2+1)^{n-1}} + \frac{2n-3}{2(n-1)} I_{n-1}.$$

Finally,

$$\int \frac{dx}{x^2+1} = \arctan x.$$

This deals with cases (IV) and (V). □

Exercises 7.8

1. Write the following in partial fractions.

   (a) $\frac{3x+4}{(x+1)(x+2)}$.

   (b) $\frac{4x^2+2x+3}{(x+2)(x^2+1)}$.

   (c) $\frac{x^3+2x^2+3x+1}{(x^2+x+1)^2}$.

   (d) $\frac{1}{x^4+1}$. *The nemesis of Leibniz.*

7.9 RADICAL SOLUTIONS

There are two ways of looking at polynomials: a practical and a theoretical. From a practical perspective, we can ask for numerical methods to calculate roots of polynomials. This can be done but we shall not do so here. From a theoretical perspective, we can ask about the nature of those roots, and it is this aspect we touch on here. We found a formula for the roots of a quadratic, but the question is whether there are similar formulae for the roots of polynomial equations of higher degree. In this section, we shall answer this question exactly for cubics and quartics, and sketch the answer for polynomial equations of higher degree. The stunning result is that there is not always a formula for the roots of a polynomial equation.[9]

---

[9]It might seem strange, but mathematicians are just as happy with negative results as they are with positive ones.

Cubics

Let

$$f(x) = a_3x^3 + a_2x^2 + a_1x + a_0$$

where $a_3 \neq 0$. We determine all the roots of $f(x)$. This problem can be simplified in two ways. First, divide through by a_3 and so, without loss of generality, we can assume that $f(x)$ is monic. That is $a_3 = 1$. Second, by means of a linear substitution, we obtain a cubic in which the coefficient of the term in x^2 is zero. To do this, put $x = y - \frac{a_2}{3}$ and check that you get a polynomial of the form

$$g(y) = y^3 + py + q.$$

We say that such a cubic is *reduced*. It follows that without loss of generality, and relabelling the variable, we need only solve the cubic

$$x^3 + px + q = 0.$$

This can be done by what looks like a miracle. Let u and v be complex unknowns and let $\omega = \cos\frac{2\pi}{3} + i\sin\frac{2\pi}{3}$ be the principal cube root of unity. Check that the cubic

$$\boxed{x^3 - 3uvx - (u^3 + v^3) = 0}$$

has the roots

$$u+v, \quad u\omega + v\omega^2, \quad u\omega^2 + v\omega.$$

Bearing this in mind, we can solve

$$\boxed{x^3 + px + q = 0}$$

if we can find u and v such that

$$p = -3uv, \quad q = -u^3 - v^3.$$

Cube the first equation above so we now have

$$-\frac{p^3}{27} = u^3v^3 \text{ and } -q = u^3 + v^3.$$

This tells us both the sum and product of the two quantities u^3 and v^3. This means that u^3 and v^3 are the roots of the quadratic equation

$$t^2 + qt - \frac{p^3}{27} = 0.$$

Solving this quadratic, we obtain

$$u^3 = \frac{1}{2}\left(-q + \sqrt{\frac{27q^2 + 4p^3}{27}}\right) \text{ and } v^3 = \frac{1}{2}\left(-q - \sqrt{\frac{27q^2 + 4p^3}{27}}\right).$$

To find u we have to take a cube root of the number u^3 and there are three possible such roots. Choose one such value for u and then choose the corresponding value of the cube root of v so that $p = -3uv$. We have values for u and v and so we can write down explicitly the values of

$$u + v, \quad u\omega + v\omega^2, \quad u\omega^2 + v\omega.$$

We have therefore solved the cubic.

Example 7.9.1. Solve the cubic $x^3 + 9x - 2 = 0$. Here $p = 9$ and $q = -2$. The quadratic equation we have to solve is therefore $t^2 - 2t - 27 = 0$. This has roots $1 \pm 2\sqrt{7}$. Put $u^3 = 1 + 2\sqrt{7}$. In this case, we can choose a real cube root to get

$$u = \sqrt[3]{1 + \sqrt{28}}.$$

We must then choose v to be the real cube root

$$v = \sqrt[3]{1 - \sqrt{28}}$$

in order that $p = -3uv$. We can now write down the three roots of our original cubic. For example

$$u + v = \sqrt[3]{1 + \sqrt{28}} + \sqrt[3]{1 - \sqrt{28}},$$

which is a nice example of a radical expression.

Example 7.9.2. The cubic $x^3 - 15x - 4 = 0$ was studied by Bombelli in 1572 and had an important influence on the development of complex numbers. It is easy to see that this equation has 4 as one of its roots. We put this fact to one side for the moment and proceed to solve the equation by the method of this section. The associated quadratic in this case is $t^2 - 4t + 125 = 0$. This has the two solutions

$$x = 2 \pm 11i.$$

Thus

$$u^3 = 2 + 11i \text{ and } v^3 = 2 - 11i.$$

There are three cube roots of $2 + 11i$ all of them complex. This is puzzling to say the least, but we continue applying the method. Write $\sqrt[3]{2 + 11i}$ to represent one of those cube roots and write $\sqrt[3]{2 - 11i}$ to represent the corresponding cube root so that their product is 5. Thus at least symbolically we can write

$$u + v = \sqrt[3]{2 + 11i} + \sqrt[3]{2 - 11i}.$$

What is surprising is that for some choices of these cube roots this value must be equal to 4. Observe that

$$(2 + i)^3 = 2 + 11i \text{ and } (2 - i)^3 = 2 - 11i.$$

If we choose $2 + i$ as one of the cube roots of $2 + 11i$ then we have to choose $2 - i$ as the corresponding cube root of $2 - 11i$. In this way, we get

$$u + v = (2 + 11i) + (2 - 11i) = 4$$

as a root. Thus we get the real root 4 by using non-real complex numbers. It was the fact that real roots arose in this way that provided the first inkling that there was a number system, the complex numbers, that extended the so-called real numbers, but had just as tangible an existence.

Quartics

Let

$$f(x) = a_4 x^4 + a_3 x^3 + a_2 x^2 + a_1 x + a_0.$$

As usual, we can assume that $a_4 = 1$. By means of a suitable linear substitution, which is left as an exercise, we can eliminate the cubic term. We therefore obtain a *reduced quartic* which it is convenient to write in the following way

$$x^4 = ax^2 + bx + c.$$

The method used to solve this equation is based on the following idea. Suppose that we could write the righthand side as a perfect square $(dx + e)^2$. Then our quartic could be written as the product of two quadratics

$$\left(x^2 - (dx+e)\right)\left(x^2 + (dx+e)\right)$$

and the roots of these two quadratics would be the four roots of our original quartic. Although it is not true that this can always be done, we shall show that by another miracle we can transform the equation into one with the same roots where it can be done. Let t be a new variable whose value will be determined later. We can write

$$(x^2 + t)^2 = (a + 2t)x^2 + bx + (c + t^2).$$

On multiplying out, this equation simplifies to the original equation and so has the same roots for any value of t. We want to choose a value of t so that the righthand side is a perfect square. This happens when the discriminant of the quadratic $(a + 2t)x^2 + bx + (c + t^2)$ is zero. That is when

$$b^2 - 4(a + 2t)(c + t^2) = 0.$$

This is a cubic in t. We use the method for solving cubics to find a specific value of t, say t_1, that solves this cubic. We then get

$$(x^2 + t_1)^2 = (a + 2t_1)\left(x + \tfrac{b}{2(a+2t_1)}\right)^2.$$

It follows that the roots of the original quartic are the roots of the following two quadratics

$$(x^2 + t_1) - \sqrt{a + 2t_1}\left(x + \tfrac{b}{2a+4t_1}\right) = 0$$

and

$$(x^2 + t_1) + \sqrt{a + 2t_1}\left(x + \tfrac{b}{2a+4t_1}\right) = 0.$$

Example 7.9.3. Solve the quartic $x^4 = 1 - 4x$. We find a value of t below

$$(x^2 + t)^2 = 2x^2 t - 4x + (1 + t^2)$$

which makes the righthand side a perfect square. By the theory above, we need a root of the cubic

$$t^3 + t - 2 = 0.$$

Here $t = 1$ works. The quartic with this value of t becomes

$$(x^2 + 1)^2 = 2(x - 1)^2.$$

Therefore the roots of the original quartic are the roots of the following two quadratics

$$(x^2 + 1) - \sqrt{2}(x - 1) = 0 \text{ and } x^2 + 1 + \sqrt{2}(x - 1) = 0.$$

The roots of our original quartic are therefore

$$\frac{1 \pm i\sqrt{\sqrt{8}+1}}{\sqrt{2}} \text{ and } \frac{-1 \pm \sqrt{\sqrt{8}-1}}{\sqrt{2}},$$

which are again nice examples of radical expressions.

Quintics and beyond

The solution of cubics and quartics relied upon two miracles. One miracle is lucky but two miracles is frankly miraculous. In mathematics, miracles are only apparent: they point towards theories waiting to be discovered that dissolve the miracles. It was Joseph-Louis Lagrange (1736–1813) who acted as the pathfinder. He wrote a book that analysed the different ways of solving cubics and quartics. Although he found a unifying approach to solving these equations and so provided a firm foundation for future work, the question of whether a formula for the roots of a quintic could be found was left open. It was left to two younger mathematicians, Paolo Ruffini (1765–1822) and Niels Henrik Abel (1802–1829), to show that the pattern that we appear to have discovered in our solutions of cubics and quartics — to solve equations of degree n use those of smaller degree — in fact breaks down when $n = 5$. Ruffini and Abel proved, with Abel's proof being acknowledged as the more complete, that there is no formula for the roots of a quintic. Quintics always have five roots, of course; we know this from the fundamental theorem of algebra, but these roots cannot in general be written down as radical expressions.

Symmetries

The theory that explains away the miracles and provides the correct setting for investigating the nature of the roots of polynomial equations was discovered by Evariste Galois, also building on the work of Lagrange. His approach was something new in algebra: to determine whether the roots of a polynomial could be expressed in algebraic terms as radical expressions, he studied the *symmetries* of the polynomial. By a symmetry of some object, we mean an operation that when applied to that object leaves it looking the same. Thus an example of a symmetry of a circle would be any rotation of that circle about its centre whereas for a square only the rotations that are multiples of a right angle will be symmetries. The fact that the circle has many more of these rotational symmetries than the square reflects the fact that the circle is a more symmetric shape than a square. What is meant by the symmeries of a polynomial is explained in a subject known as *Galois theory*. This idea of symmetry is not a mere extrapolation of existing algebraic manipulation, instead it involves working at a higher level of abstraction. As so often happens in mathematics, a development in one area led to developments in others. Sophus Lie (1842–1899) realized that symmetries could help explain the tricks used to solve differential equations. It was in this way that symmetry came to play a fundamental rôle in physics. If you hear a particle physicist talking about symmetries, they are paying unconscious tribute to Galois' work in studying the nature of the roots of polynomial equations. In this book, we cannot describe in detail what Galois did, but we can at least touch on some ideas that figure in his solution and which we shall also need later on.

Symmetric polynomials

Example 7.9.4. Let the quadratic $x^2 + ax + b = 0$ have roots r_1 and r_2. By the theory we have developed in this chapter

$$x^2 + ax + b = (x - r_1)(x - r_2).$$

Multiplying out the righthand side and equating coefficients, we get

$$a = -(r_1 + r_2) \text{ and } b = r_1 r_2.$$

We now do something that looks counterintuitive. Define

$$\sigma_1(x_1, x_2) = x_1 + x_2 \text{ and } \sigma_2(x_1, x_2) = x_1 x_2,$$

two polynomials in two variables. We can therefore express the result above as

$$a = -\sigma_1(r_1, r_2) \text{ and } b = \sigma_2(r_1, r_2).$$

We have proved that the coefficients of the quadratic can be expressed as special kinds of functions of the roots of the quadratic.

Example 7.9.5. The game we played for quadratics can also be played for cubics. If the cubic $x^3 + ax^2 + bx + c = 0$ has roots r_1, r_2, r_3 then $a = -\sigma_1(r_1, r_2, r_3)$, $b = \sigma_2(r_1, r_2, r_3)$ and $c = -\sigma_3(r_1, r_2, r_3)$ where

- $\sigma_1(x_1,x_2,x_3) = x_1 + x_2 + x_3,$

- $\sigma_2(x_1,x_2,x_3) = x_1x_2 + x_2x_3 + x_1x_3,$

- $\sigma_3(x_1,x_2,x_3) = x_1x_2x_3.$

In general, we can write down n polynomials

$$\sigma_1(x_1,\ldots,x_n),\ldots,\sigma_n(x_1,\ldots,x_n)$$

where $\sigma_i(x_1,\ldots,x_n)$ is the sum of all products of i distinct variables (and remembering that the variables commute). These polynomials are called the *elementary symmetric polynomials*. The examples above illustrate the proof of the following theorem.

Theorem 7.9.6. *Each coefficient of a monic polynomial of degree n is equal to one of the elementary symmetric polynomials $\sigma_1,\ldots,\sigma_n$ (with a possible difference in sign) evaluated at the roots of the polynomial.*

Thus locked up within the coefficients of a polynomial there is information about the roots of the polynomial.

The elementary symmetric polynomials have a property that relates directly to symmetry. Let $\sigma(x_1,\ldots,x_n)$ be any polynomial in the n variables $x_1,\ldots,x_n$. Let $f \in S_n$, the set of all permutations of the set $\{1,\ldots,n\}$. Define

$$f \cdot \sigma(x_1,\ldots,x_n) = \sigma(x_{f(1)},\ldots,x_{f(n)}).$$

Example 7.9.7. Let $\sigma(x_1,x_2,x_3) = x_1^2 + x_2x_3$ and $f = (123)$. Then $f \cdot \sigma(x_1,x_2,x_3) = x_2^2 + x_3x_1$.

Lemma 7.9.8. *Let $f,g \in S_n$ and let $\sigma = \sigma(x_1,\ldots,x_n)$ be a polynomial in the n variables $x_1,\ldots,x_n$. Then $(fg)\cdot\sigma = f\cdot(g\cdot\sigma)$.*

Proof. By definition $(fg)\cdot\sigma(x_1,\ldots,x_n) = \sigma(x_{(fg)(1)},\ldots,x_{(fg)(n)})$. But

$$\sigma(x_{(fg)(1)},\ldots,x_{(fg)(n)}) = \sigma(x_{f(g(1))},\ldots,x_{f(g(n))}) = f\cdot\sigma(x_{g(1)},\ldots,x_{g(n)})$$

which is equal to $f\cdot(g\cdot\sigma(x_1,\ldots,x_n))$. □

We say that $\sigma(x_1,\ldots,x_n)$ is *symmetric* if $f\cdot\sigma(x_1,\ldots,x_n) = \sigma(x_1,\ldots,x_n)$ for all $f \in S_n$. The elementary symmetric polynomials are indeed symmetric. Their importance stems from the following result. We prove it in the case $n=2$ in the exercises.

Theorem 7.9.9. *Let σ be a symmetric polynomial in n variables. Then there is a polynomial τ in n variables such that $\sigma = \tau(\sigma_1,\ldots,\sigma_n)$.*

Thus the elementary symmetric polynomials are the building blocks from which all symmetric polynomials can be constructed.

Example 7.9.10. The polynomial $p(x) = (x_1 - x_2)^2$ is symmetric. By Theorem 7.9.9, we should therefore be able to write it as a polynomial in the elementary symmetric polynomials σ_1 and σ_2. Observe that

$$p(x) = (x_1^2 + x_2^2) - 2x_1x_2 = (x_1 + x_2)^2 - 2x_1x_2 - 2x_1x_2 = (x_1 + x_2)^2 - 4x_1x_2.$$

Thus $p(x) = \sigma_1^2 - 4\sigma_2$ and we have written our given symmetric polynomial as a polynomial in σ_1 and σ_2.

We pursue one application of these ideas here by way of illustration. Define

$$\Delta = \Delta(x_1, \ldots, x_n) = \prod_{i<j} (x_i - x_j). \tag{7.1}$$

For example

$$\Delta(x, y) = (x - y) \text{ and } \Delta(x, y, z) = (x - y)(x - z)(y - z)$$

where we have written x, y and z instead of x_1, x_2 and x_3, respectively. Observe that if $f \in S_n$ then $f \cdot \Delta = \pm\Delta$. We shall refine this observation later, but crucially it implies that Δ^2 is a symmetric polynomial.

Example 7.9.11. The polynomial $\Delta(x_1, x_2) = (x_1 - x_2)$ is not symmetric, but the polynomial $\Delta(x_1, x_2)^2$ is. We wrote this as a polynomial in the elementary symmetric polynomials above and obtained $\sigma_1^2 - 4\sigma_2$. Suppose now that r_1 and r_2 are the roots of the polynomial $x^2 + bx + c$. Then $-b = \sigma_1$ and $c = \sigma_2$. It follows that

$$\Delta(r_1, r_2)^2 = b^2 - 4c.$$

This is nothing other than the discriminant of the polynomial as defined in Section 4.3.

The above example can be generalized. Let $p(x)$ be a polynomial of degree n with roots $r_1, \ldots, r_n$. Define the *discriminant* of $p(x)$, denoted by D, to be the number

$$D = \Delta(r_1, \ldots, r_n)^2.$$

Clearly $D = 0$ if and only if there is a repeated root. Since D is a symmetric function of the roots, the theory above implies the following.

Theorem 7.9.12. *The discriminant of a polynomial $p(x)$ can be written as a polynomial in the coeffcients of $p(x)$.*

Even in the cubic case, writing down an explicit formula for the discriminant involves quite a bit of work, but by thinking about symmetries, we know that it can be done. This hints at the way that symmetries can yield important algebraic information. See [120] for the full development of these ideas.

Odd and even permutations

The polynomial Δ in n variables defined by equation (7.1) is important in the theory of permutations introduced in Section 3.4. Let $f \in S_n$ be a permutation. There are exactly two possibilities for $f \cdot \Delta$ and we use them to make some definitions.

- We say that f is *even* if $f \cdot \Delta = \Delta$.

- We say that f is *odd* if $f \cdot \Delta = -\Delta$.

Define the *sign* of a permutation f, denoted by $\mathrm{sgn}(f)$, as follows.

$$\mathrm{sgn}(f) = \begin{cases} 1 & \text{if } f \cdot \Delta = \Delta \\ -1 & \text{if } f \cdot \Delta = -\Delta. \end{cases}$$

The proof of the following is immediate from the definition and Lemma 7.9.8.

Lemma 7.9.13. *Let $f, g \in S_n$ be permutations. Then*

$$\mathrm{sgn}(fg) = \mathrm{sgn}(f)\mathrm{sgn}(g).$$

A more convenient characterization of odd and even permutations is provided by the following.

Proposition 7.9.14.

1. *Each transposition can be written as a product of an odd number of transpositions of the form $(i, i+1)$.*

2. *A permutation is odd if and only if it can be written as a product of an odd number of transpositions, and a permutation is even if and only if it can be written as a product of an even number of transpositions.*

Proof. (1) Let (i, j) be a transposition where $i < j$. Observe that

$$(i, j) = (1, i)(1, j)(1, i).$$

Thus it is enough to prove the result for every transposition of the form $(1, j)$. If $j = 2$ there is nothing to prove. We assume therefore that $j > 2$. Observe that

$$(1, j) = (1, j-1)(j-1, j)(1, j-1).$$

The result now follows by induction.

(2) We begin with a simple calculation. Let $f = (k, k+1)$. We calculate $f \cdot \Delta$. The term $(x_k - x_{k+1})$ occurs as a factor of Δ, and the effect of f on this term yields $(x_{k+1} - x_k) = -(x_k - x_{k+1})$. The other terms of the form $(x_k - x_j)$, where $k < j \neq k+1$, will not change sign under f, nor will the terms of the form $(x_i - x_k)$ where $i < k$. Thus $f \cdot \Delta = -\Delta$.

Let g be a permutation that can be written as a product of an odd number of transpositions. Then by part (1), it can be written as a product of an odd number of transpositions of the form $(i, i+1)$. We calculate $g \cdot \Delta$ using Lemma 7.9.8 and induction together with our simple calculation above to obtain $g \cdot \Delta = -\Delta$. Now let g be a permutation that can be written as a product of an even number of transpositions. A similar calculation shows that $g \cdot \Delta = \Delta$. Conversely, suppose that g is an odd permutation. Then $g \cdot \Delta = -\Delta$. We know that g can be written as a product of transpositions. It cannot be written as a product of an even number of permutations because that would contradict what we have found above. Thus g can be written as a product of an odd number of transpositions. □

The distinction between odd and even permutations is crucial to understanding determinants. See Section 9.6.

Exercises 7.9

1. Show that the cubic
$$x^3 - 3uvx - (u^3 + v^3)$$
has the roots
$$u + v, \quad u\omega + v\omega^2, \quad u\omega^2 + v\omega,$$
where ω is the principal cube root of unity.

2. Given that
$$-\frac{p^3}{27} = u^3 v^3 \text{ and } -q = u^3 + v^3$$
prove that u^3 and v^3 are the roots of the quadratic equation
$$x^2 + qx - \frac{p^3}{27} = 0.$$

3. Solve
$$x^3 + 3x^2 - 3x - 14 = 0$$
using the methods of this section.

4. Show by means of a suitable linear substitution that the cubic term of
$$x^4 + a_3 x^3 + a_2 x^2 + a_1 x + a_0 = 0$$
can be eliminated.

5. (a) Solve
$$x^4 - x^2 - 2x - 1 = 0$$
using the methods of this section.
   (b) Solve
$$x^4 + 4x^2 + 1 = 0$$
using the methods of this section.

6. List the odd and even permutations in S_3.

7. *This question is about symmetric polynomials in two variables.[10]

   (a) Prove that

   $$x^i + y^i = (x+y)(x^{i-1} + y^{i-1}) - xy(x^{i-2} + y^{i-2})$$

   for $i \geq 2$.

   (b) Prove that

   $$x^i y^j + x^j y^i = x^i y^i (x^{j-i} + y^{j-i})$$

   for $i < j$.

   (c) Let $f(x,y)$ be a symmetric polynomial in two variables. Prove that if $ax^i y^j$, where $i \neq j$, is a term of this polynomial then $ax^j y^i$ is also a term.

   (d) Let $f(x,y)$ be a symmetric polynomial in two variables. Show that $f(x,y)$ can be written as a sum of terms of the form $a(x^i y^j + x^j y^i)$ and $ax^i y^i$.

   (e) Deduce that $f(x,y) = g(\sigma_1, \sigma_2)$ for some polynomial in two variables g.

   (f) Apply the above results to the symmetric polynomial $x^4 + 4x^3 y + 6x^2 y^2 + 4xy^3 + y^4$ and so write it as a polynomial in the elementary symmetric polynomials σ_1 and σ_2.

Box 7.2: Group Theory

We defined a group to be a set equipped with an associative binary operation that has an identity and in which every element is invertible. There are examples of groups throughout this book. A few important ones are: $(\mathbb{Z}_n, +)$ and $(\mathbb{Z}, +)$, and the group U_n of invertible elements of $(\mathbb{Z}, \times)$ from Chapter 5; the complex numbers S^1 of unit modulus from Chapter 6; the n roots of unity C_n from Chapter 7; the $n \times n$ invertible real matrices $GL_n(\mathbb{R})$ from Chapter 8; and $SO(3)$ the 3×3 orthogonal matrices with determinant 1 from Chapter 10. Historically, the most important groups were the symmetric groups $S(X)$ defined in Chapter 3, in particular the finite symmetric groups S_n. A subset of S_n closed under composition and inverses is called a *permutation group*. Whether a polynomial of degree n can be solved by radicals or not depends on the behaviour of a permutation group in S_n that can be constructed from the polynomial called its *Galois group*.

   The general question of whether there is a formula for the roots of a polynomial of degree n depends on the properties of the groups S_n themselves. These groups are algebraically different depending on whether $n \leq 4$ or $n \geq 5$: the former are examples of what are called *solvable groups* and the latter of *unsolvable groups*. This terminology arises directly from the theory of polynomials.

7.10   ALGEBRAIC AND TRANSCENDENTAL NUMBERS

In this section, we use our results on infinite numbers from Section 3.10. As a result of Galois' work we know that the roots of a polynomial with rational coefficients need not be described by radical expressions. But such roots still have an 'algebraic feel' to them even if in a more implicit than explicit way. For this reason, a root of a polynomial with rational coefficients is called an *algebraic number*. The set of algebraic numbers is denoted by $\mathbb{A}$.

---

[10] Adapted from [115].

Theorem 7.10.1. *There is a countable infinity of algebraic numbers.*

Proof. If a polynomial equation with rational coefficients is multiplied by a suitable integer it can be transformed into a polynomial with integer coefficients having the same roots. We can therefore restrict attention to such polynomials. We count the algebraic numbers that occur as roots of these polynomials of degree n. A polynomial such as this has coefficients from the set $\mathbb{Z}^n$. Each such polynomial has at most n distinct roots. Thus the number of algebraic numbers that arise in this way is at most $\aleph_0^n n = \aleph_0$. Thus the cardinality of the set of algebraic numbers is at most $\aleph_0 \aleph_0 = \aleph_0$. Since each rational number is algebraic, we have proved that the set of algebraic numbers is countably infinite. □

It can be proved, though we shall not do so here, that the sum, difference, product and quotient of algebraic numbers is also algebraic. It follows that $\mathbb{A}$ is a field.

A complex number that is not algebraic is called *transcendental*. The cardinality of the set of real numbers is c, and the cardinality of the set of complex numbers is $c^2 = c$ since each complex number can be viewed as an ordered pair of real numbers. But $c > \aleph_0$. We therefore deduce the following.

Theorem 7.10.2. *There are transcendental numbers, and in fact uncountably many.*

We have therefore proved that there are lots of transcendental numbers without producing a single example. Such are existence proofs. What these two theorems tell us is that most complex numbers cannot be described in easily intelligible algebraic ways.

Proving specific complex numbers are transcendental is hard. Here are two examples which we shall not prove here.[11] The first was proved by Charles Hermite (1822–1901) in 1873.

Theorem 7.10.3. *e is transcendental.*

The second was proved by Ferdinand von Lindemann (1852–1939) in 1882.

Theorem 7.10.4. *π is transcendental.*

The theory of transcendental numbers is such that it is still not known whether a number such as $\pi + e$ is transcendental.

7.11   MODULAR ARITHMETIC WITH POLYNOMIALS

In Chapter 5, modular arithmetic was defined with the help of the remainder theorem. We now use the remainder theorem for polynomials to define modular arithmetic for polynomials in $\mathbb{F}[x]$. Let $p(x)$ be a non-constant polynomial called the *modulus*. If $a(x), b(x) \in \mathbb{F}[x]$ we write $a(x) \equiv b(x) \pmod{p(x)}$ or just $a(x) \equiv b(x)$ if and only if $p(x) \mid a(x) - b(x)$. All the properties that we described for modular arithmetic hold good here: Lemma 5.4.2 and Theorem 5.4.9, except that irreducible polynomials play

---

[11]See [115] for proofs and more on transcendental numbers.

the rôle of prime numbers. We denote by $\mathbb{F}[x]/(p(x))$ the set of equivalence classes. If $p(x)$ is irreducible, then it will be a field. Just as in the case of the integers, we can work with representatives of the congruence classes. These will consist of all polynomials of degree strictly less than that of $p(x)$. Rather than develop the theory here, we describe three examples to whet your appetite. All three are fields.

Example 7.11.1. Consider the polynomials $\mathbb{Q}[x]$ modulo the irreducible polynomial $x^2 - 2$. We can regard this as the set of all polynomials of the form $a + bx$ where $a, b \in \mathbb{Q}$. Addition is straightforward but multiplication is interesting. We calculate $(a + bx)(c + dx)$. As a first step we get $ac + (ad + bc)x + bdx^2$. We now take the remainder when this polynomial is divided by $x^2 - 2$ to obtain $(ac + 2bd) + (ad + bc)x$. It follows that in this field

$$(a + bx)(c + dx) = (ac + 2bd) + (ad + bc)x.$$

To understand what this field is we compare it with the field whose elements are of the form $a + b\sqrt{2}$. Calculating the product $(a + b\sqrt{2})(c + d\sqrt{2})$ we get

$$(ac + 2bd) + (ad + bc)\sqrt{2}.$$

We have therefore constructed a copy of the field $\mathbb{Q}[\sqrt{2}]$ without explicitly mentioning the symbol $\sqrt{2}$.

Example 7.11.2. Consider the polynomials $\mathbb{R}[x]$ modulo the irreducible polynomial $x^2 + 1$. We can regard this as the set of all polynomials of the form $a + bx$ where $a, b \in \mathbb{R}$. As before, we need only describe multiplication. Observe that this time $x^2 = -1$. By a similar argument to the previous example we have therefore constructed a copy of $\mathbb{C}$ without explicitly using the symbol i. This construction is therefore a new way of showing that the field of complex numbers exists. Algebraically there is little difference between $\sqrt{2}$ and i in that both are the 'missing roots' of irreducible polynomials.

Example 7.11.3. Consider the polynomials $\mathbb{Z}_2[x]$ modulo the irreducible polynomial $x^2 + x + 1$. This polynomial is irreducible because neither 0 nor 1 is a root. This time

$$\mathbb{Z}_2[x]/(x^2 + x + 1) = \{0, 1, x, 1 + x\}$$

has four elements. Here is the Cayley table for multiplication.

$\times$	0	1	x	$1+x$
0	0	0	0	0
1	0	1	x	$1+x$
x	0	x	$1+x$	1
$1+x$	0	$1+x$	1	x

We therefore get a field with four elements. In fact, this is the first example of a field with four elements in this book. Every field with a finite number of elements can be constructed using methods similar to this example. It is worth observing that there is at least one element in this field, in this case x works, such that x, $x^2 = x+1$, $x^3 = 1$ are all the non-zero elements of the field. Although not proved here, such elements, called *primitive elements*, exist in every finite field. If one such element is found then constructing the Cayley table for multiplication becomes trivial using the law of exponents.

The technique we have sketched in this section is a powerful way of constructing new kinds of numbers.

Exercises 7.11

1. *Describe the multiplication of $\mathbb{R}[x]/(x^2)$, the *dual numbers*. Is it a field? [These are usually written in the form $a + b\varepsilon$ where $a, b \in \mathbb{R}$ and $\varepsilon^2 = 0$.]

2. *Describe the multiplication of $\mathbb{Z}_2[x]/(x^3 + x + 1)$. Is it a field?

3. *Describe the multiplication of $\mathbb{Z}_3[x]/(x^2 + 1)$. Is it a field?

Box 7.3: Ruler and Compass

The axioms of Euclidean geometry are based on straight lines and circles, and so it is natural to wonder what might be accomplished geometrically if the ruler, actually the straightedge, and compass were the only tools one was allowed to use. Euclid does not provide any methods to answer this question. In fact, the solution lies in algebra. The clue is to think about the sorts of numbers that can appear as the coordinates of points constructed using a ruler and compass. These are called *constructible numbers* and form a field, the elements of which can be described by means of radical expressions, but the only roots that can appear are square roots. Using this idea, two geometric problems from antiquity can be shown to be impossible: *trisection of an angle* and *doubling the cube*. Trisecting an angle means: given an angle construct using only a ruler and compass an angle one-third the size. Doubling the cube means: given a cube construct a cube of twice the volume using only a ruler and compass. A third problem from antiquity that can be handled similarly is squaring the circle. This means constructing a square of area equal to that of a circle using only a ruler and compass. This is shown to be impossible with the help of Theorem 7.10.4. Another question that can be asked in this vein is what regular polygons can be constructed using only a ruler and compass. By developing the theory of constructible numbers, Gauss proved that a regular n-gon could be constructed in this way if and only if $n = 2^m p_1 \ldots p_k$ where $p_1, \ldots, p_k$ are distinct Fermat primes (defined in Exercises 5.3). The factor 2^m is a result of the fact that angles can be bisected using a ruler and compass. The solution uses ideas that ultimately formed part of Galois theory. See [116] for a readable account of this theory and a much more precise historical discussion.

CHAPTER 8

Matrices

"What is the Matrix? The answer is out there [...] and it's looking for you, and it will find you if you want it to." – The Wachowskis

The term *matrix* was introduced by James Joseph Sylvester (1814–1897) in 1850, and the first paper on matrix algebra was published by Arthur Cayley (1821–1895) in 1858 [24]. Although the primary goal of this chapter is to describe the basics of matrix arithmetic and matrix algebra, towards the end we begin matrix theory proper with an introduction to diagonalization, an idea of fundamental importance in mathematics. This chapter, together with Chapters 9 and 10, forms the first step in the subject known as linear algebra, the importance of which is hard to overemphasize.

8.1 MATRIX ARITHMETIC

A *matrix*[1] is a rectangular array of numbers. In this book, you can usually assume that the numbers are real but on occasion complex numbers, modular arithmetic and even polynomials will be used.

Example 8.1.1. The following are all matrices.

$$\begin{pmatrix} 1 & 2 & 3 \\ 4 & 5 & 6 \end{pmatrix}, \quad \begin{pmatrix} 4 \\ 1 \end{pmatrix}, \quad \begin{pmatrix} 1 & 1 & -1 \\ 0 & 2 & 4 \\ 1 & 1 & 3 \end{pmatrix}, \quad (\, 6 \,).$$

The last example in the above list is a matrix and not just a number.

The array of numbers that comprises a matrix is enclosed in round brackets although in some books square brackets are used. Later on, we introduce determinants and these are indicated by using straight brackets. The kind of brackets you use is important and not just a matter of taste. We usually denote matrices by capital Roman letters: $A, B, C, \ldots$. The *size* of a matrix is $m \times n$ if it has m *rows* and n *columns*. The entries in a matrix or its elements are usually denoted by lower case Roman letters.

---

[1] Plural: *matrices*.

If A is an $m \times n$ matrix, and $1 \leq i \leq m$ and $1 \leq j \leq n$, then the entry in the ith row and jth column of A is denoted by $(A)_{ij}$.

Example 8.1.2. Let

$$A = \begin{pmatrix} 1 & 2 & 3 \\ 4 & 5 & 6 \end{pmatrix}.$$

Then A is a 2×3 matrix. Here $(A)_{11} = 1$, $(A)_{12} = 2$, $(A)_{13} = 3$, $(A)_{21} = 4$, $(A)_{22} = 5$, $(A)_{23} = 6$.

Matrices A and B are said to be *equal*, written $A = B$, if they have the same size, thus they are both $m \times n$ matrices, and corresponding entries are equal, meaning that $(A)_{ij} = (B)_{ij}$ for all $1 \leq i \leq m$ and $1 \leq j \leq n$.

Example 8.1.3. Given that

$$\begin{pmatrix} a & 2 & b \\ 4 & 5 & c \end{pmatrix} = \begin{pmatrix} 3 & x & -2 \\ y & z & 0 \end{pmatrix}$$

then $a = 3$, $2 = x$, $b = -2$, $4 = y$, $5 = z$ and $c = 0$.

The entries of an arbitrary matrix A are denoted by a_{ij} where i tells you the row the element lives in and j the column. For example, a typical 2×3 matrix A would be written

$$A = \begin{pmatrix} a_{11} & a_{12} & a_{13} \\ a_{21} & a_{22} & a_{23} \end{pmatrix}.$$

We now define some simple operations on matrices.

- Let A and B be two matrices *of the same size*. Then their *sum* $A + B$ is the matrix defined by

$$(A + B)_{ij} = (A)_{ij} + (B)_{ij}.$$

  That is, corresponding entries of A and B are added. If A and B are not the same size then their sum is *not defined*.

- Let A and B be two matrices *of the same size*. Then their *difference* $A - B$ is the matrix defined by

$$(A - B)_{ij} = (A)_{ij} - (B)_{ij}.$$

  That is, corresponding entries of A and B are subtracted. If A and B are not the same size then their difference is *not defined*.

- In matrix theory, numbers are called *scalars*. Let A be a matrix and λ a scalar. Then the matrix λA is defined by

$$(\lambda A)_{ij} = \lambda (A)_{ij}.$$

  That is, every entry of A is multiplied by λ.

- Let A be an $m \times n$ matrix. Then the *transpose of A*, denoted A^T, is the $n \times m$ matrix defined by $(A^T)_{ij} = (A)_{ji}$. Thus rows and columns are interchanged, with the first row of A becoming the first column of A^T, the second row of A becoming the second column of A^T, and so on.

Examples 8.1.4.

1.

$$\begin{pmatrix} 1 & 2 & -1 \\ 3 & -4 & 6 \end{pmatrix} + \begin{pmatrix} 2 & 1 & 3 \\ -5 & 2 & 1 \end{pmatrix} = \begin{pmatrix} 1+2 & 2+1 & -1+3 \\ 3+(-5) & -4+2 & 6+1 \end{pmatrix}$$

which gives

$$\begin{pmatrix} 3 & 3 & 2 \\ -2 & -2 & 7 \end{pmatrix}.$$

2.

$$\begin{pmatrix} 1 & 1 \\ 2 & 1 \end{pmatrix} - \begin{pmatrix} 3 & 3 & 2 \\ -2 & -2 & 7 \end{pmatrix}$$

is not defined since the matrices have different sizes.

3.

$$2 \begin{pmatrix} 3 & 3 & 2 \\ -2 & -2 & 7 \end{pmatrix} = \begin{pmatrix} 6 & 6 & 4 \\ -4 & -4 & 14 \end{pmatrix}.$$

4. The transposes of the following matrices

$$\begin{pmatrix} 1 & 2 & 3 \\ 4 & 5 & 6 \end{pmatrix}, \quad \begin{pmatrix} 4 \\ 1 \end{pmatrix}, \quad \begin{pmatrix} 1 & 1 & -1 \\ 0 & 2 & 4 \\ 1 & 1 & 3 \end{pmatrix}, \quad (6)$$

are, respectively,

$$\begin{pmatrix} 1 & 4 \\ 2 & 5 \\ 3 & 6 \end{pmatrix}, \quad (4 \quad 1), \quad \begin{pmatrix} 1 & 0 & 1 \\ 1 & 2 & 1 \\ -1 & 4 & 3 \end{pmatrix}, \quad (6).$$

We now turn to multiplication. You might expect this would follow the same pattern as above, but it does not. The obvious way to multiply matrices together is the wrong way.[2] Instead, matrix multiplication is defined in a more complicated but more useful way. To help define this operation, two special classes of matrix are used. A *row matrix* or *row vector* is a matrix with one row but any number of columns, and a *column matrix* or *column vector* is a matrix with one column but any number of rows. Row and column vectors are denoted by bold lower case Roman letters $\mathbf{a}, \mathbf{b}, \mathbf{c} \dots$.

---

[2] I have only once ever had to multiply matrices together the wrong way.

Examples 8.1.5. The matrix

$$(\ 1 \quad 2 \quad 3 \quad 4 \ )$$

is a row vector whilst

$$\begin{pmatrix} 1 \\ 2 \\ 3 \\ 4 \end{pmatrix}$$

is a column vector.

The full definition of matrix multiplication is reached in three stages.

Stage 1. Let **a** be a row vector and **b** a column vector, where

$$\mathbf{a} = (\ a_1 \quad a_2 \quad \ldots \quad a_n \ ) \text{ and } \mathbf{b} = \begin{pmatrix} b_1 \\ b_2 \\ \cdot \\ \cdot \\ \cdot \\ b_p \end{pmatrix}.$$

Then their product **ab** is defined if and only if the number of columns of **a** is equal to the number of rows of **b**, that is if $n = p$, in which case their product is the 1×1 matrix

$$\mathbf{ab} = (a_1 b_1 + a_2 b_2 + \ldots + a_n b_n).$$

The *number*

$$a_1 b_1 + a_2 b_2 + \ldots + a_n b_n$$

is called the *inner product* of **a** and **b** and is denoted by $\mathbf{a} \cdot \mathbf{b}$. Using this notation $\mathbf{ab} = (\mathbf{a} \cdot \mathbf{b})$. We shall meet the inner product again in Chapters 9 and 10.

Example 8.1.6. This odd way of multiplying is actually quite natural. Suppose that we have n commodities labelled by i, where $1 \leq i \leq n$. Let the price of commodity i at a specific outlet be a_i. We package this information into a single row vector **a** called the *price row vector* where $(\mathbf{a})_{1i} = a_i$. Let the quantity of commodity i we wish to buy at that outlet be b_i. We package this information into a single column vector **b** called the *quantity column vector* where $(\mathbf{b})_{i1} = b_i$. Then the total cost of all the commodities at that outlet is $\mathbf{a} \cdot \mathbf{b}$.

Stage 2. Let **a** be a $1 \times n$ row vector and let B be a $p \times q$ matrix. Then the product $\mathbf{a}B$ is defined if and only if the number of columns of **a** is equal to the number of rows of B. That is if $n = p$. To calculate the product $\mathbf{a}B$, think of B as a list of q column vectors so that $B = (\mathbf{b}_1, \ldots, \mathbf{b}_q)$. Then $\mathbf{a}B = (\ \mathbf{a} \cdot \mathbf{b}_1 \quad \ldots \quad \mathbf{a} \cdot \mathbf{b}_q \ )$.

Example 8.1.7. Let **a** be the price matrix of Example 8.1.6, and let B be the matrix whose columns list the quantity of commodities bought on different dates. Then the matrix $\mathbf{a}B$ is a row matrix that tells us how much was spent on each of those dates.

Stage 3. Let A be an $m \times n$ matrix and let B be a $p \times q$ matrix. The product AB is defined if and only if the number of columns of A is equal to the number of rows of B. That is $n = p$. If this is so then AB is an $m \times q$ matrix. To define this product we think of A as consisting of m row vectors $\mathbf{a}_1, \dots, \mathbf{a}_m$ and we think of B as consisting of q column vectors $\mathbf{b}_1, \dots, \mathbf{b}_q$. As in Stage 2 above, we multiply the first row of A into each of the columns of B and this gives us the first row of A; we then multiply the second row of A into each of the columns of B and this gives us the second row of B, and so on. It follows that

$$
AB = \begin{pmatrix} \mathbf{a}_1 \\ \cdot \\ \cdot \\ \cdot \\ \mathbf{a}_m \end{pmatrix} \begin{pmatrix} \mathbf{b}_1 \dots \mathbf{b}_q \end{pmatrix} = \begin{pmatrix} \mathbf{a}_1 \cdot \mathbf{b}_1 & \dots & \mathbf{a}_1 \cdot \mathbf{b}_q \\ \cdot & \dots & \cdot \\ \cdot & \dots & \cdot \\ \cdot & \dots & \cdot \\ \mathbf{a}_m \cdot \mathbf{b}_1 & \dots & \mathbf{a}_m \cdot \mathbf{b}_q \end{pmatrix}.
$$

Example 8.1.8. Let A be the matrix whose rows list the prices of the commodities in different outlets, and let B be the matrix whose columns list the quantity of each commodity bought on different dates. Then the matrix AB contains all the information about how much was spent at the different outlets on the different dates.

Examples 8.1.9. Here are some examples of matrix multiplication.

1.

$$
\begin{pmatrix} 1 & -1 & 0 & 2 & 1 \end{pmatrix} \begin{pmatrix} 2 \\ 3 \\ 1 \\ -1 \\ 3 \end{pmatrix} = \begin{pmatrix} 0 \end{pmatrix}.
$$

2. The product

$$
\begin{pmatrix} 1 & -1 & 2 \\ 3 & 0 & 1 \end{pmatrix} \begin{pmatrix} 0 & -2 & 3 \\ 2 & 1 & -1 \end{pmatrix}
$$

is not defined because the number of columns of the first matrix is not equal to the number of rows of the second matrix.

3. The product

$$
\begin{pmatrix} 1 & 2 & 4 \\ 2 & 6 & 0 \end{pmatrix} \begin{pmatrix} 4 & 1 & 4 & 3 \\ 0 & -1 & 3 & 1 \\ 2 & 7 & 5 & 2 \end{pmatrix}
$$

is defined because the first matrix is a 2×3 and the second is a 3×4. Thus the product is a 2×4 matrix and is

$$
\begin{pmatrix} 12 & 27 & 30 & 13 \\ 8 & -4 & 26 & 12 \end{pmatrix}.
$$

Matrix multiplication can be summarized as follows.

- Let A be an $m \times n$ matrix and B a $p \times q$ matrix. The product AB is defined if and only if $n = p$ and the result is an $m \times q$ matrix. In other words

$$(m \times n)(n \times q) = (m \times q).$$

- $(AB)_{ij}$ is the inner product of the ith row of A and the jth column of B.

- The inner product of the ith row of A and each of the columns of B in turn yields each of the elements of the ith row of AB in turn.

A matrix all of whose elements are zero is called a *zero matrix*. The $m \times n$ zero matrix is denoted $O_{m,n}$ or just O where we let the context determine its size. A *square matrix* is one in which the number of rows is equal to the number of columns. In a square matrix A the entries $(A)_{11}, (A)_{22}, \ldots, (A)_{nn}$ are called the *diagonal elements*. All the other elements of A are called the *off-diagonal elements*. A *diagonal matrix* is a square matrix in which all off-diagonal elements are zero. A *scalar matrix* is a diagonal matrix in which the diagonal elements are all the same. The $n \times n$ *identity matrix* is the scalar matrix in which all the diagonal elements are the number one. This is denoted by I_n or just I where we let the context determine its size. Thus scalar matrices are those of the form λI where λ is any scalar. It is important to observe that an identity matrix does not have every entry equal to 1. It is called an identity matrix because, as we shall see, it acts like a multiplicative identity. A matrix is *real* if all its elements are real numbers, and *complex* if all its elements are complex numbers. A matrix A is said to be *symmetric* if $A^T = A$. Symmetric matrices have important applications described in Chapter 10. We also define here a class of column vectors that we shall meet again later. Define e_i^n to be the $n \times 1$ matrix with all entries zero except $(e_i^n)_{i1} = 1$. We usually let the context determine n and omit the superscript. When $n = 3$ there are three such vectors e_1, e_2, e_3.

Examples 8.1.10.

1. The matrix

$$\begin{pmatrix} 1 & 0 & 0 \\ 0 & 2 & 0 \\ 0 & 0 & 3 \end{pmatrix}$$

    is a 3×3 diagonal matrix.

2. The matrix

$$\begin{pmatrix} 1 & 0 & 0 & 0 \\ 0 & 1 & 0 & 0 \\ 0 & 0 & 1 & 0 \\ 0 & 0 & 0 & 1 \end{pmatrix}$$

    is the 4×4 identity matrix.

3. The matrix

$$\begin{pmatrix} 42 & 0 & 0 & 0 & 0 \\ 0 & 42 & 0 & 0 & 0 \\ 0 & 0 & 42 & 0 & 0 \\ 0 & 0 & 0 & 42 & 0 \\ 0 & 0 & 0 & 0 & 42 \end{pmatrix}$$

is a 5×5 scalar matrix.

4. The matrix

$$\begin{pmatrix} 0 & 0 & 0 & 0 & 0 \\ 0 & 0 & 0 & 0 & 0 \\ 0 & 0 & 0 & 0 & 0 \\ 0 & 0 & 0 & 0 & 0 \\ 0 & 0 & 0 & 0 & 0 \\ 0 & 0 & 0 & 0 & 0 \end{pmatrix}$$

is a 6×5 zero matrix.

5. The matrix

$$\begin{pmatrix} 1 & 2 & 3 \\ 2 & 4 & 5 \\ 3 & 5 & 6 \end{pmatrix}$$

is a 3×3 symmetric matrix.

6.

$$\mathbf{e}_1 = \begin{pmatrix} 1 \\ 0 \\ 0 \end{pmatrix}, \mathbf{e}_2 = \begin{pmatrix} 0 \\ 1 \\ 0 \end{pmatrix} \text{ and } \mathbf{e}_3 = \begin{pmatrix} 0 \\ 0 \\ 1 \end{pmatrix}.$$

Matrices are useful in solving systems of linear equations. We describe the detailed method in Section 8.3. As a first step, we show here how the definition of matrix multiplication leads to a concise way of writing down such equations. This will also partly explain why matrix multiplication is defined the way it is. A *system of m linear equations in n unknowns* is a list of equations

$$a_{11}x_1 + a_{12}x_2 + \ldots + a_{1n}x_n = b_1$$
$$a_{21}x_1 + a_{22}x_2 + \ldots + a_{2n}x_n = b_2$$
$$\ldots\ldots\ldots\ldots\ldots\ldots\ldots\ldots\ldots$$
$$a_{m1}x_1 + a_{m2}x_2 + \ldots + a_{mn}x_n = b_m$$

where the a_{ij} are numbers of some kind. If there are only a few unknowns then we often use w, x, y, z rather than x_1, x_2, x_3, x_4. A *solution* is a set of values of $x_1, \ldots, x_n$ that satisfy all the equations. The set of all solutions is called the *solution set*. The equations above can be conveniently represented by matrices. Let A be the $m \times n$ matrix defined by $(A)_{ij} = a_{ij}$, let $\mathbf{b}$ be the $m \times 1$ matrix defined by $(\mathbf{b})_{i1} = b_i$, and let $\mathbf{x}$ be the $n \times 1$ matrix defined by $(\mathbf{x})_{j1} = x_j$. Then the *system* of linear equations above can be written as a *single* matrix equation $A\mathbf{x} = \mathbf{b}$. This is both efficient packaging and an important conceptual advance as we shall see. The matrix A is called the *coefficient matrix*. A *solution* is a column vector $\mathbf{x}$ that satisfies the above matrix equation.

Example 8.1.11. The following system of linear equations

$$2x + 3y = 1$$
$$x + y = 2$$

can be written in matrix form as

$$\begin{pmatrix} 2 & 3 \\ 1 & 1 \end{pmatrix} \begin{pmatrix} x \\ y \end{pmatrix} = \begin{pmatrix} 1 \\ 2 \end{pmatrix}.$$

Observe that

$$\begin{pmatrix} 5 \\ -3 \end{pmatrix}$$

is a solution to the above matrix equation and so to the original system of linear equations because

$$\begin{pmatrix} 2 & 3 \\ 1 & 1 \end{pmatrix} \begin{pmatrix} 5 \\ -3 \end{pmatrix} = \begin{pmatrix} 1 \\ 2 \end{pmatrix}.$$

Matrices are not just useful in studying linear equations. The following three examples touch on other important applications.

Example 8.1.12. We described the theory of polynomial equations in one variable in Chapter 7, and in this chapter we describe linear equations in several variables. But what about equations in several variables where both products and powers of the variables can occur? The simplest class of such equations are the *conics*

$$ax^2 + bxy + cy^2 + dx + ey + f = 0$$

where a, b, c, d, e, f are numbers. In general, the roots or *zeroes* of such equations form curves in the plane such as circles, ellipses and hyperbolas. The term conic itself arises from the way they were first defined by the Greeks as those curves obtained when a double cone is sliced by a plane. They are important in celestial mechanics since when one body orbits another under the action of gravity the path is a conic. The reason for introducing them here is that they can be represented by matrix equations of the form

$$\mathbf{x}^T A \mathbf{x} + J^T \mathbf{x} + (f) = (0)$$

where

$$A = \begin{pmatrix} a & \frac{1}{2}b \\ \frac{1}{2}b & c \end{pmatrix}, \quad \mathbf{x} = \begin{pmatrix} x \\ y \end{pmatrix} \text{ and } J = \begin{pmatrix} d \\ e \end{pmatrix}.$$

This is not just a notational convenience. The fact that the matrix A is symmetric means that powerful ideas from matrix theory, to be developed in Chapter 10, can be brought to bear on studying such conics. If we replace the $\mathbf{x}$ above by the matrix

$$\mathbf{x} = \begin{pmatrix} x \\ y \\ z \end{pmatrix}$$

and A by a 3×3 symmetric matrix and J by a 3×1 matrix then we get the matrix equation of a *quadric*. Examples are the surface of a sphere or that described by a cooling tower. Even though we are dealing with three rather than two dimensions, the matrix algebra we develop applies just as well.

Example 8.1.13. Graphs and digraphs were defined in Chapter 3. Recall that a graph consists of a set of vertices, represented here by circles, and a collection of edges, represented here by lines joining the vertices. An example of a graph is given below.

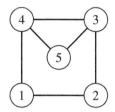

This graph can also described by means of a 5×5 symmetric matrix G, where the entry $(G)_{ij}$ is equal to the number of edges connecting vertices i and j.

$$G = \begin{pmatrix} 0 & 1 & 0 & 1 & 0 \\ 1 & 0 & 1 & 0 & 0 \\ 0 & 1 & 0 & 1 & 1 \\ 1 & 0 & 1 & 0 & 1 \\ 0 & 0 & 1 & 1 & 0 \end{pmatrix}.$$

More generally, a graph with n vertices labelled $1, 2, \ldots, n$ can be described by an $n \times n$ *adjacency matrix* A where $(A)_{ij}$ is the number of edges joining i and j. The adjacency matrices of graphs are always symmetric. We describe an application of matrix multiplication to graphs in Theorem 8.2.4.

Example 8.1.14. Most of the matrices in this book have either real or complex entries, but there is no reason why matrices should not have entries from more exotic kinds of numbers such as the modular arithmetic systems $\mathbb{Z}_n$ introduced in Section 5.4. Matrices over $\mathbb{Z}_2$, for example, play an important rôle in the theory of error-correcting codes. In this context, the elements of $\mathbb{Z}_2$ are called *bits*. We explain what is meant by such codes by means of an example. Let

$$G = \begin{pmatrix} 1 & 1 & 1 & 0 & 0 & 0 & 0 \\ 1 & 0 & 0 & 1 & 1 & 0 & 0 \\ 0 & 1 & 0 & 1 & 0 & 1 & 0 \\ 1 & 1 & 0 & 1 & 0 & 0 & 1 \end{pmatrix}$$

and

$$H = \begin{pmatrix} 0 & 0 & 0 & 1 & 1 & 1 & 1 \\ 0 & 1 & 1 & 0 & 0 & 1 & 1 \\ 1 & 0 & 1 & 0 & 1 & 0 & 1 \end{pmatrix}$$

be two matrices whose entries are taken from $\mathbb{Z}_2$. The matrix G is called the *generator matrix* and the matrix H is called the *parity check matrix*. Observe that if the ith column of H is the column vector

$$\begin{pmatrix} c'_3 \\ c'_2 \\ c'_1 \end{pmatrix}$$

then $c_3' c_2' c_1'$ is the binary representation of the number i. A row vector

$$\mathbf{m} = (\ m_1 \quad m_2 \quad m_3 \quad m_4\ )$$

with 4 entries from $\mathbb{Z}_2$ should be interpreted as some information that we wish to transmit. For example, it might be from a deep-space probe. As the message wings its way through space, errors may occur due to interference, called *noise*, turning some 0s to 1s and some 1s to 0s. Thus what is actually received could contain errors. In this example, we assume at most one error can occur. To deal with this, we send not $\mathbf{m}$ but $\mathbf{t} = \mathbf{m}G$ which is the row vector

$$(\ c_1 \quad c_2 \quad m_1 \quad c_3 \quad m_2 \quad m_3 \quad m_4\ )$$

where $c_1 = m_1 + m_2 + m_4$, $c_2 = m_1 + m_3 + m_4$, and $c_3 = m_2 + m_3 + m_4$ and all calculations are carried out in $\mathbb{Z}_2$. The Venn diagram below helps to explain what is going on.

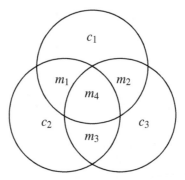

The digits c_1, c_2, c_3 are called *check digits* and they are defined so that all the numbers within one circle add up to 0. Thus $\mathbf{t}$ is transmitted and we suppose that $\mathbf{r}$ is received. To determine whether an error has occurred and if it has which bit is affected, we do the following. Calculate $H\mathbf{r}^T$ and denote the column vector that results by

$$\mathbf{c}' = \begin{pmatrix} c_3' \\ c_2' \\ c_1' \end{pmatrix}.$$

Interpret $c_3' c_2' c_1'$ as a number i in binary. Then if $i = 0$, we deduce that no error has occurred in transmission and $\mathbf{t} = \mathbf{r}$, but if $i \neq 0$, we deduce that the ith bit of $\mathbf{r}$ is in error. Once this is corrected, we reconstitute the transmitted message $\mathbf{t}$ that was sent and so the original message $\mathbf{m}$. For example, let $\mathbf{m} = (\ 1 \quad 0 \quad 1 \quad 0\ )$ be the message we wish to convey. We first encode it by defining $\mathbf{t} = \mathbf{m}G$ to get $\mathbf{t} = (\ 1 \quad 0 \quad 1 \quad 1 \quad 0 \quad 1 \quad 0\ )$. It is this which is actually transmitted. Suppose one of the bits is zapped in transit, say the 5th. What is actually received is $\mathbf{r} = (\ 1 \quad 0 \quad 1 \quad 1 \quad 1 \quad 1 \quad 0\ )$. It is important to remember that the receiver has no idea if an error has occurred and if so where it has occurred. We calculate $\mathbf{c}' = H\mathbf{r}^T$ to get $(\ 1 \quad 0 \quad 1\ )^T$. This represents the number 5 in binary and so we know the fifth bit is in error, which indeed it is. The Venn diagram can be used to explain why this ingenious procedure works. This example of an error-correcting code is called the *Hamming* $(7,4)$-*code* first described in [55] by Richard Hamming (1915–1998). The use of check digits entails a cost either in the time taken to transmit a message or by increasing the amount of space needed to store a message. Thus the goal of the theory of error-correcting codes is an optimizing one: maximize the number of errors that can be corrected whilst minimizing the number of check digits needed. By using more sophisticated codes constructed by more advanced algebra, though within hailing distance of the algebra in this book, it is possible to design efficient codes that correct many errors. Error-correcting codes have mundane as well as interplanetary applications, and will be an essential ingredient in making quantum computing work.

Exercises 8.1

1. Let $A = \begin{pmatrix} 1 & 2 \\ 1 & 0 \\ -1 & 1 \end{pmatrix}$ and $B = \begin{pmatrix} 1 & 4 \\ -1 & 1 \\ 0 & 3 \end{pmatrix}$.

   (a) Calculate $-3B$.
   (b) Calculate $A + B$.
   (c) Calculate $A - B$.

2. Let $A = \begin{pmatrix} 0 & 4 & 2 \\ -1 & 1 & 3 \\ 2 & 0 & 2 \end{pmatrix}$ and $B = \begin{pmatrix} 1 & -3 & 5 \\ 2 & 0 & -4 \\ 3 & 2 & 0 \end{pmatrix}$.

   (a) Calculate AB.
   (b) Calculate BA.

3. Let $A = \begin{pmatrix} 3 & 1 \\ 0 & -1 \end{pmatrix}$, $B = \begin{pmatrix} 0 & 1 \\ -1 & 1 \\ 3 & 1 \end{pmatrix}$ and $C = \begin{pmatrix} 1 & 0 & 3 \\ -1 & 1 & 1 \end{pmatrix}$.

   (a) Calculate BA.
   (b) Calculate AA.
   (c) Calculate CB.
   (d) Calculate AC.
   (e) Calculate BC.
   (f) Calculate $C^T A$.

4. Calculate

$$\begin{pmatrix} 1 \\ 2 \\ 3 \\ 4 \end{pmatrix} \begin{pmatrix} 1 & 2 & 3 \end{pmatrix}.$$

5. Let $A = \begin{pmatrix} 2 & 1 \\ -1 & 0 \\ 2 & 3 \end{pmatrix}$, $B = \begin{pmatrix} 3 & 0 \\ -2 & 1 \end{pmatrix}$ and $C = \begin{pmatrix} -1 & 2 & 3 \\ 4 & 0 & 1 \end{pmatrix}$.

   (a) Calculate $(AB)C$.

   (b) Calculate $A(BC)$.

6. Calculate

$$\begin{pmatrix} 2+i & 1+2i \\ i & 3+i \end{pmatrix} \begin{pmatrix} 2i & 2+i \\ 1+i & 1+2i \end{pmatrix}$$

where i is the complex number i.

7. Calculate

$$\begin{pmatrix} a & 0 & 0 \\ 0 & b & 0 \\ 0 & 0 & c \end{pmatrix} \begin{pmatrix} d & 0 & 0 \\ 0 & e & 0 \\ 0 & 0 & f \end{pmatrix}.$$

8. Calculate

   (a)

$$\begin{pmatrix} 1 & 0 & 0 \\ 0 & 1 & 0 \\ 0 & 0 & 1 \end{pmatrix} \begin{pmatrix} a & b & c \\ d & e & f \\ g & h & i \end{pmatrix}.$$

   (b)

$$\begin{pmatrix} 0 & 1 & 0 \\ 1 & 0 & 0 \\ 0 & 0 & 1 \end{pmatrix} \begin{pmatrix} a & b & c \\ d & e & f \\ g & h & i \end{pmatrix}.$$

   (c)

$$\begin{pmatrix} a & b & c \\ d & e & f \\ g & h & i \end{pmatrix} \begin{pmatrix} 0 & 1 & 0 \\ 1 & 0 & 0 \\ 0 & 0 & 1 \end{pmatrix}.$$

9. Find the transposes of each of the following matrices.

   (a) $A = \begin{pmatrix} 1 & 2 \\ 1 & 0 \\ -1 & 1 \end{pmatrix}$.

   (b) $B = \begin{pmatrix} 1 & -3 & 5 \\ 2 & 0 & -4 \\ 3 & 2 & 0 \end{pmatrix}$.

   (c) $C = \begin{pmatrix} 1 \\ 2 \\ 3 \\ 4 \end{pmatrix}$.

10. This question deals with the following four matrices with complex entries *and their negatives*: I, X, Y, Z where

$$I = \begin{pmatrix} 1 & 0 \\ 0 & 1 \end{pmatrix}, X = \begin{pmatrix} 0 & 1 \\ -1 & 0 \end{pmatrix}, Y = \begin{pmatrix} i & 0 \\ 0 & -i \end{pmatrix}, Z = \begin{pmatrix} 0 & -i \\ -i & 0 \end{pmatrix}.$$

Show that the product of any two such matrices is again a matrix of this type by completing the following Cayley table for multiplication where the entry in row A and column B is AB in that order. The entries of the Cayley table should be the symbols $\pm I$, $\pm X$, $\pm Y$ or $\pm Z$ rather than the matrices themselves.

	I	X	Y	Z	$-I$	$-X$	$-Y$	$-Z$
I								
X								
Y								
Z								
$-I$								
$-X$								
$-Y$								
$-Z$								

It is enough to calculate explicitly only the products involving the first four rows and first four columns, and then use the result from the next section that $-(AB) = A(-B) = (-A)B$ and $(-A)(-B) = AB$ to complete the rest of the table.

11. *This question refers to Example 8.1.14.

   (a) Explain why the Hamming $(7,4)$-code of Example 8.1.14 works.

   (b) Why are three check digits optimal for being able to correct 1 error in messages with 4 bits?

8.2   MATRIX ALGEBRA

In this section, we describe the properties of the algebraic operations introduced in Section 8.1. The resulting algebra is similar to that of real numbers described in Section 4.2 but differs significantly in one or two places. We use the terminology introduced in Section 4.1 despite the fact that the addition, subtraction and multiplication of matrices are not quite binary operations *because they are not always defined*. Furthermore, we should say what the entries of our matrices are, but instead we simply assume that they have enough properties to make the results true.

Properties of matrix addition

We restrict attention to the set of all $m \times n$ matrices.

(MA1) $(A+B)+C = A+(B+C)$. This is the *associative law* for matrix addition.

(MA2) $A+O=A=O+A$. The zero matrix O, the same size as A, is the *additive identity* for matrices the same size as A.

(MA3) $A+(-A)=O=(-A)+A$. The matrix $-A$ is the unique *additive inverse* of A.

(MA4) $A+B=B+A$. Matrix addition is *commutative*.

Thus matrix addition has the same properties as the addition of real numbers, apart from the fact that the sum of two matrices is only defined when they have the same size.

Properties of matrix multiplication

(MM1) The product $(AB)C$ is defined precisely when the product $A(BC)$ is defined, and when they are both defined $(AB)C=A(BC)$. This is the *associative law* for matrix multiplication.

(MM2) Let A be an $m \times n$ matrix. Then $I_m A = A = AI_n$. The matrices I_m and I_n are the *left and right multiplicative identities*, respectively. It is important to observe that for matrices that are not square different identities are needed on the left and on the right.

(MM3) $A(B+C)=AB+AC$ and $(B+C)A=BA+CA$ when the products and sums are defined. These are the *left and right distributivity laws*, respectively, for matrix multiplication over matrix addition.

Thus, apart from the fact that it is not always defined, matrix multiplication has the same properties as the multiplication of real numbers *except for* the following three major differences.

1. *Matrix multiplication is not commutative.* For example, if

$$A = \begin{pmatrix} 1 & 2 \\ 3 & 4 \end{pmatrix} \text{ and } B = \begin{pmatrix} 1 & 1 \\ -1 & 1 \end{pmatrix}$$

then $AB \neq BA$. One consequence of this is that

$$(A+B)^2 \neq A^2 + 2AB + B^2,$$

in general.

2. *The product of two matrices can be a zero matrix without either matrix being a zero matrix.* For example, if

$$A = \begin{pmatrix} 1 & 2 \\ 2 & 4 \end{pmatrix} \text{ and } B = \begin{pmatrix} -2 & -6 \\ 1 & 3 \end{pmatrix}$$

then $AB = O$.

3. *Cancellation of matrices is not allowed in general.* For example, if

$$A = \begin{pmatrix} 0 & 2 \\ 0 & 1 \end{pmatrix} \text{ and } B = \begin{pmatrix} 2 & 3 \\ 1 & 4 \end{pmatrix} \text{ and } C = \begin{pmatrix} -1 & 1 \\ 1 & 4 \end{pmatrix}$$

then $A \neq O$ and $AB = AC$ but $B \neq C$.

Example 8.2.1. Denote by $M_n(\mathbb{R})$ the set of all $n \times n$ real matrices. In this set addition, subtraction and multiplication can be performed without any restrictions. All the axioms (F1)–(F11) of Section 4.2 hold *except* (F7) and (F8). We can replace $\mathbb{R}$ by any one of $\mathbb{Z}, \mathbb{Q}, \mathbb{C}, \mathbb{Z}_n, \mathbb{F}[x]$ and obtain systems with the same properties.

Scalar multiplication causes us no problems.

Properties of scalar multiplication

(S1) $1A = A$ and $-1A = -A$.

(S2) $0A = O$.

(S3) $\lambda(A + B) = \lambda A + \lambda B$ where λ is a scalar.

(S4) $(\lambda \mu)A = \lambda(\mu A)$ where λ and μ are scalars.

(S5) $(\lambda + \mu)A = \lambda A + \mu A$ where λ and μ are scalars.

(S6) $(\lambda A)B = A(\lambda B) = \lambda(AB)$ where λ is a scalar.

The transpose is also straightforward apart from property (T4) below.

Properties of the transpose

(T1) $(A^T)^T = A$.

(T2) $(A + B)^T = A^T + B^T$.

(T3) $(\lambda A)^T = \lambda A^T$ where λ is a scalar.

(T4) $(AB)^T = B^T A^T$.

There are some important consequences of the above properties.

- Because matrix addition is associative, we can apply generalized associativity and write sums without brackets. Similarly, because matrix multiplication is associative, we can apply generalized associativity and write matrix products without brackets, though ensuring that we keep the same order.

- The left and right distributivity laws can be extended to arbitrary finite sums.

It remains to prove that the algebraic properties of matrices stated really do hold. We prove three important ones here and leave the rest as exercises. We rely heavily on the Σ-notation introduced in Section 4.2. Let $A = (a_{ij})$ be an $m \times n$ matrix and let $B = (b_{jk})$ be an $n \times p$ matrix. By definition $(AB)_{ik}$ is equal to the inner product of the ith row of A and the kth column of B. This is equal to

$$(AB)_{ik} = \sum_{j=1}^{n} a_{ij} b_{jk}.$$

Theorem 8.2.2 (Properties of matrix operations).

1. $A(BC) = (AB)C$.

2. $A(B+C) = AB + AC$.

3. $(AB)^T = B^T A^T$.

Proof. (1) Let $A = (a_{ij})$ be an $m \times n$ matrix let $B = (b_{jk})$ be an $n \times p$ matrix, and let $C = (c_{kl})$ be a $p \times q$ matrix. The matrices $A(BC)$ and $(AB)C$ are both defined and have the same size, so it remains to show that corresponding entries are the same. This means we have to show that $(A(BC))_{il} = ((AB)C)_{il}$ for every row i and every column l. By definition

$$(A(BC))_{il} = \sum_{t=1}^{n} a_{it} (BC)_{tl} \text{ and } (BC)_{tl} = \sum_{s=1}^{p} b_{ts} c_{sl}.$$

Thus

$$(A(BC))_{il} = \sum_{t=1}^{n} a_{it} \left(\sum_{s=1}^{p} b_{ts} c_{sl} \right).$$

Using generalized distributivity of multiplication over addition for real numbers this sum is equal to

$$\sum_{t=1}^{n} \sum_{s=1}^{p} a_{it} b_{ts} c_{sl}.$$

We interchange summations using the result from Section 4.2 to get

$$\sum_{s=1}^{p} \sum_{t=1}^{n} a_{it} b_{ts} c_{sl}.$$

We now use generalized distributivity again

$$\sum_{s=1}^{p} \left(\sum_{t=1}^{n} a_{it} b_{ts} \right) c_{sl}.$$

The sum within the brackets is equal to $(AB)_{is}$ and so the whole sum is

$$\sum_{s=1}^{p} (AB)_{is} c_{sl}$$

which is equal to

$$((AB)C)_{il}.$$

(2) Let A be an $m \times n$ matrix and let B and C be $n \times p$ matrices. The element in row i and column j of the lefthand side is equal to

$$\sum_{k=1}^{n} a_{ik}(b_{kj} + c_{kj})$$

where $b_{kj} + c_{kj}$ is the element in row k and column j of the sum $B + C$. The result now follows from fact that multiplication distributes over addition for real numbers.

(3) Let A be an $m \times n$ matrix and B an $n \times p$ matrix. Then AB is defined and is $m \times p$. Hence $(AB)^T$ is $p \times m$. Now B^T is $p \times n$ and A^T is $n \times m$. Thus $B^T A^T$ is defined and is $p \times m$. It follows that $(AB)^T$ and $B^T A^T$ are defined and have the same size. We show that corresponding entries are equal. By definition

$$((AB)^T)_{ij} = (AB)_{ji}.$$

This is equal to

$$\sum_{s=1}^{n} (A)_{js}(B)_{si} = \sum_{s=1}^{n} (A^T)_{sj}(B^T)_{is}.$$

But real numbers commute under multiplication and so

$$\sum_{s=1}^{n} (A^T)_{sj}(B^T)_{is} = \sum_{s=1}^{n} (B^T)_{is}(A^T)_{sj} = (B^T A^T)_{ij},$$

as required.   □

Let A be a square matrix. By associativity and the results of Section 4.1, the usual properties of exponents hold for powers of A. Thus

$$A^m A^n = A^{m+n} \text{ and } (A^m)^n = A^{mn}$$

for all possible natural numbers m and n. In particular, powers of A commute with each other so that

$$A^m A^n = A^n A^m.$$

Define A^0 to be the identity matrix the same size as A. It is a short step from this to constructing polynomials in A but we shall leave them to Section 8.6.

Example 8.2.3. Suppose that $X^2 = I$ where X is a 2×2 matrix. Then $X^2 - I = O$. We can factorize the lefthand side to get $(X - I)(X + I) = O$. But we cannot conclude from this that $X = I$ or $X = -I$ because, as we saw above, the product of two matrices can be a zero matrix without either matrix being a zero matrix. We therefore cannot deduce that the identity matrix has two square roots. In fact, it has infinitely many. All the matrices

$$\begin{pmatrix} \sqrt{1+n^2} & -n \\ n & -\sqrt{1+n^2} \end{pmatrix},$$

where n is any positive integer, are square roots of the identity matrix.

We conclude this section by giving a practical application of matrix multiplication. A *path of length n* in a graph is a sequence of n edges that can be traced out in a graph. Thus in Example 8.1.13,

$$1-4-3$$

is a path of length 2 joining vertex 1 and vertex 3, whereas

$$1-2-3-5-4-3$$

is a path of length 5 joining 1 and 3. Paths of length 1 are simply edges. There is nothing in the definition of a path that says it has to be efficient. We can use matrix multiplication to count the total number of paths between any two vertices.

Theorem 8.2.4 (Path counting in graphs). *Let A be the adjacency matrix of a graph with n vertices. Then the number of paths of length m between vertex i and vertex j is $(A^m)_{ij}$.*

Proof. We prove the result by induction on m. When $m = 1$ the result is immediate. Our induction hypothesis is that $(A^m)_{ij}$ is equal to the number of paths of length m joining i and j. We prove, assuming this hypothesis, that $(A^{m+1})_{ij}$ is equal to the number of paths of length $m + 1$ joining i and j. Every path of length $m + 1$ joining i and j is constructed from some path of length m from i to some intermediate vertex k, where $1 \leq k \leq n$, followed by an edge from k to j. It follows that the number of paths of length $m + 1$ joining i and j is given by

$$\sum_{k=1}^{n} (A^m)_{ik}(A)_{kj}$$

using the induction hypothesis. By the definition of matrix multiplication this is equal to $(A^m A)_{ij} = (A^{m+1})_{ij}$, as required. □

Box 8.1: Quantum Mechanics

Quantum mechanics is one of the fundamental theories of physics. At its heart are matrices. We have defined the transpose of a matrix but for matrices with complex entries there is another, related, operation. Given any complex matrix A define the matrix $A^\dagger$ to be the one obtained by transposing A and then taking the complex conjugate of all entries. It is therefore the *conjugate-transpose* of A. A matrix A is called *Hermitian* if $A^\dagger = A$. Observe that a real matrix is Hermitian precisely when it is symmetric. It turns out that quantum mechanics is based on Hermitian matrices and their generalizations. The fact that matrix multiplication is not commutative is one of the reasons that quantum mechanics is so different from classical mechanics. The theory of quantum computing makes heavy use of Hermitian matrices and their properties.

Exercises 8.2

1. Calculate

$$\begin{pmatrix} 2 & 0 \\ 7 & -1 \end{pmatrix} + \begin{pmatrix} 1 & 1 \\ 1 & 0 \end{pmatrix} + \begin{pmatrix} 0 & 1 \\ 1 & 1 \end{pmatrix} + \begin{pmatrix} 2 & 2 \\ 3 & 3 \end{pmatrix}.$$

2. Calculate

$$\begin{pmatrix} 1 \\ 2 \\ 3 \end{pmatrix} (3 \quad 2 \quad 1) \begin{pmatrix} 1 \\ -1 \\ -4 \end{pmatrix} (3 \quad 1 \quad 5).$$

3. Calculate A^2, A^3 and A^4 where

$$A = \begin{pmatrix} 1 & -1 \\ 1 & 2 \end{pmatrix}.$$

4. Calculate $A\mathbf{x}, A^2\mathbf{x}, A^3\mathbf{x}, A^4\mathbf{x}$ and $A^5\mathbf{x}$ where

$$A = \begin{pmatrix} 1 & 1 \\ 1 & 0 \end{pmatrix} \text{ and } \mathbf{x} = \begin{pmatrix} 1 \\ 0 \end{pmatrix}.$$

What do you notice?

5. Calculate

$$\begin{pmatrix} \cos\theta & \sin\theta \\ -\sin\theta & \cos\theta \end{pmatrix} \begin{pmatrix} \cos\phi & \sin\phi \\ -\sin\phi & \cos\phi \end{pmatrix}$$

and simplify.

6. Calculate $A^3 - 5A^2 + 8A - 4I$ where

$$A = \begin{pmatrix} 3 & 1 & -1 \\ 2 & 2 & -1 \\ 2 & 2 & 0 \end{pmatrix}$$

and I is the 3×3 identity matrix.

7. Under what circumstances can you use the binomial theorem to expand

$$(A + B)^n$$

where A and B are square matrices of the same size?

8. This question refers to the following graph

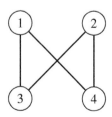

(a) Write down its adjacency matrix.

(b) How many paths of length 2 are there joining vertex 1 and vertex 2?

(c) How many paths of length 3 are there joining vertex 1 and vertex 4?

9. *Let A, B and C be square matrices of the same size. Define $[A,B] = AB - BA$. Calculate

$$[[A,B],C] + [[B,C],A] + [[C,A],B].$$

10. *Recall that a matrix A is *symmetric* if $A^T = A$.

    (a) Show that a symmetric matrix must be square.

    (b) Show that if A is any matrix then AA^T is defined and symmetric.

    (c) Let A and B be symmetric matrices of the same size. Prove that AB is symmetric if and only if $AB = BA$.

11. *A matrix A is said to be *skew-symmetric* if $A^T = -A$.

    (a) Show that the diagonal entries of a skew-symmetric matrix are all zero.

    (b) If B is any $n \times n$-matrix, show that $B + B^T$ is symmetric and that $B - B^T$ is skew-symmetric.

    (c) Deduce that every square matrix can be expressed as the sum of a symmetric matrix and a skew-symmetric matrix.

12. *Let A be an $n \times n$ matrix such that $AB = BA$ for all $n \times n$ matrices B. Show that A is a scalar matrix. That is, $A = \lambda I$ for some scalar λ. Hint: It is a good idea to start with the special case $n = 2$ to get a sense of what is involved in this question.

8.3 SOLVING SYSTEMS OF LINEAR EQUATIONS

The goal of this section is to use matrices to help us solve systems of linear equations. We begin by proving some general results on linear equations, and then describe Gaussian elimination, an algorithm for solving such equations. We write our equations in matrix form as described in Section 8.1. Let $Ax = b$ be a system of equations. If b is a zero vector, the equations are said to be *homogeneous*; otherwise they are said to be *inhomogeneous*. A system of linear equations that has no solution is said to be *inconsistent*; otherwise it is said to be *consistent*. We begin with two major results that tell us what to expect when solving systems of linear equations.

Theorem 8.3.1 (Structure of solutions).

1. *Homogeneous equations $Ax = 0$ are always consistent, because $x = 0$ is always a solution. In addition, the sum of any two solutions is again a solution, and the scalar multiple of any solution is again a solution.*

2. *Let $Ax = b$ be a consistent system of linear equations. Let p be any solution. Then every solution of the equation is of the form $p + h$ for some solution h of $Ax = 0$.*

Proof. (1) Let $Ax = 0$ be a homogeneous system of equations. If $\mathbf{a}$ and $\mathbf{b}$ are solutions then $A\mathbf{a} = 0$ and $A\mathbf{b} = 0$. We calculate $A(\mathbf{a} + \mathbf{b})$. To do this we use the fact that matrix multiplication satisfies the left distributivity law

$$A(\mathbf{a} + \mathbf{b}) = A\mathbf{a} + A\mathbf{b} = 0 + 0 = 0.$$

Thus $\mathbf{a} + \mathbf{b}$ is also a solution. Now let $\mathbf{a}$ be a solution and λ any scalar. Then

$$A(\lambda \mathbf{a}) = \lambda A\mathbf{a} = \lambda 0 = 0$$

where we use the fact that scalars pass freely through products of matrices. Thus $\lambda \mathbf{a}$ is a solution.

(2) Let $\mathbf{a}$ be any solution to $Ax = \mathbf{b}$. Let $\mathbf{h} = \mathbf{a} - \mathbf{p}$. Then $A\mathbf{h} = 0$. It follows that $\mathbf{a} = \mathbf{p} + \mathbf{h}$. □

Theorem 8.3.1 tells us that homogeneous systems of linear equations are important even when our system is inhomogeneous.

Theorem 8.3.2 (Fundamental theorem of linear equations). *We assume that the scalars are the rationals, the reals or the complexes. For a system of linear equations $Ax = \mathbf{b}$ exactly one of the following holds.*

1. *There are no solutions.*

2. *There is exactly one solution.*

3. *There are infinitely many solutions.*

Proof. We prove that if we can find two different solutions we can in fact find infinitely many solutions. Let $\mathbf{u}$ and $\mathbf{v}$ be two distinct solutions to the equation. Then $A\mathbf{u} = \mathbf{b}$ and $A\mathbf{v} = \mathbf{b}$. Consider now the column vector $\mathbf{w} = \mathbf{u} - \mathbf{v}$. Then

$$A\mathbf{w} = A(\mathbf{u} - \mathbf{v}) = A\mathbf{u} - A\mathbf{v} = 0$$

using the distributive law. Thus $\mathbf{w}$ is a non-zero column vector that satisfies the equation $Ax = 0$. Consider now the infinite set of column vectors of the form $\mathbf{u} + \lambda \mathbf{w}$ where λ is any scalar. We calculate

$$A(\mathbf{u} + \lambda \mathbf{w}) = A\mathbf{u} + \lambda A\mathbf{w} = \mathbf{b}$$

using the distributive law and properties of scalars. It follows that all the column vectors $\mathbf{u} + \lambda \mathbf{w}$ are solutions to the equation $Ax = \mathbf{b}$. □

> *It is important to observe that solving a system of linear equations does not just mean solving them when there is a unique solution. If the system has infinitely many solutions then all solutions must be found.*

This is all we can say about the theory of linear equations at this stage. We now turn to the practice. We develop an algorithm that will take as input a system of linear equations and produce as output the following: if the system has no solutions it will tell us; on the other hand if it has solutions then it will determine them all. Our method is based on three simple ideas.

1. Certain systems of linear equations have a shape that makes them easy to solve.

2. Certain operations can be carried out on systems of linear equations which simplify them but do not change the solution set.

3. Everything can be done using matrices.

Here are examples of each of these ideas.

Examples 8.3.3.

1. The system of equations

$$
\begin{aligned}
2x + 3y &= 1 \\
y &= -3
\end{aligned}
$$

is easy to solve. From the second equation we get $y = -3$. Substituting this value back into the first equation gives us $x = 5$. This process is called *back-substitution*.

2. The system of equations

$$
\begin{aligned}
2x + 3y &= 1 \\
x + y &= 2
\end{aligned}
$$

can be converted into a system with the same solutions but which is easier to solve. Multiply the second equation by 2. This gives us the new equations

$$
\begin{aligned}
2x + 3y &= 1 \\
2x + 2y &= 4
\end{aligned}
$$

which have the same solutions as the original ones. Next, subtract the first equation from the second equation to get

$$
\begin{aligned}
2x + 3y &= 1 \\
-y &= 3.
\end{aligned}
$$

Finally, multiply the last equation by -1. The resulting equations have the same solutions as the original ones, but they can now be easily solved as we showed above.

3. The system of equations

$$
\begin{aligned}
2x + 3y &= 1 \\
x + y &= 2
\end{aligned}
$$

can be written in matrix form as

$$
\begin{pmatrix} 2 & 3 \\ 1 & 1 \end{pmatrix} \begin{pmatrix} x \\ y \end{pmatrix} = \begin{pmatrix} 1 \\ 2 \end{pmatrix}.
$$

For the purposes of our algorithm, however, we rewrite this equation in terms of what is called an *augmented matrix*

$$\left(\begin{array}{cc|c} 2 & 3 & 1 \\ 1 & 1 & 2 \end{array} \right).$$

The operations carried out in the previous example can be applied directly to the augmented matrix.

$$\left(\begin{array}{cc|c} 2 & 3 & 1 \\ 1 & 1 & 2 \end{array} \right) \rightarrow \left(\begin{array}{cc|c} 2 & 3 & 1 \\ 2 & 2 & 4 \end{array} \right) \rightarrow \left(\begin{array}{cc|c} 2 & 3 & 1 \\ 0 & -1 & 3 \end{array} \right) \rightarrow \left(\begin{array}{cc|c} 2 & 3 & 1 \\ 0 & 1 & -3 \end{array} \right).$$

The resulting augmented matrix can then be converted back into the usual matrix form and solved by back-substitution.

To formalize the above ideas we need two definitions.

- A matrix is called a *row echelon matrix* or is said to be in *row echelon form* if it satisfies the following three conditions.

  1. Any zero rows are at the bottom of the matrix below all the non-zero rows.

  2. If there are non-zero rows then they begin with the number 1, called the *leading 1*.

  3. In the column beneath a leading 1, all the entries are zero.

The following is a picture of a typical row echelon matrix where the asterisks can be any elements.

$$\left(\begin{array}{cccccc} 0 & 1 & * & * & * & * \\ 0 & 0 & 1 & * & * & * \\ 0 & 0 & 0 & 0 & 1 & * \\ 0 & 0 & 0 & 0 & 0 & 0 \\ 0 & 0 & 0 & 0 & 0 & 0 \end{array} \right).$$

Echelon means arranged in a step-like manner. The row echelon matrices are precisely those which have a good shape, and a system of linear equations that have this step-like pattern is easy to solve.

- The following operations on a matrix are called *elementary row operations*.

  1. Multiply row i by a non-zero scalar λ. Denote this operation by $R_i \leftarrow \lambda R_i$. This means that the lefthand side is replaced by the righthand side.

  2. Interchange rows i and j. Denote this operation by $R_i \leftrightarrow R_j$.

  3. Add a multiple λ of row i to another row j. Denote this operation by $R_j \leftarrow R_j + \lambda R_i$.

The following result is not hard to prove and is left as an exercise. It explains the significance of elementary row operations.

Proposition 8.3.4. *Applying elementary row operations to a system of linear equations does not change their solution set.*

Given a system of linear equations $A\mathbf{x} = \mathbf{b}$ the matrix $(A|\mathbf{b})$ is called the *augmented matrix*.

Theorem 8.3.5 (Gaussian elimination). *This is an algorithm for solving a system $A\mathbf{x} = \mathbf{b}$ of linear equations. In outline, the algorithm runs as follows.*

(Step 1). *Form the augmented matrix $(A|\mathbf{b})$.*

(Step 2). *By using elementary row operations, convert $(A|\mathbf{b})$ into a row echelon matrix $(A'|\mathbf{b}')$.*

(Step 3). *Solve the equations obtained from $(A'|\mathbf{b}')$ by back-substitution. These are also the solutions to the original equations.*

The proof of the above theorem will be described in stages.

- The procedure in step (2) has to be carried out systematically to avoid going around in circles.

- Elementary row operations applied to a set of linear equations do not change the solution set by Proposition 8.3.4. Thus the solution sets of $A\mathbf{x} = \mathbf{b}$ and $A'\mathbf{x} = \mathbf{b}'$ are the same.

- Solving systems of linear equations where the associated augmented matrix is a row echelon matrix is easy and can be accomplished by back-substitution.

Here is a more detailed description of step (2) of the algorithm where the input is a matrix B and the output is a matrix B' which is a row echelon matrix.

1. Locate the leftmost column that does not consist entirely of zeros.

2. Interchange the top row with another row if necessary to bring a non-zero entry to the top of the column found in step 1.

3. If the entry now at the top of the column found in step 1 is a, then multiply the first row by $\frac{1}{a}$ in order to introduce a leading 1.

$$\begin{pmatrix} 0 & 1 & * & * & * & * \\ 0 & * & * & * & * & * \\ 0 & * & * & * & * & * \\ 0 & * & * & * & * & * \\ 0 & * & * & * & * & * \end{pmatrix}.$$

4. Add suitable multiples of the top row to the rows below so that all entries below the leading 1 become zeros.

$$\begin{pmatrix} 0 & 1 & * & * & * & * \\ 0 & 0 & * & * & * & * \\ 0 & 0 & * & * & * & * \\ 0 & 0 & * & * & * & * \\ 0 & 0 & * & * & * & * \end{pmatrix}.$$

5. Now cover up the top row, and begin again with step 1 applied to the matrix that remains.

$$\begin{pmatrix} 0 & 1 & * & * & * & * \\ 0 & 0 & 1 & * & * & * \\ 0 & 0 & 0 & * & * & * \\ 0 & 0 & 0 & * & * & * \\ 0 & 0 & 0 & * & * & * \end{pmatrix}.$$

Continue in this way until the matrix is a row echelon matrix.

$$\begin{pmatrix} 0 & 1 & * & * & * & * \\ 0 & 0 & 1 & * & * & * \\ 0 & 0 & 0 & 0 & 1 & * \\ 0 & 0 & 0 & 0 & 0 & 0 \\ 0 & 0 & 0 & 0 & 0 & 0 \end{pmatrix}.$$

In applying this algorithm it is important to remember to start at the top and work downwards.

We now consider in more detail part (3) of the algorithm where the set of equations $A'\mathbf{x} = \mathbf{b}'$ is derived from an augmented matrix which is a row echelon matrix. There are three possible outcomes.

1. In the augmented matrix $(A' \mid \mathbf{b}')$, there is a zero row to the left of the $\mid$ but a corresponding non-zero entry to the right of the $\mid$. This means that the equation corresponding to this row is of the form 'zero equal to something non-zero'. This is impossible and so the original equations are inconsistent.

2. Assume that the equations are consistent, so that outcome (1) does not occur. Suppose that in the augmented matrix $(A' \mid \mathbf{b}')$, the number of leading 1s of A' is equal to the number of unknowns. This means the original equations have a unique solution.

3. Assume that the equations are consistent, so that outcome (1) does not occur and that there is more than one solution, so that outcome (2) does not occur. The variables are divided into two groups: those variables corresponding to the columns of A' containing leading 1s are called *leading variables*, and the rest are called *free variables*.

   - The free variables can be assigned arbitrary values independently of each other.
   - We solve for the leading variables in terms of the free variables.

A few examples will clarify the above algorithm and illustrate the three possible outcomes.

Examples 8.3.6.

1. We show that the following system of equations is inconsistent.

$$\begin{aligned} x+2y-3z &= -1 \\ 3x-y+2z &= 7 \\ 5x+3y-4z &= 2. \end{aligned}$$

The first step is to write down the augmented matrix of the system.

$$\left(\begin{array}{ccc|c} 1 & 2 & -3 & -1 \\ 3 & -1 & 2 & 7 \\ 5 & 3 & -4 & 2 \end{array} \right).$$

Carry out the elementary row operations $R_2 \leftarrow R_2 - 3R_1$ and $R_3 \leftarrow R_3 - 5R_1$. This gives

$$\left(\begin{array}{ccc|c} 1 & 2 & -3 & -1 \\ 0 & -7 & 11 & 10 \\ 0 & -7 & 11 & 7 \end{array} \right).$$

Now carry out the elementary row operation $R_3 \leftarrow R_3 - R_2$ which yields

$$\left(\begin{array}{ccc|c} 1 & 2 & -3 & -1 \\ 0 & -7 & 11 & 10 \\ \mathbf{0} & \mathbf{0} & \mathbf{0} & -3 \end{array} \right).$$

There is no need to continue. The equation corresponding to the last line of the augmented matrix is $0x + 0y + 0z = -3$. Clearly, this equation has no solutions because it is zero on the left of the equals sign and non-zero on the right. It follows that the original set of equations has no solutions.

2. We show that the following system of equations has exactly one solution, and we shall also check it.

$$\begin{aligned} x+2y+3z &= 4 \\ 2x+2y+4z &= 0 \\ 3x+4y+5z &= 2. \end{aligned}$$

We first write down the augmented matrix

$$\left(\begin{array}{ccc|c} 1 & 2 & 3 & 4 \\ 2 & 2 & 4 & 0 \\ 3 & 4 & 5 & 2 \end{array} \right).$$

Carry out the elementary row operations $R_2 \leftarrow R_2 - 2R_1$ and $R_3 \leftarrow R_3 - 3R_1$ to get

$$\left(\begin{array}{ccc|c} 1 & 2 & 3 & 4 \\ 0 & -2 & -2 & -8 \\ 0 & -2 & -4 & -10 \end{array} \right).$$

Now carry out the elementary row operations $R_2 \leftarrow -\frac{1}{2}R_2$ and $R_3 \leftarrow -\frac{1}{2}R_3$ that yield

$$\begin{pmatrix} 1 & 2 & 3 & 4 \\ 0 & 1 & 1 & 4 \\ 0 & 1 & 2 & 5 \end{pmatrix}.$$

Finally, carry out the elementary row operation $R_3 \leftarrow R_3 - R_2$

$$\begin{pmatrix} 1 & 2 & 3 & 4 \\ 0 & 1 & 1 & 4 \\ 0 & 0 & 1 & 1 \end{pmatrix}.$$

This is a row echelon matrix and there are no free variables. Write down the corresponding set of equations

$$\begin{aligned} x + 2y + 3z &= 4 \\ y + z &= 4 \\ z &= 1. \end{aligned}$$

Solve by back-substitution to get $x = -5$, $y = 3$ and $z = 1$. We check that

$$\begin{pmatrix} 1 & 2 & 3 \\ 2 & 2 & 4 \\ 3 & 4 & 5 \end{pmatrix} \begin{pmatrix} -5 \\ 3 \\ 1 \end{pmatrix} = \begin{pmatrix} 4 \\ 0 \\ 2 \end{pmatrix}.$$

3. We show that the following system of equations has infinitely many solutions, and we shall check them.

$$\begin{aligned} x + 2y - 3z &= 6 \\ 2x - y + 4z &= 2 \\ 4x + 3y - 2z &= 14. \end{aligned}$$

The augmented matrix for this system is

$$\begin{pmatrix} 1 & 2 & -3 & 6 \\ 2 & -1 & 4 & 2 \\ 4 & 3 & -2 & 14 \end{pmatrix}.$$

Carry out the following elementary row operations $R_2 \leftarrow R_2 - 2R_1$, $R_3 \leftarrow R_3 - 4R_1$, $R_2 \leftarrow -\frac{1}{5}R_2$, $R_3 \leftarrow -\frac{1}{5}R_3$ and $R_3 \leftarrow R_3 - R_2$. This yields

$$\begin{pmatrix} 1 & 2 & -3 & 6 \\ 0 & 1 & -2 & 2 \\ \mathbf{0} & \mathbf{0} & \mathbf{0} & \mathbf{0} \end{pmatrix}.$$

Because the bottom row consists entirely of zeros, this means that there are only two equations

$$\begin{aligned} x + 2y - 3z &= 6 \\ y - 2z &= 2. \end{aligned}$$

The variables x and y are leading variables and z is the only free variable. The variable z can be assigned any value $z = \lambda$ where $\lambda \in \mathbb{R}$. By back-substitution, both x and y can be expressed in terms of λ. The solution set can be written in the form

$$\begin{pmatrix} x \\ y \\ z \end{pmatrix} = \begin{pmatrix} 2 \\ 2 \\ 0 \end{pmatrix} + \lambda \begin{pmatrix} -1 \\ 2 \\ 1 \end{pmatrix}.$$

We now check that these solutions work

$$\begin{pmatrix} 1 & 2 & -3 \\ 2 & -1 & 4 \\ 4 & 3 & -2 \end{pmatrix} \begin{pmatrix} 2-\lambda \\ 2+2\lambda \\ \lambda \end{pmatrix} = \begin{pmatrix} 6 \\ 2 \\ 14 \end{pmatrix}$$

as required. Observe that the λs cancel out.

We finish off with a simple but important consequence of the theory described in this section. The proof is immediate by thinking about the echelon forms A can assume and observing that there will always be free variables.

Theorem 8.3.7 (Existence of solutions). *Let* $Ax = 0$ *be a homogeneous system of* m *equations in* n *unknowns. Suppose that* $n > m$. *Then there is a non-zero solution.*

Exercises 8.3

1. In each case, determine whether the system of equations is consistent or not. When consistent, find all solutions and show that they work.

(a)

$$\begin{aligned} 5x + 4y - 2z &= 3 \\ 3x + 3y - z &= 2 \\ 2x + 4y + 0z &= 2. \end{aligned}$$

(b)

$$\begin{aligned} 5x + 4y - 2z &= 3 \\ 3x + 3y - z &= 2 \\ 2x + 4y + 0z &= 3. \end{aligned}$$

(c)

$$\begin{aligned} 4x + 4y - 8z &= 20 \\ 3x + 2y + 2z &= 1 \\ 5x + 4y + 3z &= 4. \end{aligned}$$

(d)

$$\begin{aligned} 8x + 11y + 16z &= 78 \\ 6x + 9y + 13z &= 63 \\ 10x + 13y + 24z &= 108. \end{aligned}$$

(e)

$$\begin{aligned} x + 2y + 3z &= 1 \\ x + 3y + 6z &= 3 \\ 3x + 9y + 19z &= 8. \end{aligned}$$

(f)

$$\begin{aligned} 3x + 7y - 11z &= 5 \\ 2x + 5y - 8z &= 4 \\ 3x + 8y - 13z &= 7. \end{aligned}$$

2.

$$\begin{aligned} 2x + 2y + 2z + 2w &= 0 \\ 4x + 5y + 3z + 3w &= 1 \\ 2x + 3y + z + w &= 1 \\ 7x + 10y + 4z + 4w &= 3. \end{aligned}$$

3. Prove Proposition 8.3.4.

8.4   DETERMINANTS

In this section, all matrices are square. We show how to compute one number, called the determinant of the matrix, that contains important information about that matrix. It is a quirk of history that determinants were studied long before matrices.[3] The usefulness of determinants will become apparent in Section 8.5 and what they mean will be clarified in Sections 9.3 and 9.6. The next example helps motivate the definition.

Example 8.4.1. Here is the augmented matrix of two linear equations in two unknowns.

$$\left(\begin{array}{cc|c} a & b & e \\ c & d & f \end{array} \right).$$

Throwing caution to the wind, multiply the top row by c and the bottom row by a and then subtract the top row from the bottom row to get

$$\left(\begin{array}{cc|c} ac & bc & ec \\ 0 & ad - bc & af - ce \end{array} \right).$$

---

[3]Don't believe anything bad anyone tells you about determinants: they are important.

The nature of the solutions to the equations hinges on the value of the number

$$ad - bc.$$

This is a determinant.

The *determinant* of A is denoted by $\det(A)$ or by replacing the round brackets of the matrix A with straight brackets. It is defined recursively: this means that we define an $n \times n$ determinant in terms of $(n-1) \times (n-1)$ determinants when $n > 2$ and we define determinants explicitly in the cases where $n = 1$ and $n = 2$.

- The determinant of the 1×1 matrix $\begin{pmatrix} a \end{pmatrix}$ is a.

- The determinant of the 2×2 matrix

$$A = \begin{pmatrix} a & b \\ c & d \end{pmatrix}$$

denoted by

$$\begin{vmatrix} a & b \\ c & d \end{vmatrix}$$

is the number $ad - bc$.

- The determinant of the 3×3 matrix

$$\begin{pmatrix} a & b & c \\ d & e & f \\ g & h & i \end{pmatrix}$$

denoted by

$$\begin{vmatrix} a & b & c \\ d & e & f \\ g & h & i \end{vmatrix}$$

is the number

$$a \begin{vmatrix} e & f \\ h & i \end{vmatrix} - b \begin{vmatrix} d & f \\ g & i \end{vmatrix} + c \begin{vmatrix} d & e \\ g & h \end{vmatrix}.$$

We could in fact define the determinant of any square matrix by continuing in the same way but we limit ourselves to calculating the determinants of 3×3 matrices, at least by hand. It is important to pay attention to the signs in the definition. You multiply alternately by plus one and minus one

$$+ \quad - \quad + \quad - \quad \dots$$

More generally, the signs are found by computing $(-1)^{i+j}$ where i is the row of the element and j the column. In the above definition, we are taking $i = 1$ throughout. We have defined determinants by expansion along the first row but they can also be expanded along any row and along any column. The fact that the same answers arise however you expand them is a first indication of their remarkable properties.

Examples 8.4.2.

1.
$$\begin{vmatrix} 2 & 3 \\ 4 & 5 \end{vmatrix} = 2 \times 5 - 3 \times 4 = -2.$$

2.
$$\begin{vmatrix} 1 & 2 & 1 \\ 3 & 1 & 0 \\ 2 & 0 & 1 \end{vmatrix} = 1 \begin{vmatrix} 1 & 0 \\ 0 & 1 \end{vmatrix} - 2 \begin{vmatrix} 3 & 0 \\ 2 & 1 \end{vmatrix} + \begin{vmatrix} 3 & 1 \\ 2 & 0 \end{vmatrix} = -7.$$

Theorem 8.4.3 (Properties of determinants). *The following properties hold.*

1. *Let A be a square matrix. Then* $\det(A^T) = \det(A)$.

2. *Let A and B be square matrices having the same size. Then* $\det(AB) = \det(A)\det(B)$.

3. *Let A be a square matrix and let B be obtained from A by interchanging any two columns. Then* $\det(B) = -\det(A)$.

4. *If two columns of a determinant are equal then the determinant is zero.*

5. *If B is the matrix obtained from the matrix A by multiplying* one column *of A by the scalar* λ *then* $\det(B) = \lambda \det(A)$.

6. *Let* $A = (\mathbf{a}_1, \ldots, \mathbf{a}_i + \mathbf{b}_i, \ldots, \mathbf{a}_n)$ *be regarded as a list of column vectors. Then*

$$\det(A) = \det(\mathbf{a}_1, \ldots, \mathbf{a}_i, \ldots, \mathbf{a}_n) + \det(\mathbf{a}_1, \ldots, \mathbf{b}_i, \ldots, \mathbf{a}_n).$$

Proof. The results are true for arbitrary determinants, but we only prove them in the case of 2×2 matrices. If you are feeling energetic you can check the case of 3×3 matrices yourself. The general case is proved by using a different, but equivalent, definition of the determinant that is derived in Section 9.6.

(1) Let
$$A = \begin{pmatrix} a & b \\ c & d \end{pmatrix}.$$

We calculate $\det(A^T)$

$$\begin{vmatrix} a & c \\ b & d \end{vmatrix} = ad - cb = \begin{vmatrix} a & b \\ c & d \end{vmatrix}$$

as claimed.

(2) Let
$$A = \begin{pmatrix} a & b \\ c & d \end{pmatrix} \text{ and } B = \begin{pmatrix} e & f \\ g & h \end{pmatrix}.$$

We prove directly that $\det(AB) = \det(A)\det(B)$. First

$$AB = \begin{pmatrix} ae + bg & af + bh \\ ce + dg & cf + dh \end{pmatrix}.$$

Thus
$$\det(AB) = (ae + bg)(cf + dh) - (af + bh)(ce + dg).$$

The first bracket multiplies out as
$$acef + adeh + bcgf + bdgh$$

and the second as
$$acef + adfg + bceh + bdgh.$$

Subtracting these two expressions we get
$$adeh + bcgf - adfg - bceh.$$

Now we calculate $\det(A)\det(B)$. This is equal to
$$(ad - bc)(eh - fg)$$

which multiplies out to give
$$adeh + bcfg - adfg - bceh.$$

Thus the two sides are equal, and we have proved the result.

(3) Let
$$A = \begin{pmatrix} a & b \\ c & d \end{pmatrix} \text{ and } B = \begin{pmatrix} b & a \\ d & c \end{pmatrix}.$$

So B is obtained from A by interchanging two columns. It is easy to check that $\det(B) = -\det(A)$.

(4) Let A be a matrix with two columns equal. If those two columns are swapped the matrix remains unchanged, but by part (3) above $\det(A) = -\det(A)$. It follows that $\det A = 0$.

(5) Let
$$A = \begin{pmatrix} a & b \\ c & d \end{pmatrix} \text{ and } B = \begin{pmatrix} \lambda a & b \\ \lambda c & d \end{pmatrix}.$$

So we have multiplied one column of A by the scalar λ. It is easy to check that $\det(B) = \lambda \det(A)$. It is important to observe here that we only multiply a column of A by λ and not every element of A.

(6) Calculate
$$\begin{vmatrix} a & b+x \\ c & d+y \end{vmatrix}$$

to get
$$\begin{vmatrix} a & b \\ c & d \end{vmatrix} + \begin{vmatrix} a & x \\ c & y \end{vmatrix}.$$

□

Property (1) above tells us something that is not obvious from the definition: the properties of determinants are the same with respect to rows and columns. Property (2) will be fundamental in our study of invertible matrices. Properties (3)–(6) will be explained in more detail in Chapter 9 where we discuss determinants further. Of these, property (6) is the most complex. Here is another concrete special case.

$$\begin{vmatrix} a & b+x & c \\ d & e+y & f \\ g & h+z & i \end{vmatrix} = \begin{vmatrix} a & b & c \\ d & e & f \\ g & h & i \end{vmatrix} + \begin{vmatrix} a & x & c \\ d & y & f \\ g & z & i \end{vmatrix}.$$

The evaluation of determinants can be simplified using these properties. To understand how, we need to define two important classes of matrices. A square matrix is said to be *upper triangular* if all the entries below the main diagonal are zero, and it is said to be *lower triangular* if all the entries above the main diagonal are zero. The transpose of an upper triangular matrix is lower triangular and vice versa.

Example 8.4.4. The matrix

$$\begin{pmatrix} 1 & 2 & 3 \\ 0 & 4 & 5 \\ 0 & 0 & 6 \end{pmatrix}$$

is upper triangular.

We state the following result only for 3×3 matrices but it is true in general.

Lemma 8.4.5. *The determinant of the 3×3 upper triangular matrix*

$$\begin{pmatrix} \mathbf{a} & b & c \\ 0 & \mathbf{d} & e \\ 0 & 0 & \mathbf{f} \end{pmatrix}$$

is just adf.

Proof. The result is immediately true for lower triangular matrices using our definition of determinant. But determinants do not change when we take transposes. □

Lemma 8.4.5 tells us that the determinant of an upper triangular matrix can be written down on sight.

Lemma 8.4.6. *Let B be the matrix obtained from the square matrix A by adding a scalar multiple of one row of A to another row of A. Then $\det(B) = \det(A)$.*

Proof. The matrix B^T is obtained from the matrix A^T by adding a scalar multiple of one column of A^T to another column of A^T. Thus B^T is equal to

$$(\mathbf{a}_1, \ldots, \mathbf{a}_i + \lambda \mathbf{a}_j, \ldots, \mathbf{a}_n)$$

where $i \neq j$ and the $\mathbf{a}_i$ are the columns of A^T. Thus by parts (6), (5) and (4) of Theorem 8.4.3, we deduce that $\det(B^T) = \det(A^T)$. It follows by part (1) of Theorem 8.4.3 that $\det(B) = \det(A)$. □

Example 8.4.7. By part (3) of Lemma 8.4.3 and Lemma 8.4.6, we can calculate

$$\begin{vmatrix} 2 & 1 & 3 \\ -1 & 0 & 2 \\ 2 & 1 & 0 \end{vmatrix}$$

using two of the three elementary row operations, where multiplication of a row by a scalar cannot, of course, be used, and where the sign changes introduced by row interchanges are recorded. The goal is to transform the matrix into an upper triangular matrix. The determinant can then be written down on sight by Lemma 8.4.5. Apply $R_3 \leftarrow R_3 - R_1$ to get

$$\begin{vmatrix} 2 & 1 & 3 \\ -1 & 0 & 2 \\ 0 & 0 & -3 \end{vmatrix}.$$

This does not change the determinant. Now apply $R_1 \leftrightarrow R_2$ which changes the determinant by -1. Thus the determinant is equal to

$$- \begin{vmatrix} -1 & 0 & 2 \\ 2 & 1 & 3 \\ 0 & 0 & -3 \end{vmatrix}.$$

Finally, apply $R_2 \leftarrow R_2 + 2R_1$ to get

$$- \begin{vmatrix} -1 & 0 & 2 \\ 0 & 1 & 7 \\ 0 & 0 & -3 \end{vmatrix}.$$

The determinant is therefore $-(-1 \times 1 \times -3) = -3$. This method of calculating determinants is called *reduction to an upper triangular matrix*.

Exercises 8.4

1. Compute the following determinants from the definition. Then check your calculations using reduction to an upper triangular matrix.

    (a)

    $$\begin{vmatrix} 1 & -1 \\ 2 & 3 \end{vmatrix}.$$

    (b)

    $$\begin{vmatrix} 3 & 2 \\ 6 & 4 \end{vmatrix}.$$

    (c)

    $$\begin{vmatrix} 1 & -1 & 1 \\ 2 & 3 & 4 \\ 0 & 0 & 1 \end{vmatrix}.$$

(d)

$$\begin{vmatrix} 1 & 2 & 0 \\ 0 & 1 & 1 \\ 2 & 3 & 1 \end{vmatrix}.$$

(e)

$$\begin{vmatrix} 2 & 2 & 2 \\ 1 & 0 & 5 \\ 100 & 200 & 300 \end{vmatrix}.$$

(f)

$$\begin{vmatrix} 1 & 3 & 5 \\ 102 & 303 & 504 \\ 1000 & 3005 & 4999 \end{vmatrix}.$$

(g)

$$\begin{vmatrix} 1 & 1 & 2 \\ 2 & 1 & 1 \\ 1 & 2 & 1 \end{vmatrix}.$$

(h)

$$\begin{vmatrix} 15 & 16 & 17 \\ 18 & 19 & 20 \\ 21 & 22 & 23 \end{vmatrix}.$$

2. Solve $\begin{vmatrix} 1-x & 4 \\ 2 & 3-x \end{vmatrix} = 0.$

3. Calculate

$$\begin{vmatrix} x & \cos x & \sin x \\ 1 & -\sin x & \cos x \\ 0 & -\cos x & -\sin x \end{vmatrix}.$$

4. *Prove that

$$\begin{vmatrix} a & b \\ c & d \end{vmatrix} = 0$$

if and only if one column is a scalar multiple of the other. Hint: Consider two cases: (1) $ad = bc \neq 0$ and (2) $ad = bc = 0$ where you will need to consider various possibilities.

8.5 INVERTIBLE MATRICES

The simplest kind of linear equation is $ax = b$ where $a \neq 0$ and b are scalars. We recall how such an equation can be solved. Multiply both sides of the equation by a^{-1} to get $a^{-1}(ax) = a^{-1}b$; use associativity to shift the brackets $(a^{-1}a)x = a^{-1}b$; finally, use $a^{-1}a = 1$ to obtain $1x = a^{-1}b$ which yields the solution $x = a^{-1}b$. The number a^{-1} is the multiplicative inverse of the non-zero number a. These equations were described first in the Introduction and then in Section 4.2.

Systems of linear equations in several variables cannot apparently be solved in such a straightforward way. But by using matrices, the system can be packaged into a single matrix equation $A\mathbf{x} = \mathbf{b}$ in one vector variable. We now explore the circumstances under which such an equation can be solved using the same ideas as in the one variable case. Assume that A is a square matrix and that there is a matrix B such that $BA = I$.

1. Multiply both sides of the equation $A\mathbf{x} = \mathbf{b}$ on the left by B to get $B(A\mathbf{x}) = B\mathbf{b}$. Because order matters when matrices are multiplied, which side you multiply on also matters.

2. By associativity of matrix multiplication $B(A\mathbf{x}) = (BA)\mathbf{x}$.

3. By assumption $BA = I$, and $I\mathbf{x} = \mathbf{x}$ since I is an identity matrix.

4. It follows that $\mathbf{x} = B\mathbf{b}$.

We appear to have solved our equation, but we need to check it. We calculate $A(B\mathbf{b})$. By associativity this is $(AB)\mathbf{b}$. But at this point we are snookered unless we also assume that $AB = I$. If we do so then $I\mathbf{b} = \mathbf{b}$, as required. We conclude that in order to copy for matrix equations the method used for solving a linear equation in one unknown, the coefficient matrix A must have the property that there is a matrix B such that $AB = I = BA$. We take this as the basis of the following definition.

A matrix A is said to be *invertible* or *non-singular* if there is a matrix B such that $AB = I = BA$. The matrix B is called an *inverse* of A. Observe that A has to be square.[4] A matrix that is not invertible is said to be *singular*.

Example 8.5.1. A real number r regarded as a 1×1 matrix is invertible if and only if it is non-zero, in which case its multiplicative inverse is its reciprocal. Thus the definition of a matrix inverse directly generalizes what we mean by the inverse of a number.

It is clear that if A is a zero matrix, then it cannot be invertible just as in the case of real numbers. The next example shows that even if A is not a zero matrix, it need not be invertible.

Example 8.5.2. Let A be the matrix

$$\begin{pmatrix} 1 & 1 \\ 0 & 0 \end{pmatrix}.$$

We show there is no matrix B such that $AB = I = BA$. Let B be the matrix

$$\begin{pmatrix} a & b \\ c & d \end{pmatrix}.$$

From $BA = I$ we get $a = 1$ and $a = 0$. It is impossible to meet both these conditions at the same time and so B does not exist.

---

[4]There are rectangular invertible matrices but their entries need to be more exotic creatures than real numbers. See [12].

On the other hand here is an example of a matrix that is invertible.

Example 8.5.3. Let

$$A = \begin{pmatrix} 1 & 2 & 3 \\ 0 & 1 & 4 \\ 0 & 0 & 1 \end{pmatrix} \text{ and } B = \begin{pmatrix} 1 & -2 & 5 \\ 0 & 1 & -4 \\ 0 & 0 & 1 \end{pmatrix}.$$

Check that $AB = I = BA$. We deduce that A is invertible with inverse B.

The set $M_n(\mathbb{R})$ of all $n \times n$ real matrices is equipped with an associative binary operation and an identity. The following is therefore a special case of Lemma 4.1.10 and is a nice example of applying abstract results to special cases.

Lemma 8.5.4. *Let A be invertible and suppose that B and C are matrices such that $AB = I = BA$ and $AC = I = CA$. Then $B = C$.*

If a matrix A is invertible then there is only one matrix B such that $AB = I = BA$. We call the matrix B *the* inverse of A and it is denoted by A^{-1}. It is important to remember that we can only write A^{-1} when we know that A is invertible. In the following, we describe some important properties of the inverse of a matrix. They all follow by Lemma 4.1.11.

Lemma 8.5.5.

1. *If A is invertible then A^{-1} is invertible and its inverse is A.*

2. *If A and B are both invertible and AB is defined then AB is invertible with inverse $B^{-1}A^{-1}$.*

3. *If $A_1, \ldots, A_n$ are all invertible and $A_1 \ldots A_n$ is defined then $A_1 \ldots A_n$ is invertible and its inverse is $A_n^{-1} \ldots A_1^{-1}$.*

Example 8.5.6. The set of all invertible $n \times n$ real matrices is denoted by $GL_n(\mathbb{R})$ and is called the *general linear group*. By Lemma 8.5.5, this set is closed under matrix multiplication and inverses. The table below illustrates the parallels with similar results in other sections.

Elements	Invertible elements	Section
$T(X)$	$S(X)$	3.4
$\mathbb{Z}_n$	U_n	5.4
$M_n(\mathbb{R})$	$GL_n(\mathbb{R})$	8.5

We can use inverses to solve certain kinds of linear equations as suggested by the motivation to this section.

Theorem 8.5.7 (Matrix inverse method). *If the matrix A is invertible then the system $Ax = b$ has the unique solution $x = A^{-1}b$.*

Proof. We verify that $A^{-1}\mathbf{b}$ is a solution by calculating

$$A(A^{-1}\mathbf{b}) = (AA^{-1})\mathbf{b} = I\mathbf{b} = \mathbf{b}.$$

It is unique because if $\mathbf{x}'$ is any solution, then $A\mathbf{x}' = \mathbf{b}$ and $A^{-1}(A\mathbf{x}') = A^{-1}\mathbf{b}$ and so $\mathbf{x}' = A^{-1}\mathbf{b}$. □

Example 8.5.8. We solve the following system of equations using the matrix inverse method

$$
\begin{aligned}
x + 2y &= 1 \\
3x + y &= 2.
\end{aligned}
$$

Write the equations in matrix form.

$$\begin{pmatrix} 1 & 2 \\ 3 & 1 \end{pmatrix} \begin{pmatrix} x \\ y \end{pmatrix} = \begin{pmatrix} 1 \\ 2 \end{pmatrix}.$$

It can be checked that the inverse of the coefficient matrix is

$$A^{-1} = -\frac{1}{5} \begin{pmatrix} 1 & -2 \\ -3 & 1 \end{pmatrix}.$$

Now we can solve the equations. From $A\mathbf{x} = \mathbf{b}$ we obtain $\mathbf{x} = A^{-1}\mathbf{b}$. Thus

$$\mathbf{x} = -\frac{1}{5} \begin{pmatrix} 1 & -2 \\ -3 & 1 \end{pmatrix} \begin{pmatrix} 1 \\ 2 \end{pmatrix} = \begin{pmatrix} \frac{3}{5} \\ \frac{1}{5} \end{pmatrix}$$

and so $x = \frac{3}{5}$ and $y = \frac{1}{5}$.

The definition of an invertible matrix does not provide a practical method for deciding whether a matrix is invertible or not. We now describe such a method using determinants.

The results that follow assume that the scalars are real or complex. A flavour of what happens when other kinds of scalars are allowed is described in Example 8.5.17.

Recall from Theorem 8.4.3 that $\det(AB) = \det(A)\det(B)$. This property is used below to obtain a necessary condition for a matrix to be invertible.

Lemma 8.5.9. *If A is invertible then $\det(A) \neq 0$.*

Proof. From the definition of invertibility, there is a matrix B such that $AB = I$. Take determinants of both sides of this equation $\det(AB) = \det(I)$. But $\det(I) = 1$ and $\det(AB) = \det(A)\det(B)$. Thus $\det(A)\det(B) = 1$. In particular, $\det(A) \neq 0$. □

It is natural to ask if there are any other properties that a matrix must satisfy in order to have an inverse. The answer is, surprisingly, no. To motivate the general case, we start with a 2×2 matrix A where everything is easy to calculate. Let

$$A = \begin{pmatrix} a & b \\ c & d \end{pmatrix}.$$

We construct a new matrix as follows.

- Replace each entry a_{ij} of A by the element remaining when the ith row and jth column are crossed out

$$\begin{pmatrix} d & c \\ b & a \end{pmatrix}.$$

- Use the following matrix of signs

$$\begin{pmatrix} + & - \\ - & + \end{pmatrix},$$

where the entry in row i and column j is the sign of $(-1)^{i+j}$, to get

$$\begin{pmatrix} d & -c \\ -b & a \end{pmatrix}.$$

- Take the transpose of this matrix to get the matrix we call the *adjugate* of A

$$\mathrm{adj}(A) = \begin{pmatrix} d & -b \\ -c & a \end{pmatrix}.$$

- Observe that

$$A\,\mathrm{adj}(A) = \det(A)I = \mathrm{adj}(A)\,A.$$

- We deduce that if $\det(A) \neq 0$ then

$$\boxed{A^{-1} = \tfrac{1}{\det(A)} \begin{pmatrix} d & -b \\ -c & a \end{pmatrix}.}$$

We have therefore proved the following.

Proposition 8.5.10. *A 2×2 matrix is invertible if and only if its determinant is non-zero.*

Example 8.5.11. Let

$$A = \begin{pmatrix} 1 & 2 \\ 3 & 1 \end{pmatrix}.$$

Its determinant is -5 and so the matrix is invertible. The adjugate of A is

$$\mathrm{adj}(A) = \begin{pmatrix} 1 & -2 \\ -3 & 1 \end{pmatrix}.$$

Thus the inverse of A is

$$A^{-1} = -\frac{1}{5} \begin{pmatrix} 1 & -2 \\ -3 & 1 \end{pmatrix}.$$

We now describe the general case. Let A be an $n \times n$ matrix with entries a_{ij}. We define its adjugate as the result of the following sequence of operations.

- Choose an entry a_{ij} in the matrix A.

- Crossing out the entries in row i and column j, an $(n-1) \times (n-1)$ matrix is constructed, denoted by $M(A)_{ij}$, and called a *submatrix*.

- The determinant $\det(M(A)_{ij})$ is called the *minor* of the element a_{ij}.

- If $\det(M(A)_{ij})$ is multiplied by the corresponding sign, we get the *cofactor* $c_{ij} = (-1)^{i+j} \det(M(A)_{ij})$ of the element a_{ij}.

- Replace each element a_{ij} by its cofactor to obtain the matrix $C(A)$ of cofactors of A.

- The *transpose* of the matrix of cofactors $C(A)$, denoted adj(A), is called the *adjugate*[5] *matrix of A*. Thus the adjugate is the transpose of the matrix of signed minors.

The defining property of the adjugate is described in the next result.

Theorem 8.5.12. *For any square matrix A,*

$$A\,(\mathrm{adj}(A)) = \det(A)I = (\mathrm{adj}(A))A.$$

Proof. We have verified the result in the 2×2 case. We now prove the 3×3 case by means of an argument that generalizes. Let $A = (a_{ij})$ and write

$$B = \mathrm{adj}(A) = \begin{pmatrix} c_{11} & c_{21} & c_{31} \\ c_{12} & c_{22} & c_{32} \\ c_{13} & c_{23} & c_{33} \end{pmatrix}.$$

Compute AB. We have that

$$(AB)_{11} = a_{11}c_{11} + a_{12}c_{12} + a_{13}c_{13} = \det A$$

by expanding the determinant along the top row. The next element is

$$(AB)_{12} = a_{11}c_{21} + a_{12}c_{22} + a_{13}c_{23}.$$

But this is the determinant of the matrix

$$\begin{pmatrix} a_{11} & a_{12} & a_{13} \\ a_{11} & a_{12} & a_{13} \\ a_{31} & a_{32} & a_{33} \end{pmatrix}$$

[5]This odd word comes from Latin and means 'yoked together'.

which, having two rows equal, is zero by Theorem 8.4.3. This pattern now continues with all the off-diagonal entries being zero for similar reasons and the diagonal entries being the determinant. □

We can now prove the main theorem of this section.

Theorem 8.5.13 (Existence of inverses). *Let A be a square matrix. Then A is invertible if and only if* $\det(A) \neq 0$. *When A is invertible, its inverse is given by*

$$A^{-1} = \frac{1}{\det(A)} \operatorname{adj}(A).$$

Proof. Let A be invertible. Then $\det(A) \neq 0$ by Lemma 8.5.9. We can therefore form the matrix

$$\frac{1}{\det(A)} \operatorname{adj}(A)$$

and then

$$A \frac{1}{\det(A)} \operatorname{adj}(A) = \frac{1}{\det(A)} A \operatorname{adj}(A) = I$$

by Theorem 8.5.12. Thus A has the advertised inverse.

Conversely, suppose that $\det(A) \neq 0$. Then again we can form the matrix

$$\frac{1}{\det(A)} \operatorname{adj}(A)$$

and verify that this is the inverse of A and so A is invertible. □

Theorem 8.5.13 tells us that the determinant is the arbiter of whether a matrix is invertible or not. Algebraically there is nothing more to be said, but no insight is provided as to why this works. Geometry does provide the insight and just how is explained in Example 9.6.4.

Example 8.5.14. Let

$$A = \begin{pmatrix} 1 & 2 & 3 \\ 2 & 0 & 1 \\ -1 & 1 & 2 \end{pmatrix}.$$

Its determinant is -5 and so the matrix is invertible. We calculate its inverse. The matrix of minors is

$$\begin{pmatrix} -1 & 5 & 2 \\ 1 & 5 & 3 \\ 2 & -5 & -4 \end{pmatrix}.$$

The matrix of cofactors is

$$\begin{pmatrix} -1 & -5 & 2 \\ -1 & 5 & -3 \\ 2 & 5 & -4 \end{pmatrix}.$$

The adjugate is the transpose of the matrix of cofactors

$$\begin{pmatrix} -1 & -1 & 2 \\ -5 & 5 & 5 \\ 2 & -3 & -4 \end{pmatrix}.$$

The inverse of A is the adjugate with each entry divided by the determinant of A

$$A^{-1} = -\frac{1}{5}\begin{pmatrix} -1 & -1 & 2 \\ -5 & 5 & 5 \\ 2 & -3 & -4 \end{pmatrix}.$$

Box 8.2: The Moore-Penrose Inverse

We have proved that a square matrix has an inverse if and only if it has a non-zero determinant. For rectangular matrices, the existence of an inverse does not even come up for discussion. However, in later applications of matrix theory it is very convenient if every matrix have an 'inverse'. Let A be any matrix. We say that A^+ is its *Moore-Penrose inverse* if the following conditions hold:

1. $A = AA^+A$.
2. $A^+ = A^+AA^+$.
3. $(A^+A)^T = A^+A$.
4. $(AA^+)^T = AA^+$.

It is not obvious, but every matrix A has a Moore-Penrose inverse A^+ and, in fact, such an inverse is uniquely determined by the above four conditions. In the case where A is invertible in the vanilla-sense, its Moore-Penrose inverse is just its inverse. But even singular matrices have Moore-Penrose inverses. You can check that the matrix defined below satisfies the four conditions above

$$\begin{pmatrix} 1 & 2 \\ 3 & 6 \end{pmatrix}^+ = \begin{pmatrix} 0\cdot002 & 0\cdot006 \\ 0\cdot04 & 0\cdot12 \end{pmatrix}.$$

The Moore-Penrose inverse can be used to find approximate solutions to systems of linear equations that might otherwise have no solution.

The adjugate is an important theoretical tool and leads to a method for calculating inverses. But for larger matrices this method is onerous. A more practical method is to use elementary row operations. We describe the method and then prove it works. Let A be an $n \times n$ matrix. We want to determine whether it is invertible and, if it is, calculate its inverse. We do this at the same time and we shall not need to calculate a determinant. We write down a new kind of augmented matrix, this time of the form $B = (A \mid I)$ where I is the $n \times n$ identity matrix. The first part of the algorithm is to carry out elementary row operations on B guided by A. Our goal is to convert A into a row echelon matrix. This will have zeros below the leading diagonal. We are interested in what entries lie on the leading diagonal. If one of them is zero we stop and say that A is not invertible. If all of them are 1 then the algorithm continues.

$$\begin{pmatrix} 1 & * & * & * & * & * \\ 0 & 1 & * & * & * & * \\ 0 & 0 & 1 & * & * & * \end{pmatrix}.$$

We now use the 1s that lie on the leading diagonal to remove the elements above. First we obtain

$$\begin{pmatrix} 1 & * & 0 & | & * & * & * \\ 0 & 1 & 0 & | & * & * & * \\ 0 & 0 & 1 & | & * & * & * \end{pmatrix}$$

and then

$$\begin{pmatrix} 1 & 0 & 0 & | & * & * & * \\ 0 & 1 & 0 & | & * & * & * \\ 0 & 0 & 1 & | & * & * & * \end{pmatrix}.$$

Our original matrix B now has the following form $(I \mid A')$. We prove that in fact $A' = A^{-1}$. This method is illustrated by means of an example.

Example 8.5.15. Let

$$A = \begin{pmatrix} -1 & 1 & 1 \\ 1 & -1 & 1 \\ 1 & 1 & -1 \end{pmatrix}.$$

We show that A is invertible and calculate its inverse. We first write down the augmented matrix

$$\begin{pmatrix} -1 & 1 & 1 & | & 1 & 0 & 0 \\ 1 & -1 & 1 & | & 0 & 1 & 0 \\ 1 & 1 & -1 & | & 0 & 0 & 1 \end{pmatrix}.$$

We now carry out a sequence of elementary row operations to obtain the following where the matrix on the left of the partition is an echelon matrix.

$$\begin{pmatrix} 1 & -1 & -1 & | & -1 & 0 & 0 \\ 0 & 1 & 0 & | & \frac{1}{2} & 0 & \frac{1}{2} \\ 0 & 0 & 1 & | & \frac{1}{2} & \frac{1}{2} & 0 \end{pmatrix}.$$

The leading diagonal contains only 1s and so our original matrix is invertible. We now use each such 1 on the leading diagonal to make all entries above it equal to zero by means of a suitable elementary row operation of the third type

$$\begin{pmatrix} 1 & 0 & 0 & | & 0 & \frac{1}{2} & \frac{1}{2} \\ 0 & 1 & 0 & | & \frac{1}{2} & 0 & \frac{1}{2} \\ 0 & 0 & 1 & | & \frac{1}{2} & \frac{1}{2} & 0 \end{pmatrix}.$$

We deduce that the inverse of A is

$$A^{-1} = \begin{pmatrix} 0 & \frac{1}{2} & \frac{1}{2} \\ \frac{1}{2} & 0 & \frac{1}{2} \\ \frac{1}{2} & \frac{1}{2} & 0 \end{pmatrix}.$$

We now explain why this method works in the 3×3 case though the argument works in general. We start with a lemma whose proof is left as an exercise.

Lemma 8.5.16. *Let A be a square matrix. Then A is invertible if and only if there is a square matrix B such that $AB = I$.*

Thus to show the matrix A is invertible we need only find a 3×3 matrix B such that $AB = I$. Regard $B = (\mathbf{b}_1, \mathbf{b}_2, \mathbf{b}_3)$ as a list of its column vectors. Then $AB = (A\mathbf{b}_1, A\mathbf{b}_2, A\mathbf{b}_3)$. It follows that $AB = I$ if and only if the following three equations hold

$$A\mathbf{b}_1 = \mathbf{e}_1, \quad A\mathbf{b}_2 = \mathbf{e}_2 \text{ and } A\mathbf{b}_3 = \mathbf{e}_3.$$

To find $\mathbf{b}_i$, we solve the equation $A\mathbf{x} = \mathbf{e}_i$. Thus our method for finding the inverse of A using elementary row operations is nothing other than a method for solving these three equations at the same time.

Example 8.5.17. We have considered invertible matrices only with reference to real or complex numbers. But we can also consider matrices whose entries come from $\mathbb{Z}_n$. The same results apply as long as we interpret our calculations correctly. For concreteness, we consider matrices whose entries are from $\mathbb{Z}_{26}$ since this leads to a nice application to cryptography. Let

$$A = \begin{pmatrix} a & b \\ c & d \end{pmatrix}$$

be a matrix with entries from $\mathbb{Z}_{26}$. It is invertible if there is another such matrix B such that $AB = I = BA$. As before, we take determinants and obtain $\det(A)\det(B) = 1$. It is at this point that we have to interpret our calculations correctly. The equality $\det(A)\det(B) = 1$ in $\mathbb{Z}_{26}$ says that $\det(A)$ has multiplicative inverse $\det(B)$. Here is the table from Section 5.4 which lists the invertible elements of $\mathbb{Z}_{26}$ and their corresponding multiplicative inverses.

a	1	3	5	7	9	11	15	17	19	21	23	25
a^{-1}	1	9	21	15	3	19	7	23	11	5	17	25

Thus the inverse of A exists only if $\det(A) \in U_{26}$, in which case we can use the adjugate formula as before to obtain A^{-1}. More generally, if A is a square matrix with entries from $\mathbb{Z}_n$, it is invertible if and only if $\det(A) \in U_n$. Here is an example. The matrix

$$A = \begin{pmatrix} 1 & 3 \\ 3 & 4 \end{pmatrix}$$

has entries from $\mathbb{Z}_{26}$. Its determinant $4 - 9 = -5 \equiv 21$ when working modulo 26. This is an invertible element of $\mathbb{Z}_{26}$, and consulting the above table, we find that

$$21^{-1} = 5.$$

The inverse of A is therefore

$$5 \begin{pmatrix} 4 & -3 \\ -3 & 1 \end{pmatrix} = \begin{pmatrix} 20 & -15 \\ -15 & 5 \end{pmatrix} = \begin{pmatrix} 20 & 11 \\ 11 & 5 \end{pmatrix}.$$

This works because

$$\begin{pmatrix} 1 & 3 \\ 3 & 4 \end{pmatrix} \begin{pmatrix} 20 & 11 \\ 11 & 5 \end{pmatrix} = \begin{pmatrix} 1 & 0 \\ 0 & 1 \end{pmatrix},$$

remembering that all calculations are modulo 26. Invertible matrices with entries from $\mathbb{Z}_{26}$ can be used in cryptography. We give an example using the invertible matrix A above. A message is written in upper case letters without punctuation. We assume the message has even length and if not add a 'Z' to the end as padding. Replace each letter by its numerical equivalent according to the recipe $A \leftrightarrow 0, \ldots, Z \leftrightarrow 25$. Group the sequence of numbers into pairs regarded as column vectors. Encrypt

$$\begin{pmatrix} x \\ y \end{pmatrix} \text{ by } \begin{pmatrix} 1 & 3 \\ 3 & 4 \end{pmatrix} \begin{pmatrix} x \\ y \end{pmatrix}$$

and then replace by the letter equivalents. Thus the message SPY is first padded out to give SPYZ. This consists of two blocks of two letters SP and YZ. We replace each of these by their numerical equivalents

$$\begin{pmatrix} 18 \\ 15 \end{pmatrix} \text{ and } \begin{pmatrix} 24 \\ 25 \end{pmatrix}$$

which are encrypted as

$$\begin{pmatrix} 11 \\ 10 \end{pmatrix} \text{ and } \begin{pmatrix} 17 \\ 16 \end{pmatrix},$$

respectively. Thus the original message, the *cleartext*, is encrypted as LKRQ, the *ciphertext*. Decrypting such a message involves repeating the above steps but using the matrix A^{-1} instead. Thus, for example,

$$\begin{pmatrix} 20 & 11 \\ 11 & 5 \end{pmatrix} \begin{pmatrix} 11 \\ 10 \end{pmatrix} = \begin{pmatrix} 18 \\ 15 \end{pmatrix}$$

as expected. This is an example of a *Hill cipher* [65].

We now make two fundamental definitions.

- Let $\mathbf{v}_1, \ldots, \mathbf{v}_m$ be a list of m column vectors each having n rows. A sum such as

$$\lambda_1 \mathbf{v}_1 + \ldots + \lambda_m \mathbf{v}_m$$

is called a *linear combination* of the column vectors. This is nothing other than nomenclature.

- The equation

$$x_1 \mathbf{v}_1 + \ldots + x_m \mathbf{v}_m = \mathbf{0}$$

always has the trivial solution where $x_1 = \ldots = x_m = 0$. If this is the only solution we say that $\mathbf{v}_1, \ldots, \mathbf{v}_m$ are *linearly independent* whereas if the equation has a non-trivial solution we say that $\mathbf{v}_1, \ldots, \mathbf{v}_m$ are *linearly dependent*.

The concept of linear (in)dependence turns out to be indispensable in matrix theory particularly in a geometric context. This will become clear in Chapter 10. We now explain the simple idea that lies behind the definition of linear dependence. Assume that the column vectors $\mathbf{v}_1, \ldots, \mathbf{v}_m$ are linearly dependent. Then

$$x_1 \mathbf{v}_1 + \ldots + x_m \mathbf{v}_m = \mathbf{0}$$

for some x_i where not all the $x_i = 0$. By relabelling if necessary, we can assume that $x_1 \neq 0$. We can therefore write

$$\mathbf{v}_1 = -\frac{x_2}{x_1}\mathbf{v}_2 - \ldots - \frac{x_m}{x_1}\mathbf{v}_m.$$

We have shown that $\mathbf{v}_1$ can be written as a linear combination of $\mathbf{v}_2, \ldots, \mathbf{v}_m$. Thus if a list of vectors is linearly dependent one of the vectors can be written as a linear combination of the others. Conversely, if a vector $\mathbf{v}_1$ can be written as a linear combination of $\mathbf{v}_2, \ldots, \mathbf{v}_m$ then $\mathbf{v}_1, \ldots, \mathbf{v}_m$ are linearly dependent.

> *The concept of linear dependence is simply a neat way of saying that one vector from a list can be written as a linear combination of the remaining vectors without having to specify which vector we are talking about.*

Example 8.5.18. These concepts may also be applied[6] to row vectors and they shed light on the nature of Gaussian elimination. When a zero row is obtained in an augmented matrix as a result of applying elementary row operations, we have actually shown that one of the original equations is a linear combination of the others. Such an equation is redundant since it provides no new information and so can be discarded. Thus Gaussian elimination is a procedure for discarding redundant information in a systematic way.

The notion of linear dependence can be related to solving linear equations.

Lemma 8.5.19. *Let $\mathbf{v}_1, \ldots, \mathbf{v}_n$ be n column vectors each with n entries. Let A be the square matrix whose columns are exactly $\mathbf{v}_1, \ldots, \mathbf{v}_n$. Then the column vectors $\mathbf{v}_1, \ldots, \mathbf{v}_n$ are linearly dependent if and only if the system of equations $A\mathbf{x} = \mathbf{0}$ has a non-zero solution.*

Proof. In the case where $n = 3$ this follows from

$$\begin{pmatrix} a_{11} & a_{12} & a_{13} \\ a_{21} & a_{22} & a_{23} \\ a_{31} & a_{32} & a_{33} \end{pmatrix} \begin{pmatrix} \lambda_1 \\ \lambda_2 \\ \lambda_3 \end{pmatrix} = \lambda_1 \begin{pmatrix} a_{11} \\ a_{21} \\ a_{31} \end{pmatrix} + \lambda_2 \begin{pmatrix} a_{12} \\ a_{22} \\ a_{32} \end{pmatrix} + \lambda_3 \begin{pmatrix} a_{13} \\ a_{23} \\ a_{33} \end{pmatrix}.$$

---

[6]Mutatis mutandis.

The proof of the general case is similar.   □

We now relate the notion of linear independence to invertibility of matrices. The proof of the following result is immediate by parts (3) and (5) of Theorem 8.4.3, Lemma 8.4.6 and Theorem 8.5.13.

Lemma 8.5.20. *Let A be a square matrix and let A' be any matrix obtained from A by applying a sequence of elementary row operations. Then A is invertible if and only if A' is invertible.*

Thus if a sequence of elementary row operations is applied to a matrix A then the matrix that results is invertible or singular precisely when A is.

Theorem 8.5.21 (Determinants and linear dependence). *Let A be an $n \times n$ matrix with column vectors $\mathbf{a}_i$ where $1 \le i \le n$. Then $\det(A) = 0$ if and only if the column vectors $\mathbf{a}_1, \dots, \mathbf{a}_n$ are linearly dependent.*

Proof. Suppose that the columns of A are linearly dependent. Then one of the columns can be written as a linear combination of the others. It follows by Theorem 8.4.3 that $\det(A) = 0$. To see what is going on here, consider a special case. Regard A as a list of its column vectors and for simplicity assume that it is a 3×3 matrix and that the first column is a linear combination of the other two. Then

$$\det(\lambda \mathbf{a}_2 + \mu \mathbf{a}_3, \mathbf{a}_2, \mathbf{a}_3) = \det(\lambda \mathbf{a}_2, \mathbf{a}_2, \mathbf{a}_3) + \det(\mu \mathbf{a}_3, \mathbf{a}_2, \mathbf{a}_3)$$

which is equal to

$$\lambda \det(\mathbf{a}_2, \mathbf{a}_2, \mathbf{a}_3) + \mu \det(\mathbf{a}_3, \mathbf{a}_2, \mathbf{a}_3).$$

This sum is zero because in both terms the determinants have repeated columns.

To prove the converse, let $\det(A) = 0$. Suppose that the columns of A are linearly independent. Then the system of equations $A\mathbf{x} = \mathbf{0}$ has exactly one solution, the zero column vector, by Lemma 8.5.19. Thus the augmented matrix $(A \mid \mathbf{0})$ can be transformed by means of elementary row operations into $(A' \mid \mathbf{0})$ where A' is a row echelon matrix with no zero rows. In particular, A' is an upper triangular matrix and so $\det(A') \ne 0$ by Lemma 8.4.5. But then $\det(A) \ne 0$ by Lemma 8.5.20, which is a contradiction.   □

The proof of Theorem 8.5.21 is complex when viewed algebraically, but becomes obvious when viewed geometrically. This is explained in Example 9.3.4. We conclude this chapter with a portmanteau theorem which describes when a matrix is invertible from a number of different points of view.

Theorem 8.5.22 (Characterizations of invertibility). *Let A be an $n \times n$ matrix over the real or complex numbers. Then the following are equivalent.*

1. A is invertible.

2. $\det(A) \ne 0$.

3. The system $A\mathbf{x} = \mathbf{b}$ has a unique solution for every column vector $\mathbf{b}$.

4. *The homogeneous system* $A\mathbf{x} = \mathbf{0}$ *has exactly one solution.*

5. *The columns of A are linearly independent.*

6. *The function* $\mathbf{x} \mapsto A\mathbf{x}$ *from* $\mathbb{R}^n$ *to itself is bijective.*

Proof. The proof that (1)⇔(2) follows from Theorem 8.5.13. The proof that (1)⇒(3) is immediate. The proof that (3)⇒(1) follows by choosing for **b**, in turn, the column vectors $\mathbf{e}_1, \ldots, \mathbf{e}_n$. The solutions provide the columns of a matrix B such that $AB = I$ which means that A is invertible. Clearly (1)⇒(4). The proof that (4)⇔(5) follows from Lemma 8.5.19. To prove that (4)⇒(1), it is enough to prove that (5)⇒(2) but this follows by Theorem 8.5.21. The proof that (1)⇒(6) is left as an exercise, and the proof that (6)⇒(4) is immediate. This is a tangled web of implications but it can be checked that each of the six statements is equivalent to each of the other six statements. □

Exercises 8.5

1. Use the adjugate method and then elementary row operations to compute the inverses of the following matrices. In each case, check that your solution works.

   (a) $\begin{pmatrix} 1 & 0 \\ 0 & 2 \end{pmatrix}$.

   (b) $\begin{pmatrix} 1 & 1 \\ 1 & 2 \end{pmatrix}$.

   (c) $\begin{pmatrix} 1 & 0 & 0 \\ 0 & 2 & 0 \\ 0 & 0 & 3 \end{pmatrix}$.

   (d) $\begin{pmatrix} 1 & 2 & 3 \\ 2 & 0 & 1 \\ -1 & 1 & 2 \end{pmatrix}$.

   (e) $\begin{pmatrix} 1 & 2 & 3 \\ 1 & 3 & 3 \\ 1 & 2 & 4 \end{pmatrix}$.

   (f) $\begin{pmatrix} 2 & 2 & 1 \\ -2 & 1 & 2 \\ 1 & -2 & 2 \end{pmatrix}$.

2. Prove Lemma 8.5.20.

3. (a) Prove that if A is invertible then A^{-1} is invertible and $\det(A^{-1}) = \det(A)^{-1}$.

   (b) Prove that if A is invertible then A^T is invertible and $(A^T)^{-1} = (A^{-1})^T$.

4. Prove 8.5.16.

5. (a) Find the inverse of the matrix

$$\begin{pmatrix} 2 & 3 \\ 3 & 2 \end{pmatrix}$$

with entries from $\mathbb{Z}_{26}$.

(b) The coded message EBNRYR is received. It is known to have been encrypted using the matrix

$$\begin{pmatrix} 3 & 2 \\ 12 & 9 \end{pmatrix}$$

with entries from $\mathbb{Z}_{26}$. Decrypt the message.

6. Let A be an $n \times n$ matrix. Define the function $f: \mathbb{R}^n \to \mathbb{R}^n$ by $\mathbf{x} \mapsto A\mathbf{x}$. Prove that if A is invertible then f is bijective.

7. For each triple of column vectors below, determine whether they are linearly independent or linearly dependent. For those that are linearly dependent, show that one vector can be written as a linear combination of the other two.

(a)

$$\begin{pmatrix} 1 \\ 2 \\ 0 \end{pmatrix}, \begin{pmatrix} 2 \\ 7 \\ 6 \end{pmatrix}, \begin{pmatrix} 0 \\ 1 \\ 2 \end{pmatrix}.$$

(b)

$$\begin{pmatrix} 2 \\ 3 \\ 2 \end{pmatrix}, \begin{pmatrix} -1 \\ 1 \\ -1 \end{pmatrix}, \begin{pmatrix} -1 \\ 11 \\ -1 \end{pmatrix}.$$

(c)

$$\begin{pmatrix} 5 \\ 3 \\ 3 \end{pmatrix}, \begin{pmatrix} -2 \\ 1 \\ 6 \end{pmatrix}, \begin{pmatrix} 2 \\ -1 \\ 3 \end{pmatrix}.$$

8. Prove that any four column vectors in $\mathbb{R}^3$ are linearly dependent.

8.6 DIAGONALIZATION

In this section, we take the first steps in matrix theory to demonstrate the real power of matrix methods. The first theorem we prove, the Cayley-Hamilton theorem,[7] will show that the theory of matrices is intimately bound up with the theory of polynomials described in Chapter 7. In Section 8.2, we observed that we can form powers of matrices, multiply them by scalars and add them together. We can therefore form sums like

$$A^3 + 3A^2 + A + 4I.$$

[7] For readers in Ireland, this should be read as the Hamilton-Cayley theorem throughout.

This is a polynomial in A or, viewed from another perspective, we can substitute A into the polynomial

$$x^3 + 3x^2 + x + 4,$$

remembering that $4 = 4x^0$ and so has to be replaced by $4I$. This opens up the following possibility. Let $f(x)$ be a polynomial and A a square matrix. We say that A is a *(matrix) root* of $f(x)$ if $f(A)$ is the zero matrix.

The next definition is conjured out of thin air, but we explain how it arises later. Let A be an $n \times n$ matrix. Define

$$\chi_A(x) = \det(A - xI).$$

A little thought reveals that $\chi_A(x)$ is a polynomial of degree n in x. It is called the *characteristic polynomial* of A. Observe that $\chi_A(0) = \det(A)$, when $x = 0$, and so the determinant arises as the constant term.[8]

The following theorem was proved by Cayley in [24]. In fact, he proved the 2×2 and 3×3 cases explicitly, and then wrote the following

> "...but I have not thought it necessary to undertake the labour of a formal proof of the theorem in the general case of a matrix of any degree".

Such were the days.

Theorem 8.6.1 (Cayley-Hamilton). *Every square matrix is a root of its characteristic polynomial.*

Proof. We prove the theorem in the 2×2 case by direct calculation, and then in the 3×3 case by means of a method that generalizes. Let

$$A = \begin{pmatrix} a & b \\ c & d \end{pmatrix}.$$

By definition the characteristic polynomial is

$$\begin{vmatrix} a-x & b \\ c & d-x \end{vmatrix}.$$

Thus

$$\chi_A(x) = x^2 - (a+d)x + (ad - bc).$$

We calculate $\chi_A(A)$ which is equal to

$$\begin{pmatrix} a & b \\ c & d \end{pmatrix}^2 - (a+d) \begin{pmatrix} a & b \\ c & d \end{pmatrix} + \begin{pmatrix} ad - bc & 0 \\ 0 & ad - bc \end{pmatrix}.$$

This simplifies to

$$\begin{pmatrix} 0 & 0 \\ 0 & 0 \end{pmatrix}$$

---

[8]The characteristic polynomial is sometimes defined to be $\det(xI - A)$ which has the advantage that this is a monic polynomial without sign issues.

which proves the theorem in this case. As an aside, the number $a + d$ is called the *trace* of the matrix A. We shall meet it again, in greater generality, in Chapter 10.

Let A be a 3×3 matrix and let

$$\chi_A(x) = -x^3 + a_2 x^2 + a_1 x + a_0.$$

Observe that $\mathrm{adj}(A - xI)$ is a 3×3 matrix whose entries are polynomials of degree 2. We can therefore write

$$\mathrm{adj}(A - xI) = x^2 B_2 + x B_1 + B_0$$

where B_0, B_1, B_2 are 3×3 matrices with scalar entries. The defining property of the adjugate is that

$$(A - xI)\mathrm{adj}(A - xI) = \chi_A(x)I.$$

Multiplying out both sides of the above equation, we obtain

$$-x^3 B_2 + x^2(AB_2 - B_1) + x(AB_1 - B_0) + AB_0 = -x^3 I + x^2 a_2 I + x a_1 I + a_0 I.$$

Equating like terms on each side gives rise to the following matrix equations.

- $B_2 = I.$

- $AB_2 - B_1 = a_2 I.$

- $AB_1 - B_0 = a_1 I.$

- $AB_0 = a_0 I.$

Multiply each of these equations on the left by $-A^3$, A^2, A and I, respectively. On adding up both sides and equating, we obtain

$$-A^3 + a_2 A^2 + a_1 A + a_0 I = O$$

which is equal to $\chi_A(A)$ as claimed. □

Example 8.6.2. We illustrate the proof of the 3×3 case of the Cayley-Hamilton theorem by means of an example. Let

$$A = \begin{pmatrix} 1 & 1 & 1 \\ 0 & 1 & 1 \\ 0 & 0 & 1 \end{pmatrix}.$$

By definition $\chi_A(x) = \det(A - xI) = (1 - x)^3$. Put $B = \mathrm{adj}(A - xI)$. Then

$$B = \begin{pmatrix} (1-x)^2 & (x-1) & x \\ 0 & (1-x)^2 & (x-1) \\ 0 & 0 & (1-x)^2 \end{pmatrix}.$$

This can be written as a quadratic polynomial with matrix coefficients.

$$B = x^2 \begin{pmatrix} 1 & 0 & 0 \\ 0 & 1 & 0 \\ 0 & 0 & 1 \end{pmatrix} + x \begin{pmatrix} -2 & 1 & 1 \\ 0 & -2 & 1 \\ 0 & 0 & -2 \end{pmatrix} + \begin{pmatrix} 1 & -1 & 0 \\ 0 & 1 & -1 \\ 0 & 0 & 1 \end{pmatrix}.$$

Direct calculation shows that

$$(A - xI)B = (1 - x)^3 I.$$

There is an immediate application of Theorem 8.6.1.

Proposition 8.6.3. *Let A be an invertible $n \times n$ matrix. Then the inverse of A can be written as a polynomial in A of degree $n - 1$.*

Proof. We can write $\chi_A(x) = f(x) + \det(A)$ where $f(x)$ is a polynomial with constant term zero. Thus $f(x) = xg(x)$ for some polynomial $g(x)$ of degree $n - 1$. By the Cayley-Hamilton theorem, $O = Ag(A) + \det(A)I$. Thus $Ag(A) = -\det(A)I$. Define $B = -\frac{1}{\det(A)}g(A)$. Then $AB = I$ and so by Lemma 8.5.16, the matrix B is the inverse of A. □

Example 8.6.4. Let

$$A = \begin{pmatrix} 3 & 1 & -1 \\ 2 & 2 & -1 \\ 2 & 2 & 0 \end{pmatrix}.$$

Its characteristic polynomial is

$$\chi_A(x) = -x^3 + 5x^2 - 8x + 4 = -(x - 1)(x - 2)^2.$$

In order to illustrate the Cayley-Hamilton theorem in this case, we could, of course, calculate

$$- \left(A^3 - 5A^2 + 8A - 4I \right)$$

and check that this is equal to the 3×3 zero matrix. But it is possible to simplify this calculation, which is onerous by hand, using the factorization of the polynomial. We then need only calculate $(A - I)(A - 2I)^2$. You can easily check that this also yields the zero matrix. It is important to understand why this works. If $(A - I)(A - 2I)^2$ is expanded using the rules of matrix algebra then the sum

$$A^3 - 5A^2 + 8A - 4I$$

really is obtained. The reason is that A commutes with all powers of itself and any scalar multiples of these powers. In addition, the identity matrix and any scalar multiple of the identity matrix commute with all matrices. Thus the usual rules of algebra apply in this special case.

Finally, we express the inverse of A as a polynomial in A. By the Cayley-Hamilton theorem

$$A^3 - 5A^2 + 8A - 4I = O.$$

Thus

$$I = \frac{1}{4}\left(A^3 - 5A^2 + 8A\right) = A\frac{1}{4}\left(A^2 - 5A + 8I\right).$$

It follows that

$$A^{-1} = \frac{1}{4}\left(A^2 - 5A + 8I\right).$$

This can, of course, be calculated explicitly

$$A^{-1} = \frac{1}{4}\begin{pmatrix} 2 & -2 & 1 \\ -2 & 2 & 1 \\ 0 & -4 & 1 \end{pmatrix}.$$

The goal of the remainder of this section is to explain where the definition of the characteristic polynomial comes from and why it is important. The simplest matrices to work with are the scalar ones λI but the set $\{\lambda I : \lambda \in \mathbb{C}\}$ is nothing other than a disguised version of the complex numbers and so do not get us off the ground. The next simplest matrices are the diagonal ones. A 3×3 diagonal matrix looks like this

$$A = \begin{pmatrix} \lambda_1 & 0 & 0 \\ 0 & \lambda_2 & 0 \\ 0 & 0 & \lambda_3 \end{pmatrix}.$$

It turns out that they are general enough to be useful but not so general that they are hard to compute with. They are certainly easy to compute with. The determinant of A is $\lambda_1\lambda_2\lambda_3$ and if this is non-zero the inverse of A is

$$\begin{pmatrix} \frac{1}{\lambda_1} & 0 & 0 \\ 0 & \frac{1}{\lambda_2} & 0 \\ 0 & 0 & \frac{1}{\lambda_3} \end{pmatrix}.$$

Calculating powers of A is also easy because

$$A^n = \begin{pmatrix} \lambda_1^n & 0 & 0 \\ 0 & \lambda_2^n & 0 \\ 0 & 0 & \lambda_3^n \end{pmatrix}.$$

In fact, anything you want to know about a diagonal matrix can be computed easily and quickly which contrasts with the difficulty of computing with arbitrary matrices. To make things concrete, we focus on the problem of computing powers of a matrix, which has important applications. For example, Theorem 8.2.4 uses powers of a matrix to count the number of paths in a graph. Calculating powers of a matrix takes a lot of work in general, but of course is easy for diagonal matrices. If only all matrices were diagonal. They are not, of course, but this is the germ of an idea that leads to the following definition. It is the most important one of this section.

We say that a square matrix A is *diagonalizable* if there is an invertible matrix P and a diagonal matrix D such that

$$P^{-1}AP = D.$$

We say that P *diagonalizes* A. We claim that for diagonalizable matrices the labour in computing powers can be drastically reduced. To see why, multiply $P^{-1}AP = D$ on the left by P and on the right by P^{-1} to get, after a little simplification, $A = PDP^{-1}$. If both sides of this equation are squared something interesting happens. We have that

$$A^2 = (PDP^{-1})(PDP^{-1}) = PDDP^{-1} = PD^2P^{-1}$$

because the P^{-1} and P cancel each other out. By induction

$$A^n = PD^nP^{-1}.$$

Thus to calculate a large power of A, we simply calculate that power of D, which is easy, and then carry out a couple of matrix multiplications. We have shown that if a matrix is diagonalizable then computing its powers can be simplified.

This looks wonderful in theory but will only be wonderful in practice if we can show that many matrices are diagonalizable and if we can develop simple methods for diagonalizing a matrix if it is indeed diagonalizable. The good news is this can all be done but to do so requires the development of some matrix theory. The characteristic polynomial emerges as the star of the show. It is convenient to make one further definition here. A matrix B is *similar* to a matrix A, denoted by $A \equiv B$, if $B = P^{-1}AP$ for some invertible matrix P. Thus a matrix is diagonalizable if it is similar to a diagonal matrix.

Lemma 8.6.5. *Similar matrices have the same characteristic polynomial.*

Proof. Let $B = P^{-1}AP$. By definition

$$\chi_B(x) = \det(B - xI) = \det(P^{-1}AP - xI)$$

and this is equal to

$$\det\left(P^{-1}(A - xI)P\right) = \det(P^{-1})\det(A - xI)\det(P) = \chi_A(x)$$

using the fact that $I = P^{-1}IP$ and x is a scalar. $\qquad\square$

Suppose that a 3×3 matrix A is similar to the diagonal matrix

$$\begin{pmatrix} \lambda_1 & 0 & 0 \\ 0 & \lambda_2 & 0 \\ 0 & 0 & \lambda_3 \end{pmatrix}.$$

Then by Lemma 8.6.5

$$\chi_A = (\lambda_1 - x)(\lambda_2 - x)(\lambda_3 - x).$$

Thus the numbers $\lambda_1, \lambda_2, \lambda_3$ that appear along the diagonal of the associated diagonal matrix are precisely the roots of the characteristic polynomial of A. The roots of the characteristic polynomial of a matrix A are given a special name: they are called *eigenvalues*.[9] We have therefore proved the following.

---

[9]The word *eigenvalue* is a rare example of an Anglo-German hybrid where *eigen* means *characteristic*. This theory was developed in Germany.

Theorem 8.6.6. *If a matrix A is similar to a diagonal matrix D then the diagonal entries of D are the eigenvalues of A occurring according to their multiplicity.*

Let λ be an eigenvalue of A. Then an *eigenvector* belonging to the eigenvalue λ is any *non-zero* column vector $\mathbf{v}$ such that $A\mathbf{v} = \lambda\mathbf{v}$. Our next result will at last explain where our definition of the characteristic polynomial came from.

Theorem 8.6.7 (Eigenvalues and eigenvectors). *Let A be a square matrix. The following are equivalent for a scalar λ.*

1. *The equation $A\mathbf{x} = \lambda\mathbf{x}$ has non-trivial solutions which are precisely the eigenvectors of A belonging to λ.*

2. *The system $(A - \lambda I)\mathbf{x} = \mathbf{0}$ of homogeneous equations has a non-trivial solution.*

3. *λ is a root of the characteristic polynomial of A.*

Proof. (1)$\Leftrightarrow$(2). The equation $A\mathbf{v} = \lambda\mathbf{v}$ can be written $A\mathbf{v} = \lambda I\mathbf{v}$. This is equivalent to $A\mathbf{v} - \lambda I\mathbf{v} = \mathbf{0}$ which in turn is equivalent to $(A - \lambda I)\mathbf{v} = \mathbf{0}$.

(2)$\Leftrightarrow$(3). The system $(A - \lambda I)\mathbf{x} = \mathbf{0}$ of homogeneous equations has a non-trivial solution if and only if $\det(A - \lambda I) = 0$ by Theorem 8.5.22. But this is equivalent to saying that λ is a root of the characteristic polynomial of A. $\qquad\square$

We now have what we need to give a criterion for a matrix to be diagonalizable.

Theorem 8.6.8 (Diagonalization criterion). *Let A and P be square matrices of the same size. Then A is diagonalized by P if and only if P is an invertible matrix whose columns are eigenvectors of A.*

Proof. We just consider the 3×3 case here, but the argument generalizes and also clearly applies to the 2×2 case. Suppose that the matrix A is diagonalized by P. Then $P^{-1}AP = D$ for some diagonal matrix D with diagonal entries $\lambda_1, \lambda_2, \lambda_3$. Thus $AP = PD$. Regard P as consisting of its columns so $P = (\mathbf{p}_1, \mathbf{p}_2, \mathbf{p}_3)$. We can therefore regard the product AP as being $(A\mathbf{p}_1, A\mathbf{p}_2, A\mathbf{p}_3)$. If we calculate PD we get $(\lambda_1\mathbf{p}_1, \lambda_2\mathbf{p}_2, \lambda_3\mathbf{p}_3)$. Since $AP = PD$, we have that $A\mathbf{p}_i = \lambda_i\mathbf{p}_i$ for $i = 1, 2, 3$. Hence the matrix P consists of column vectors which are eigenvectors. These eigenvectors are linearly independent because P is invertible. The converse is proved by reversing these calculations and using again the fact that a square matrix is invertible if and only if its columns are linearly independent by Theorem 8.5.22. $\qquad\square$

Theorem 8.6.8 raises the question of what properties a matrix A should have in order that we can construct an invertible matrix whose columns are eigenvectors of A. To answer this question fully requires more theory than we describe here. Instead, we derive two useful sufficient conditions for diagonalizability. Another important special case is described in Chapter 10. The following theorem is a key step.

Theorem 8.6.9. *Let A be a square matrix. Let $\lambda_1, \ldots, \lambda_r$ be distinct eigenvalues of A, and let $\mathbf{v}_1, \ldots, \mathbf{v}_r$ be corresponding eigenvectors. Then $\mathbf{v}_1, \ldots, \mathbf{v}_r$ are linearly independent.*

Proof. If $r = 1$ there is nothing to prove. We assume the result is true for $r - 1$ eigenvectors. Let

$$\mu_1 \mathbf{v}_1 + \ldots + \mu_r \mathbf{v}_r = \mathbf{0}. \tag{8.1}$$

Multiply both sides of equation (8.1) by A to get

$$\mu_1 \lambda_1 \mathbf{v}_1 + \ldots + \mu_r \lambda_r \mathbf{v}_r = \mathbf{0}. \tag{8.2}$$

Multiply both sides of equation (8.1) by λ_1 to get

$$\mu_1 \lambda_1 \mathbf{v}_1 + \ldots + \mu_r \lambda_1 \mathbf{v}_r = \mathbf{0}. \tag{8.3}$$

Subtract equation (8.3) from equation (8.2) to get

$$\mu_2 (\lambda_2 - \lambda_1) \mathbf{v}_2 + \ldots + \mu_r (\lambda_r - \lambda_1) \mathbf{v}_r = \mathbf{0}.$$

By the induction hypothesis, the coefficients are zero. This implies that $\mu_2 = \ldots = \mu_r = 0$ because the eigenvalues are distinct. Hence $\mu_1 = 0$ and we have proved that the eigenvectors are linearly independent. □

We can now state two sufficient conditions for diagonalizability. The first is immediate by Theorem 8.6.9 and by Theorem 8.6.8.

Theorem 8.6.10. *Let A be an $n \times n$ matrix. If the characteristic polynomial of A has n distinct roots then A is diagonalizable.*

The second condition also follows by Theorem 8.6.9 and Theorem 8.6.8 and a small extra calculation. The next result is immediate by part (2) of Theorem 8.6.7 and Theorem 8.3.1.

Lemma 8.6.11. *Let A be a square matrix and let λ be an eigenvalue. Any non-zero linear combination of eigenvectors belonging to λ is an eigenvector belonging to λ.*

Thus by Lemma 8.6.11 the set of all eigenvectors belonging to λ together with the zero vector is closed under the taking of arbitrary linear combinations. This set is called the *eigenspace* of λ.

Theorem 8.6.12. *Let A be a 3×3 matrix. Suppose that the eigenvalue λ_1 occurs once and the eigenvalue λ_2 occurs twice. Then A is diagonalizable if and only if there are two linearly independent eigenvectors belonging to λ_2.*

Proof. Let $\mathbf{a}$ be an eigenvector belonging to λ_1 and let $\mathbf{b}, \mathbf{c}$ be linearly independent eigenvectors belonging to λ_2. We prove that $\mathbf{a}, \mathbf{b}, \mathbf{c}$ are linearly independent which, by Theorem 8.6.8, implies that A is diagonalizable. Suppose that $\mu_1 \mathbf{a} + \mu_2 \mathbf{b} + \mu_3 \mathbf{c} = \mathbf{0}$. Consider the linear combination $\mu_2 \mathbf{b} + \mu_3 \mathbf{c}$. If it is the zero vector then $\mu_1 = \mu_2 = 0$ by assumption and therefore $\mu_1 = 0$. If $\mu_2 \mathbf{b} + \mu_3 \mathbf{c}$ is non-zero then it is an eigenvector belonging to λ_2 by Lemma 8.6.11. But then $\mu_1 = 0$ by Theorem 8.6.9, and $\mu_2 \mathbf{b} + \mu_3 \mathbf{c} = \mathbf{0}$. In either event, we deduce that $\mu_1 = \mu_2 = \mu_3 = 0$ and the vectors are linearly independent. □

The following examples illustrate the methods needed to determine whether a 2×2 or 3×3 real matrix is diagonalizable and, if it is, how to find an invertible matrix that does the diagonalizing.

Examples 8.6.13.

1. Let
$$A = \begin{pmatrix} -3 & 5 \\ -2 & 4 \end{pmatrix}.$$

Its characteristic polynomial is $(x-2)(x+1)$. It has two distinct real eigenvalues 2 and -1 and so the matrix can be diagonalized. To find an eigenvector belonging to 2, we need to find a non-zero solution to the system of homogeneous equations $(A - 2I)\mathbf{x} = \mathbf{0}$. To do this, we use Gaussian elimination. The column vector
$$\begin{pmatrix} 1 \\ 1 \end{pmatrix}$$

works. Observe that fractions can always be cleared out since eigenvectors are only determined up to scalar multiples. We repeat the above procedure for the eigenvalue -1 and obtain the corresponding eigenvector
$$\begin{pmatrix} 5 \\ 2 \end{pmatrix}.$$

We accordingly define
$$P = \begin{pmatrix} 1 & 5 \\ 1 & 2 \end{pmatrix}.$$

The order in which the eigenvectors are written only affects the order in which the eigenvalues appear in the diagonalized matrix. Because the eigenvalues are distinct the matrix P is invertible. From the theory
$$P^{-1}AP = \begin{pmatrix} 2 & 0 \\ 0 & -1 \end{pmatrix}.$$

2. Let
$$A = \begin{pmatrix} 1 & -1 \\ 1 & 1 \end{pmatrix}$$

be a real matrix. Its characteristic polynomial is $x^2 - 2x + 2$. This has no real roots and so the matrix cannot be diagonalized.

3. Let
$$A = \begin{pmatrix} 2 & 1 \\ 0 & 2 \end{pmatrix}$$

be a real matrix. Its characteristic polynomial is $(x-2)^2$ and so has the eigenvalue 2 (twice). Any eigenvector belonging to 2 is a scalar multiple of
$$\begin{pmatrix} 1 \\ 0 \end{pmatrix}.$$

We therefore cannot find two linearly independent eigenvectors belonging to 2 and so the matrix is not diagonalizable.

Examples 8.6.14.

1. Let
$$A = \begin{pmatrix} 4 & -8 & -6 \\ 0 & 1 & 0 \\ 1 & 8 & 9 \end{pmatrix}.$$

Its characteristic polynomial is $-x^3 + 14x^2 - 55x + 42$. There are three distinct eigenvalues: 6, 7 and 1. Corresponding eigenvectors are

$$\begin{pmatrix} 3 \\ 0 \\ -1 \end{pmatrix}, \quad \begin{pmatrix} 2 \\ 0 \\ -1 \end{pmatrix} \text{ and } \begin{pmatrix} 8 \\ 15 \\ -16 \end{pmatrix}.$$

The matrix P having these three column vectors as its columns is invertible because the eigenvalues are distinct and so A is diagonalizable.

2. Let
$$A = \begin{pmatrix} -5 & 0 & 6 \\ -3 & 1 & 3 \\ -3 & 0 & 4 \end{pmatrix}.$$

The characteristic polynomial is $-x^3 + 3x - 2$. There are three eigenvalues: 1 (twice) and -2. The eigenvectors belonging to 1 are the non-zero elements of the set

$$\begin{pmatrix} \lambda \\ \mu \\ \lambda \end{pmatrix}$$

where $\lambda, \mu \in \mathbb{R}$. The fact that there are two free variables means that we can find two linearly independent eigenvectors belonging to 1. Corresponding eigenvectors are

$$\begin{pmatrix} 1 \\ 0 \\ 1 \end{pmatrix}, \quad \begin{pmatrix} 0 \\ 1 \\ 0 \end{pmatrix} \text{ and } \begin{pmatrix} 2 \\ 1 \\ 1 \end{pmatrix}$$

where the two eigenvectors belonging to 1 are indeed linearly independent. It follows that the matrix P having these three column vectors as its columns is invertible, and A is diagonalizable.

3. Let
$$A = \begin{pmatrix} 0 & -1 & 1 \\ 1 & 2 & 0 \\ 3 & 1 & 1 \end{pmatrix}.$$

The characteristic polynomial is $-x^3 + 3x^2 - 4$. There are three eigenvalues: -1 and 2 (twice). The eigenvectors belonging to 2 are

$$\begin{pmatrix} 0 \\ \lambda \\ \lambda \end{pmatrix}$$

where $\lambda \in \mathbb{R}$. The fact that there is only one free variable means that we cannot find two linearly independent eigenvectors belonging to 2. Corresponding eigenvectors are

$$\begin{pmatrix} 3 \\ -1 \\ -4 \end{pmatrix} \text{ and } \begin{pmatrix} 0 \\ 1 \\ 1 \end{pmatrix}.$$

We therefore do not have enough linearly independent eigenvectors and so the matrix A is not diagonalizable.

A real matrix can fail to be diagonalizable in two ways. First, it might have complex eigenvalues. This is not really a problem and, as we learnt in Chapter 6, we should be working with complex numbers anyway. Second, even if all its eigenvalues are real, we might not be able to find enough linearly independent eigenvectors. This is a problem with the matrix and means that we need to find an alternative to diagonalization to simplify our calculations. This can be done but just how is left to a course in linear algebra.

Exercises 8.6

1. Let

$$A = \begin{pmatrix} 1 & 2 & 2 \\ 2 & 3 & -2 \\ -5 & 3 & 8 \end{pmatrix} \text{ and } C = \begin{pmatrix} 1 & 2 & 0 \\ 0 & 2 & 1 \\ 1 & 1 & -1 \end{pmatrix}.$$

   (a) Show that C is invertible and calculate its inverse.

   (b) Calculate $C^{-1}AC$.

2. Calculate the characteristic polynomials of the following matrices and determine their eigenvalues.

   (a) $\begin{pmatrix} 3 & 1 \\ 2 & 2 \end{pmatrix}$.

   (b) $\begin{pmatrix} 1 & 2 & 0 \\ 2 & 2 & 2 \\ 0 & 2 & 3 \end{pmatrix}$.

3. Determine which of the following matrices are diagonalizable and, for those that are, diagonalize them.

(a)
$$\begin{pmatrix} 7 & 3 & -9 \\ 0 & 0 & 1 \\ 2 & 0 & -1 \end{pmatrix}.$$

(b)
$$\begin{pmatrix} 2 & 1 & 1 \\ 0 & 0 & -2 \\ 0 & 1 & 3 \end{pmatrix}.$$

(c)
$$\begin{pmatrix} 2 & 2 & -1 \\ 1 & 3 & -1 \\ 2 & 2 & 0 \end{pmatrix}.$$

4. Let $M_n(\mathbb{R})$ be the set of all $n \times n$ real matrices. Prove that the relation of similarity $\equiv$ is an equivalence relation on this set.

5. *This question deals with the properties of the following innocuous looking graph.

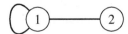

(a) Find the adjacency matrix A of this graph.
(b) Calculate the characteristic polynomial and eigenvalues of A.
(c) Find a matrix P such that $P^{-1}AP = D$, a diagonal matrix.
(d) Hence find a formula for the number of paths from vertex 1 to itself of length n.

6. *Calculate the characteristic polynomial of the following matrix
$$\begin{pmatrix} 0 & 0 & a_0 \\ 1 & 0 & a_1 \\ 0 & 1 & a_2 \end{pmatrix}.$$

What do you deduce?

7. *Let A be a square matrix. Let $A\mathbf{v} = \lambda\mathbf{v}$ and let $f(x)$ be any polynomial. Then $f(A)\mathbf{v} = f(\lambda)\mathbf{v}$.

8. *In this question, we show that the complex numbers could be defined via (real) matrices. Let $\mathscr{C}$ be the set of all real matrices that have the following shape
$$\begin{pmatrix} a & -b \\ b & a \end{pmatrix}.$$

(a) Prove that $\mathscr{C}$ is closed under sum, difference and product.

(b) Prove that matrix multiplication is commutative in $\mathscr{C}$.

(c) Prove that every non-zero matrix in $\mathscr{C}$ is invertible and that its inverse also belongs to $\mathscr{C}$.

(d) Deduce that the set $\mathscr{C}$ with these operations satisfies the axioms (F1)–(F11) from Section 4.3 and so is a field.

(e) It remains to show how $\mathscr{C}$ is related to $\mathbb{C}$. Define

$$\mathbf{1} = \begin{pmatrix} 1 & 0 \\ 0 & 1 \end{pmatrix} \text{ and } \mathbf{i} = \begin{pmatrix} 0 & -1 \\ 1 & 0 \end{pmatrix}.$$

Show that we may write our matrices in the form

$$a\mathbf{1} + b\mathbf{i} \text{ where } \mathbf{i}^2 = -1.$$

8.7 BLANKINSHIP'S ALGORITHM

The ideas of this chapter lead to an alternative, and better, procedure to the one described in Section 5.2 for calculating both the gcd of numbers a and b and the integers x and y such that $\gcd(a,b) = ax + by$. It was first described by W. A. Blankinship [16]. We begin by writing down the following simple equation

$$\begin{pmatrix} 1 & 0 \\ 0 & 1 \end{pmatrix} \begin{pmatrix} a \\ b \end{pmatrix} = \begin{pmatrix} a \\ b \end{pmatrix}$$

and its corresponding augmented matrix

$$\left(\begin{array}{cc|c} 1 & 0 & a \\ 0 & 1 & b \end{array} \right).$$

If we carry out elementary row operations on this matrix, we will not change the solution. The particular elementary row operations used are determined by Euclid's algorithm. Without loss of generality, assume that $a > b$. Then we may write $a = qb + r$ where $r < b$. Accordingly, we carry out the elementary row operation $R_1 \leftarrow R_1 - qR_2$. This leads to the augmented matrix

$$\left(\begin{array}{cc|c} 1 & -q & r \\ 0 & 1 & b \end{array} \right).$$

But we know that $\gcd(a,b) = \gcd(b,r)$ so our next elementary row operation will involve subtracting a multiple of the first row from the second row. In this way, we implement Euclid's algorithm on the numbers to the right of the | whilst at the same time recording what we do in the entries to the left of the |. By Euclid's algorithm one of the numbers on the right will ultimately become zero, and the non-zero number that remains will be $\gcd(a,b)$. For the sake of argument, we shall suppose that we have the following situation

$$\left(\begin{array}{cc|c} u & v & 0 \\ x & y & \gcd(a,b) \end{array} \right).$$

We now recall that because we have been applying elementary row operations the solution space has not changed and this solution space contains only one element. Thus

$$\begin{pmatrix} u & v \\ x & y \end{pmatrix} \begin{pmatrix} a \\ b \end{pmatrix} = \begin{pmatrix} 0 \\ \gcd(a,b) \end{pmatrix}.$$

It follows that

$$ax + by = \gcd(a,b).$$

Example 8.7.1. Calculate x, y such that

$$\gcd(2520, 154) = 2520x + 154y.$$

We start with the matrix

$$\begin{pmatrix} 1 & 0 & | & 2520 \\ 0 & 1 & | & 154 \end{pmatrix}.$$

If we divide 154 into 2520 it goes 16 times plus a remainder. Thus we subtract 16 times the second row from the first to get

$$\begin{pmatrix} 1 & -16 & | & 56 \\ 0 & 1 & | & 154 \end{pmatrix}.$$

We now repeat the process but, since the larger number, 154, is on the bottom, we have to subtract some multiple of the first row from the second. This time we subtract twice the first row from the second to get

$$\begin{pmatrix} 1 & -16 & | & 56 \\ -2 & 33 & | & 42 \end{pmatrix}.$$

Now repeat this procedure to get

$$\begin{pmatrix} 3 & -49 & | & 14 \\ -2 & 33 & | & 42 \end{pmatrix}.$$

And again

$$\begin{pmatrix} 3 & -49 & | & 14 \\ -11 & 180 & | & 0 \end{pmatrix}.$$

The process now terminates because there is a zero in the rightmost column. The non-zero entry in the rightmost column is $\gcd(2520, 154)$. We also know that

$$\begin{pmatrix} 3 & -49 \\ -11 & 180 \end{pmatrix} \begin{pmatrix} 2520 \\ 154 \end{pmatrix} = \begin{pmatrix} 14 \\ 0 \end{pmatrix}.$$

This matrix equation corresponds to two equations. It is the one corresponding to the non-zero value that says

$$14 = 3 \times 2520 - 49 \times 154$$

as required.

Exercises 8.7

1. Use Blankinship's algorithm to find integers x and y such that $\gcd(a,b) = ax + by$ for each of the following pairs of numbers.

   (a) 112, 267.

   (b) 242, 1870.

   (c) 1079, 741.

CHAPTER 9

Vectors

"Let no one ignorant of geometry enter." – Welcome sign to Plato's Academy

Euclid's *Elements* codified what was known about geometry into a handful of axioms and then showed that all of geometry could be deduced from them. His achievement was impressive, but even proving simple results, like Pythagoras' theorem, can take dozens of intermediate results. The *Elements* can be compared to a low-level programming language in which everything has to be spelt out. It was not until the nineteenth century that a more high-level approach to three-dimensional geometry was developed. On the basis of the work carried out by Hamilton on quaternions, about which we say more in Section 9.7, the theory of vectors was developed by Josiah Willard Gibbs (1839–1903) and Oliver Heaviside (1850–1925). Their theory is introduced in this chapter. Pythagoras' theorem can now be proved in a couple of lines. But it should be remembered that the geometry described by Euclid and the geometry described in this chapter are one and the same: only the form of the descriptions differ. We have not attempted to develop the subject in this chapter completely rigorously, so we often make appeals to geometric intuition in setting up the algebraic theory of vectors. If you have a firm grasp of this intuitive theory then a rigorous development can be found in the first four chapters of [103], for example. On the other hand, the theory of vectors developed in this chapter itself provides an alternative axiomatization of geometry.

9.1 VECTORS GEOMETRICALLY

The following ideas in three dimensions will be assumed.

- The notion of a *point*.

- The notion of a *line* and of a *line segment*. Lines are of infinite extent whereas line segments are of finite extent.

- The notion of the *length* of a line segment and the *angle* between two line segments. The angle between two line segments that begin at the same point is the smallest angle between them and so is always between 0 and π.

- The notion of a *directed line segment*. This amounts to drawing an arrow on the line segment pointing in one of two directions.

- The notion of *parallel lines*. This is fundamental to Euclidean geometry and was used in Chapter 2 to prove that the angles in a triangle add up to two right angles.

- The notion of a *plane*.

This chapter is based entirely on the following definition. Two directed line segments which are parallel, have the same length and point in the same direction are said to represent the same *vector*.[1] The word 'vector' means carrier in Latin and what a vector carries is information about length and direction, and nothing else. Because vectors stay the same when they move parallel to themselves, they also preserve information about angles. But crucially they do not contain information about place. This lack will be rectified later when we define position vectors. Vectors are denoted by bold letters $\mathbf{a}, \mathbf{b}, \ldots$. If P and Q are points then the directed line segment from P to Q is written $\overrightarrow{PQ}$. The *zero vector* $\mathbf{0}$ is represented by any degenerate directed line segment $\overrightarrow{PP}$ which is just a point. It has zero length and, exceptionally, no uniquely defined direction. Vectors are denoted by arrows: the vector starts at the base of the arrow (where the feathers would be), we call this the *initial point* of the vector, and ends at the tip (where the arrowhead is), which we call the *terminal point* of the vector. All the directed line segments below represent the same vector.

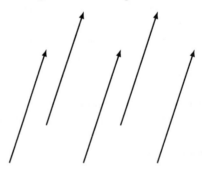

The set of all vectors is denoted by $\mathbb{E}^3$. We now define a binary operation on this set. Let $\mathbf{a}$ and $\mathbf{b}$ be vectors. Their *sum* is defined as follows: slide the vectors parallel to themselves so that the terminal point of $\mathbf{a}$ touches the initial point of $\mathbf{b}$. The directed line segment from the initial point of $\mathbf{a}$ to the terminal point of $\mathbf{b}$ represents the vector $\mathbf{a} + \mathbf{b}$.

---

[1] A vector is really defined to be an equivalence class as in Section 3.5 though this will not be emphasized here. The corresponding equivalence relation is defined on the set of all directed line segments and says that two such segments are related if they are parallel, point in the same direction and have the same length. It amounts to the same thing to regard a vector as being a directed line segment that can be moved parallel to itself but this viewpoint is more intuitively appealing.

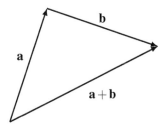

If **a** is a vector, then −**a** is defined to be the vector with the same length as **a** but pointing in the opposite direction.

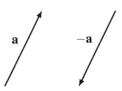

Theorem 9.1.1 (Properties of vector addition).

    1. $\mathbf{a} + (\mathbf{b} + \mathbf{c}) = (\mathbf{a} + \mathbf{b}) + \mathbf{c}$.

    2. $\mathbf{0} + \mathbf{a} = \mathbf{a} = \mathbf{a} + \mathbf{0}$.

    3. $\mathbf{a} + (-\mathbf{a}) = \mathbf{0} = (-\mathbf{a}) + \mathbf{a}$.

    4. $\mathbf{a} + \mathbf{b} = \mathbf{b} + \mathbf{a}$.

Proof. We show first that the definition of vector addition makes sense.[2] With reference to the diagram below, let A, B and C be three points in space, and let A', B' and C' be three other points. Suppose in addition that AB is parallel to and the same length as $A'B'$, and that BC is parallel to and the same length as $B'C'$. Then it is intuitively clear that AC is parallel to and the same length as $A'C'$ and both line segments point in the same direction.

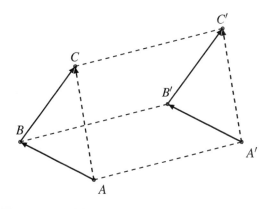

---

[2]Because a vector is actually an equivalence class and so we have a choice of representatives.

(1) The idea behind the proof of associativity is illustrated in the diagram below. We choose directed line segments such that $\mathbf{a} = \overrightarrow{AB}$, $\mathbf{b} = \overrightarrow{BC}$ and $\mathbf{c} = \overrightarrow{CD}$. The directed line segment $\overrightarrow{AD}$ represents both $(\mathbf{a} + \mathbf{b}) + \mathbf{c}$, and $\mathbf{a} + (\mathbf{b} + \mathbf{c})$.

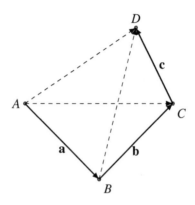

(4) The idea behind the proof of commutativity is illustrated in the diagram below. The directed line segments $\overrightarrow{AB}$ and $\overrightarrow{DC}$ represent $\mathbf{a}$ and the directed line segments $\overrightarrow{BC}$ and $\overrightarrow{AD}$ represent $\mathbf{b}$. The directed line segment $\overrightarrow{AC}$ represents both $\mathbf{a} + \mathbf{b}$ and $\mathbf{b} + \mathbf{a}$.

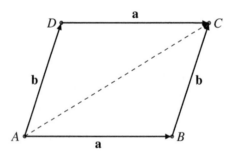

The proofs of (2) and (3) are straightforward. ☐

As usual, subtraction is defined in terms of addition

$$\mathbf{a} - \mathbf{b} = \mathbf{a} + (-\mathbf{b}).$$

Example 9.1.2. Consider the following square, though any other polygon would work in a similar way, and choose vectors as shown.

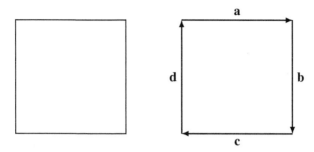

Then $\mathbf{a} + \mathbf{b} + \mathbf{c} + \mathbf{d} = \mathbf{0}$. This is a property useful in solving a number of geometric problems.

Let $\mathbf{a}$ be a vector. Denote by $\|\mathbf{a}\|$ the *length* of the vector. If $\|\mathbf{a}\| = 1$ then $\mathbf{a}$ is called a *unit vector*. We always have $\|\mathbf{a}\| \geq 0$ with $\|\mathbf{a}\| = 0$ if and only if $\mathbf{a} = \mathbf{0}$. By results on triangles

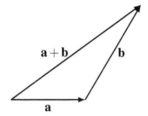

the *triangle inequality* $\|\mathbf{a} + \mathbf{b}\| \leq \|\mathbf{a}\| + \|\mathbf{b}\|$ holds.

We now define *multiplication of a vector by a scalar*. Let λ be a scalar and $\mathbf{a}$ a vector. If $\lambda = 0$ define $\lambda \mathbf{a} = \mathbf{0}$. If $\lambda > 0$ define $\lambda \mathbf{a}$ to have the same direction as $\mathbf{a}$ and length $\lambda \|\mathbf{a}\|$. If $\lambda < 0$ define $\lambda \mathbf{a}$ to have the opposite direction to $\mathbf{a}$ and length $(-\lambda) \|\mathbf{a}\|$. Observe that in all cases

$$\|\lambda \mathbf{a}\| = |\lambda| \, \|\mathbf{a}\| \,.$$

If $\mathbf{a}$ is non-zero, define

$$\hat{\mathbf{a}} = \frac{\mathbf{a}}{\|\mathbf{a}\|},$$

a unit vector in the same direction as $\mathbf{a}$. The process of constructing $\hat{\mathbf{a}}$ from $\mathbf{a}$ is called *normalization*. This will be important in Chapter 10. Vectors that differ by a scalar multiple are said to be *parallel*. The following are all intuitively plausible and the proofs are omitted.

Theorem 9.1.3 (Properties of scalar multiplication). *Let λ and μ be scalars.*

1. $0\mathbf{a} = \mathbf{0}$.

2. $1\mathbf{a} = \mathbf{a}$.

3. $(-1)\mathbf{a} = -\mathbf{a}$.

4. $(\lambda + \mu)\mathbf{a} = \lambda\mathbf{a} + \mu\mathbf{a}$.

5. $\lambda(\mathbf{a} + \mathbf{b}) = \lambda\mathbf{a} + \lambda\mathbf{b}$.

6. $\lambda(\mu\mathbf{a}) = (\lambda\mu)\mathbf{a}$.

Even with the little we have so far introduced, we can prove some simple geometric theorems.

Example 9.1.4. We prove that if the midpoints of the consecutive sides of any quadrilateral are joined by line segments, then the resulting quadrilateral is a parallelogram. We refer to the picture below.

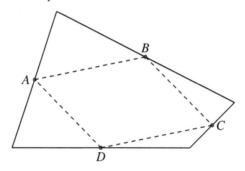

Choose vectors as shown in the diagram below.

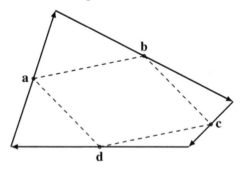

Then $\mathbf{a} + \mathbf{b} + \mathbf{c} + \mathbf{d} = \mathbf{0}$. Thus $\mathbf{a} + \mathbf{b} = -(\mathbf{c} + \mathbf{d})$. Now

$$\overrightarrow{AB} = \tfrac{1}{2}\mathbf{a} + \tfrac{1}{2}\mathbf{b} \text{ and } \overrightarrow{CD} = \tfrac{1}{2}\mathbf{c} + \tfrac{1}{2}\mathbf{d}.$$

It follows that $\overrightarrow{AB} = -\overrightarrow{CD}$. Hence the line segment AB is parallel to the line segment CD and they have the same length. Similarly, BC is parallel to AD and they have the same length.

We now introduce a notion that will enable us to measure angles. It is based on the idea of the perpendicular projection of a line segment onto another line. Let $\mathbf{a}$ and $\mathbf{b}$ be two vectors. If $\mathbf{a}, \mathbf{b} \neq \mathbf{0}$ denote the angle between them by θ and define

$$\mathbf{a} \cdot \mathbf{b} = \|\mathbf{a}\| \|\mathbf{b}\| \cos\theta.$$

If either **a** or **b** is zero, define $\mathbf{a} \cdot \mathbf{b} = 0$. We call $\mathbf{a} \cdot \mathbf{b}$ the *inner product* of **a** and **b**. It is important to remember that it is a scalar and not a vector. The inner product $\mathbf{a} \cdot \mathbf{a}$ is abbreviated $\mathbf{a}^2$. We say that non-zero vectors **a** and **b** are *orthogonal* if the angle between them is a right angle. The key property of the inner product is that for non-zero **a** and **b** we have that $\mathbf{a} \cdot \mathbf{b} = 0$ if and only if **a** and **b** are orthogonal.

Theorem 9.1.5 (Properties of the inner product).

1. $\mathbf{a} \cdot \mathbf{b} = \mathbf{b} \cdot \mathbf{a}$.

2. $\mathbf{a} \cdot \mathbf{a} = \|\mathbf{a}\|^2$.

3. $\lambda(\mathbf{a} \cdot \mathbf{b}) = (\lambda \mathbf{a}) \cdot \mathbf{b} = \mathbf{a} \cdot (\lambda \mathbf{b})$ *for any scalar* λ.

4. $\mathbf{a} \cdot (\mathbf{b} + \mathbf{c}) = \mathbf{a} \cdot \mathbf{b} + \mathbf{a} \cdot \mathbf{c}$.

Proof. The proofs of (1), (2) and (3) are straightforward. We outline the proof of (4). Let **x** and **y** be a pair of vectors. Define *the component of* **x** *in the direction of* **y**, written $\mathrm{comp}(\mathbf{x}, \mathbf{y})$, to be the number $\|\mathbf{x}\| \cos \theta$ where θ is the angle between **x** and **y**.

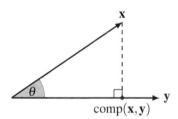

Clearly $\mathbf{x} \cdot \mathbf{y} = \|\mathbf{y}\| \mathrm{comp}(\mathbf{x}, \mathbf{y})$. The inner product of two vectors is therefore a symmetrized version of this component operation. Geometry shows that

$$\mathrm{comp}(\mathbf{b} + \mathbf{c}, \mathbf{a}) = \mathrm{comp}(\mathbf{b}, \mathbf{a}) + \mathrm{comp}(\mathbf{c}, \mathbf{a}).$$

We therefore have

$$
\begin{aligned}
(\mathbf{b} + \mathbf{c}) \cdot \mathbf{a} &= \|\mathbf{a}\| \mathrm{comp}(\mathbf{b} + \mathbf{c}, \mathbf{a}) \\
&= \|\mathbf{a}\| \mathrm{comp}(\mathbf{b}, \mathbf{a}) + \|\mathbf{a}\| \mathrm{comp}(\mathbf{c}, \mathbf{a}) \\
&= \mathbf{b} \cdot \mathbf{a} + \mathbf{c} \cdot \mathbf{a}.
\end{aligned}
$$

$\square$

The inner product enables us to prove more interesting theorems.

Example 9.1.6. Draw a semicircle. Choose any point on the circumference of the semicircle and join it to the points at either end of the diameter of the semicircle. We claim that the resulting triangle is right-angled.

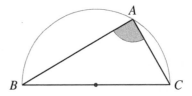

We are interested in the angle formed by AB and AC. We choose vectors as shown in the diagram below.

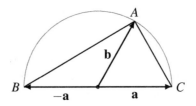

Observe that $\overrightarrow{AB} = -(\mathbf{a}+\mathbf{b})$ and $\overrightarrow{AC} = \mathbf{a}-\mathbf{b}$. Thus

$$
\begin{aligned}
\overrightarrow{AB} \cdot \overrightarrow{AC} &= -(\mathbf{a}+\mathbf{b}) \cdot (\mathbf{a}-\mathbf{b}) \\
&= -(\mathbf{a}^2 - \mathbf{a}\cdot\mathbf{b} + \mathbf{b}\cdot\mathbf{a} - \mathbf{b}^2) \\
&= -(\mathbf{a}^2 - \mathbf{b}^2) \\
&= 0
\end{aligned}
$$

where we have used the properties given in Theorem 9.1.5, such as $\mathbf{a}\cdot\mathbf{b} = \mathbf{b}\cdot\mathbf{a}$, and the fact that $\|\mathbf{a}\| = \|\mathbf{b}\|$, because this is just the radius of the semicircle. It follows that the angle $B\hat{A}C$ is a right angle, as claimed.

Example 9.1.7. We prove Pythagoras' theorem, that $a^2 + b^2 = c^2$ in the triangle below, using vectors in just a few lines.

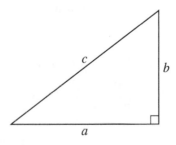

Choose vectors as shown in the diagram below.

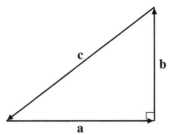

Then $\mathbf{a} + \mathbf{b} + \mathbf{c} = \mathbf{0}$. Thus $\mathbf{a} + \mathbf{b} = -\mathbf{c}$. Now

$$(\mathbf{a} + \mathbf{b})^2 = (-\mathbf{c}) \cdot (-\mathbf{c}) = \|\mathbf{c}\|^2$$

and

$$(\mathbf{a} + \mathbf{b})^2 = \|\mathbf{a}\|^2 + 2\mathbf{a} \cdot \mathbf{b} + \|\mathbf{b}\|^2.$$

This is equal to $\|\mathbf{a}\|^2 + \|\mathbf{b}\|^2$ because $\mathbf{a} \cdot \mathbf{b} = 0$. It follows that

$$\|\mathbf{a}\|^2 + \|\mathbf{b}\|^2 = \|\mathbf{c}\|^2.$$

We define another binary operation on the set of vectors. Let $\mathbf{a}$ and $\mathbf{b}$ be vectors. A unique vector $\mathbf{a} \times \mathbf{b}$ is defined in terms of these two which contains information about the area enclosed by $\mathbf{a}$ and $\mathbf{b}$ and about the orientation in space of the plane determined by $\mathbf{a}$ and $\mathbf{b}$. If either $\mathbf{a} = \mathbf{0}$ or $\mathbf{b} = \mathbf{0}$, define $\mathbf{a} \times \mathbf{b} = \mathbf{0}$. Let θ be the angle between $\mathbf{a}$ and $\mathbf{b}$. If $\theta = 0$ define $\mathbf{a} \times \mathbf{b} = \mathbf{0}$. There is no loss of generality in assuming that $\mathbf{a}$ and $\mathbf{b}$ lie in the plane of the page.

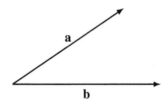

These two vectors determine a unique parallelogram.

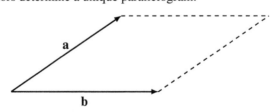

With reference to the diagram below

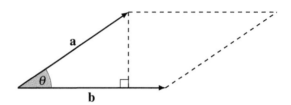

the area enclosed is $\|\mathbf{a}\| \|\mathbf{b}\| \sin\theta$ where θ is the angle between $\mathbf{a}$ and $\mathbf{b}$. For the unit vector $\mathbf{n}$, we need a vector orthogonal to both $\mathbf{a}$ and $\mathbf{b}$. There are only two choices: either we choose the vector pointing out of the page or we choose the vector pointing into the page. We choose the vector pointing into the page. With all this in mind, define

$$\mathbf{a} \times \mathbf{b} = (\|\mathbf{a}\| \|\mathbf{b}\| \sin\theta)\,\mathbf{n}.$$

We call $\mathbf{a} \times \mathbf{b}$ the *vector product* of $\mathbf{a}$ and $\mathbf{b}$. The key property of the vector product is that for non-zero vectors $\mathbf{a} \times \mathbf{b} = \mathbf{0}$ if and only if $\mathbf{a}$ and $\mathbf{b}$ are parallel.

Theorem 9.1.8 (Properties of the vector product).

1. $\mathbf{a} \times \mathbf{b} = -\mathbf{b} \times \mathbf{a}$.

2. $\lambda(\mathbf{a} \times \mathbf{b}) = (\lambda\mathbf{a}) \times \mathbf{b} = \mathbf{a} \times (\lambda\mathbf{b})$ *for any scalar* λ.

3. $\mathbf{a} \times (\mathbf{b} + \mathbf{c}) = \mathbf{a} \times \mathbf{b} + \mathbf{a} \times \mathbf{c}$.

Proof. The proofs of (1) and (2) are straightforward. We prove (3). The vector product was defined in terms of geometry and so we have to prove this property by means of geometry.

Fix the vector $\mathbf{a}$. It is no loss of generality in assuming that it is orthogonal to the page and pointing at you gentle reader. By part (2), we can also assume that it has unit length. Let $\mathbf{x}$ be any vector parallel to the page. Then $\mathbf{a} \times \mathbf{x}$ is obtained from $\mathbf{x}$ by an anticlockwise rotation by a right angle. Suppose now that $\mathbf{x}$ and $\mathbf{y}$ are two vectors parallel to the page. Recall that their sum is constructed from the parallelogram they determine. Thus $\mathbf{a} \times (\mathbf{x} + \mathbf{y}) = \mathbf{a} \times \mathbf{x} + \mathbf{a} \times \mathbf{y}$, since the parallelogram is simply rotated.

Let $\mathbf{u}$ be any vector. Position it so its base rests on the plane. We project the tip straight down onto the plane. In this way, we obtain a vector parallel to the plane, denoted by $\mathbf{u}'$, such that $\mathbf{u} = \lambda\mathbf{a} + \mathbf{u}'$ for some scalar λ. We say that $\mathbf{u}'$ is obtained from $\mathbf{u}$ by orthogonal projection. It is clear on geometrical grounds that $(\mathbf{u} + \mathbf{v})' = \mathbf{u}' + \mathbf{v}'$ since parallelograms are projected to parallelograms. From the definition of the vector product, $\mathbf{a} \times \mathbf{u} = \mathbf{a} \times \mathbf{u}'$.

We now have enough to prove the result by assembling these separate observations. We have that

$$\mathbf{a} \times (\mathbf{b} + \mathbf{c}) = \mathbf{a} \times (\mathbf{b} + \mathbf{c})' = \mathbf{a} \times (\mathbf{b}' + \mathbf{c}') = \mathbf{a} \times \mathbf{b}' + \mathbf{a} \times \mathbf{c}' = \mathbf{a} \times \mathbf{b} + \mathbf{a} \times \mathbf{c}.$$

$\square$

Observe that $\mathbf{a} \times (\mathbf{b} \times \mathbf{c}) \neq (\mathbf{a} \times \mathbf{b}) \times \mathbf{c}$ in general. In other words, the vector product is not associative. This is shown in the exercises by means of a counter-example.

Example 9.1.9. We prove the *law of sines* for triangles using the vector product. With reference to the diagram below

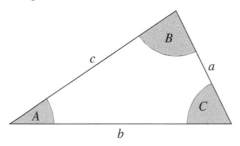

this states that

$$\frac{\sin A}{a} = \frac{\sin B}{b} = \frac{\sin C}{c}.$$

Choose vectors as shown.

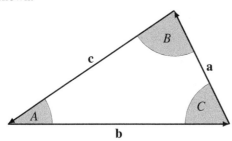

Then $\mathbf{a} + \mathbf{b} + \mathbf{c} = \mathbf{0}$ and so $\mathbf{a} + \mathbf{b} = -\mathbf{c}$. Take the vector product of this equation on both sides on the left with $\mathbf{a}$ and $\mathbf{b}$ in turn to obtain:

1. $\mathbf{a} \times \mathbf{b} = \mathbf{a} \times (-\mathbf{c})$.

2. $\mathbf{b} \times \mathbf{a} = \mathbf{b} \times (-\mathbf{c})$.

From (1), we get

$$\|(-\mathbf{b}) \times \mathbf{a}\| = \|(-\mathbf{a}) \times \mathbf{c}\|$$

which simplifies to

$$\|\mathbf{b}\| \sin C = \|\mathbf{c}\| \sin B$$

which yields the equation

$$\frac{\sin B}{b} = \frac{\sin C}{c}.$$

From (2), we obtain in a similar way the second equation

$$\frac{\sin A}{a} = \frac{\sin C}{c}.$$

The set of vectors $\mathbb{E}^3$ equipped with the operations defined in this section is called *three-dimensional Euclidean space*. It is the space of the *Elements* in modern dress.

Exercises 9.1

1. Consider the following diagram.

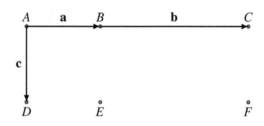

Now answer the following questions.

(a) Write the vector $\overrightarrow{BD}$ in terms of **a** and **c**.

(b) Write the vector $\overrightarrow{AE}$ in terms of **a** and **c**.

(c) What is the vector $\overrightarrow{DE}$?

(d) What is the vector $\overrightarrow{CF}$?

(e) What is the vector $\overrightarrow{AC}$?

(f) What is the vector $\overrightarrow{BF}$?

2. If **a**, **b**, **c** and **d** represent the consecutive sides of a quadrilateral, show that the quadrilateral is a parallelogram if and only if $\mathbf{a} + \mathbf{c} = \mathbf{0}$.

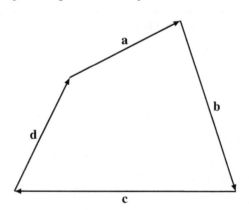

3. In the regular pentagon $ABCDE$, let $\overrightarrow{AB} = \mathbf{a}$, $\overrightarrow{BC} = \mathbf{b}$, $\overrightarrow{CD} = \mathbf{c}$ and $\overrightarrow{DE} = \mathbf{d}$. Express $\overrightarrow{EA}$, $\overrightarrow{DA}$, $\overrightarrow{DB}$, $\overrightarrow{CA}$, $\overrightarrow{EC}$, $\overrightarrow{BE}$ in terms of **a**, **b**, **c** and **d**.

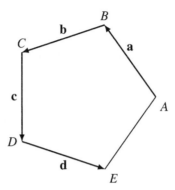

4. Let **a** and **b** represent adjacent sides of a regular hexagon so that the initial point of **b** is the terminal point of **a**. Represent the remaining sides by means of vectors expressed in terms of **a** and **b**.

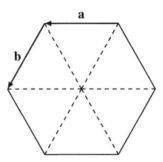

5. Prove that $\|\mathbf{a}\|\,\mathbf{b} + \|\mathbf{b}\|\,\mathbf{a}$ is orthogonal to $\|\mathbf{a}\|\,\mathbf{b} - \|\mathbf{b}\|\,\mathbf{a}$ for all vectors **a** and **b**.

6. Let **a** and **b** be two non-zero vectors. Define

$$\mathbf{u} = \left(\frac{\mathbf{a}\cdot\mathbf{b}}{\mathbf{a}\cdot\mathbf{a}}\right)\mathbf{a}.$$

Show that $\mathbf{b} - \mathbf{u}$ is orthogonal to **a**.

7. (a) Simplify $(\mathbf{u}+\mathbf{v})\times(\mathbf{u}+\mathbf{v})$.

   (b) Simplify $(\mathbf{u}+\mathbf{v})\times(\mathbf{u}-\mathbf{v})$.

8. Let **a** and **b** be two unit vectors, the angle between them being $\frac{\pi}{3}$. Show that $2\mathbf{b}-\mathbf{a}$ and **a** are orthogonal.

9. Prove that
$$\|\mathbf{u}-\mathbf{v}\|^2 + \|\mathbf{u}+\mathbf{v}\|^2 = 2(\|\mathbf{u}\|^2 + \|\mathbf{v}\|^2).$$

Deduce that the sum of the squares of the diagonals of a parallelogram is equal to the sum of the squares of all four sides.

9.2 VECTORS ALGEBRAICALLY

The theory introduced in Section 9.1 is useful for proving general results about geometry, but does not enable us to calculate with specific vectors. To do this we need coordinates. Set up a *cartesian coordinate system* consisting of x-, y- and z-axes. We orient the system so that in rotating the x-axis clockwise to the y-axis, we are looking in the direction of the positive z-axis. Let $\mathbf{i}$, $\mathbf{j}$ and $\mathbf{k}$ be unit vectors parallel to the x-, y- and z-axes, respectively, pointing in the positive directions.

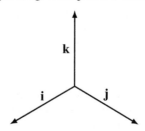

Every vector $\mathbf{a}$ can now be uniquely written in the form

$$\mathbf{a} = a_1\mathbf{i} + a_2\mathbf{j} + a_3\mathbf{k}$$

for some scalars a_1, a_2, a_3. This is achieved by *orthogonal projection* of the vector $\mathbf{a}$ (moved so that it starts at the origin) onto each of the three coordinate axes. The numbers a_i are called the *components* of $\mathbf{a}$ in each of the three directions. The proofs of the following can be deduced from the properties of vector addition and scalar multiplication.

Theorem 9.2.1.

1. *If* $\mathbf{a} = a_1\mathbf{i} + a_2\mathbf{j} + a_3\mathbf{k}$ *and* $\mathbf{b} = b_1\mathbf{i} + b_2\mathbf{j} + b_3\mathbf{k}$ *then* $\mathbf{a} = \mathbf{b}$ *if and only if* $a_i = b_i$ *for* $1 \leq i \leq 3$. *That is, corresponding components are equal.*

2. $\mathbf{0} = 0\mathbf{i} + 0\mathbf{j} + 0\mathbf{k}$.

3. *If* $\mathbf{a} = a_1\mathbf{i} + a_2\mathbf{j} + a_3\mathbf{k}$ *and* $\mathbf{b} = b_1\mathbf{i} + b_2\mathbf{j} + b_3\mathbf{k}$ *then*

$$\mathbf{a} + \mathbf{b} = (a_1 + b_1)\mathbf{i} + (a_2 + b_2)\mathbf{j} + (a_3 + b_3)\mathbf{k}.$$

4. *If* $\mathbf{a} = a_1\mathbf{i} + a_2\mathbf{j} + a_3\mathbf{k}$ *then* $\lambda\mathbf{a} = \lambda a_1\mathbf{i} + \lambda a_2\mathbf{j} + \lambda a_3\mathbf{k}$ *for any scalar* λ.

The vector $\mathbf{a} = a_1\mathbf{i} + a_2\mathbf{j} + a_3\mathbf{k}$ can equally well be represented by the column vector

$$\begin{pmatrix} a_1 \\ a_2 \\ a_3 \end{pmatrix}.$$

In this notation, $\mathbf{i}$, $\mathbf{j}$ and $\mathbf{k}$ can be represented by the column vectors

$$\mathbf{e}_1 = \begin{pmatrix} 1 \\ 0 \\ 0 \end{pmatrix}, \quad \mathbf{e}_2 = \begin{pmatrix} 0 \\ 1 \\ 0 \end{pmatrix} \text{ and } \mathbf{e}_3 = \begin{pmatrix} 0 \\ 0 \\ 1 \end{pmatrix},$$

respectively. The choice of notation is sometimes down to taste and sometimes down to context. In vector analysis, for example, the $\mathbf{i}, \mathbf{j}, \mathbf{k}$ notation is used. When we come to discuss matrices and vectors, however, the column vector notation is more convenient.[3]

The form taken by the inner and vector products in coordinates can now be revealed.

Theorem 9.2.2 (Inner products). *Let* $\mathbf{a} = a_1\mathbf{i} + a_2\mathbf{j} + a_3\mathbf{k}$ *and* $\mathbf{b} = b_1\mathbf{i} + b_2\mathbf{j} + b_3\mathbf{k}$. *Then*

$$\mathbf{a} \cdot \mathbf{b} = a_1b_1 + a_2b_2 + a_3b_3.$$

Proof. This is proved using part (4) of Theorem 9.1.5 and the following table

·	**i**	**j**	**k**
i	1	0	0
j	0	1	0
k	0	0	1

computed from the definition of the inner product. We have that

$$\mathbf{a} \cdot \mathbf{b} = \mathbf{a} \cdot (b_1\mathbf{i} + b_2\mathbf{j} + b_3\mathbf{k}) = b_1(\mathbf{a} \cdot \mathbf{i}) + b_2(\mathbf{a} \cdot \mathbf{j}) + b_3(\mathbf{a} \cdot \mathbf{k}).$$

Compute $\mathbf{a} \cdot \mathbf{i}$, $\mathbf{a} \cdot \mathbf{j}$ and $\mathbf{a} \cdot \mathbf{k}$ in turn to get a_1, a_2 and a_3 respectively. Putting everything together we obtain

$$\mathbf{a} \cdot \mathbf{b} = a_1b_1 + a_2b_2 + a_3b_3,$$

as required. □

If $\mathbf{a} = a_1\mathbf{i} + a_2\mathbf{j} + a_3\mathbf{k}$ then $\|\mathbf{a}\| = \sqrt{a_1^2 + a_2^2 + a_3^2}$.

Theorem 9.2.3 (Vector products). *Let* $\mathbf{a} = a_1\mathbf{i} + a_2\mathbf{j} + a_3\mathbf{k}$ *and* $\mathbf{b} = b_1\mathbf{i} + b_2\mathbf{j} + b_3\mathbf{k}$. *Then*

$$\mathbf{a} \times \mathbf{b} = \begin{vmatrix} \mathbf{i} & \mathbf{j} & \mathbf{k} \\ a_1 & a_2 & a_3 \\ b_1 & b_2 & b_3 \end{vmatrix}.$$

It is important to note that this 'determinant' can only be expanded along the first row.

Proof. This follows by part (3) of Theorem 9.1.8 and the following table

×	**i**	**j**	**k**
i	**0**	**k**	**−j**
j	**−k**	**0**	**i**
k	**j**	**−i**	**0**

---

[3]Some books use $\mathbf{i}, \mathbf{j}, \mathbf{k}$ for free vectors and the column vector notation for bound vectors. I don't. I shall explain the difference between free and bound vectors later.

computed from the definition of the vector product. By distributivity

$$\mathbf{a} \times \mathbf{b} = \mathbf{a} \times (b_1\mathbf{i} + b_2\mathbf{j} + b_3\mathbf{k}) = b_1(\mathbf{a} \times \mathbf{i}) + b_2(\mathbf{a} \times \mathbf{j}) + b_3(\mathbf{a} \times \mathbf{k}).$$

Compute $\mathbf{a} \times \mathbf{i}$, $\mathbf{a} \times \mathbf{j}$ and $\mathbf{a} \times \mathbf{k}$ in turn to get

- $\mathbf{a} \times \mathbf{i} = -a_2\mathbf{k} + a_3\mathbf{j}$.
- $\mathbf{a} \times \mathbf{j} = a_1\mathbf{k} - a_3\mathbf{i}$.
- $\mathbf{a} \times \mathbf{k} = -a_1\mathbf{j} + a_2\mathbf{i}$.

Putting everything together we obtain

$$\mathbf{a} \times \mathbf{b} = (a_2b_3 - a_3b_2)\mathbf{i} - (a_1b_3 - a_3b_1)\mathbf{j} + (a_1b_2 - a_2b_1)\mathbf{k}$$

which is equal to the given determinant. □

In Section 9.1, we defined vectors and vector operations geometrically whereas in Section 9.2, we showed that once a coordinate system had been chosen vectors and vector operations could be described algebraically. The important point to remember in what follows is that the two approaches give the same answers.

Exercises 9.2

1. Let $\mathbf{a} = 2\mathbf{i} + \mathbf{j} + \mathbf{k}$ and $\mathbf{b} = \mathbf{i} + 2\mathbf{j} + 3\mathbf{k}$.

    (a) Calculate $\|\mathbf{a}\|$ and $\|\mathbf{b}\|$.
    (b) Calculate $\mathbf{a} + \mathbf{b}$.
    (c) Calculate $\mathbf{a} - \mathbf{b}$.
    (d) Calculate $\mathbf{a} \cdot \mathbf{b}$.
    (e) Calculate the angle between $\mathbf{a}$ and $\mathbf{b}$ to the nearest degree.
    (f) Calculate $\mathbf{a} \times \mathbf{b}$.
    (g) Find a unit vector orthogonal to both $\mathbf{a}$ and $\mathbf{b}$.

2. Calculate $\mathbf{i} \times (\mathbf{i} \times \mathbf{k})$ and $(\mathbf{i} \times \mathbf{i}) \times \mathbf{k}$. What do you deduce as a result of this?

3. Calculate $\mathbf{a} \cdot (\mathbf{a} \times \mathbf{b})$ and $\mathbf{b} \cdot (\mathbf{a} \times \mathbf{b})$.

4. The unit cube is determined by the three vectors $\mathbf{i}$, $\mathbf{j}$ and $\mathbf{k}$. Find the angle between the long diagonal of the unit cube and one of its edges.

5. Calculate $\mathbf{u} \cdot (\mathbf{v} \times \mathbf{w})$ where $\mathbf{u} = 3\mathbf{i} - 2\mathbf{j} - 5\mathbf{k}$, $\mathbf{v} = \mathbf{i} + 4\mathbf{j} - 4\mathbf{k}$ and $\mathbf{w} = 3\mathbf{j} + 2\mathbf{k}$.

9.3 GEOMETRIC MEANING OF DETERMINANTS

We begin by defining a third product that is simply a combination of inner and vector products. Let $\mathbf{a}$, $\mathbf{b}$ and $\mathbf{c}$ be three vectors. Then $\mathbf{b} \times \mathbf{c}$ is a vector and so $\mathbf{a} \cdot (\mathbf{b} \times \mathbf{c})$ is a scalar. Define $[\mathbf{a}, \mathbf{b}, \mathbf{c}] = \mathbf{a} \cdot (\mathbf{b} \times \mathbf{c})$. This is called the *scalar triple product*. As we now show, it is the determinant wearing a false beard.

Theorem 9.3.1 (Scalar triple products and determinants). *Let*

$$\mathbf{a} = a_1\mathbf{i} + a_2\mathbf{j} + a_3\mathbf{k},\ \mathbf{b} = b_1\mathbf{i} + b_2\mathbf{j} + b_3\mathbf{k} \text{ and } \mathbf{c} = c_1\mathbf{i} + c_2\mathbf{j} + c_3\mathbf{k}.$$

Then

$$[\mathbf{a}, \mathbf{b}, \mathbf{c}] = \begin{vmatrix} a_1 & a_2 & a_3 \\ b_1 & b_2 & b_3 \\ c_1 & c_2 & c_3 \end{vmatrix} = \begin{vmatrix} a_1 & b_1 & c_1 \\ a_2 & b_2 & c_2 \\ a_3 & b_3 & c_3 \end{vmatrix}.$$

Thus the properties of scalar triple products are the same as the properties of 3×3 determinants.

Proof. Calculate $\mathbf{a} \cdot (\mathbf{b} \times \mathbf{c})$. By definition this is

$$(a_1\mathbf{i} + a_2\mathbf{j} + a_3\mathbf{k}) \cdot [(b_2c_3 - b_3c_2)\mathbf{i} - (b_1c_3 - b_3c_1)\mathbf{j} + (b_1c_2 - b_2c_1)\mathbf{k}].$$

This is equal to

$$a_1(b_2c_3 - b_3c_2) - a_2(b_1c_3 - b_3c_1) + a_3(b_1c_2 - b_2c_1)$$

which is nothing other than

$$\begin{vmatrix} a_1 & a_2 & a_3 \\ b_1 & b_2 & b_3 \\ c_1 & c_2 & c_3 \end{vmatrix}.$$

The second equality follows from the fact that the determinant of the transpose of a matrix and the determinant of the original matrix are the same. □

The connection between determinants and scalar triple products will enable us to describe the geometric meaning of determinants. Start with 1×1 matrices. The determinant of the matrix (a) is just a. The *length* of a is $|a|$, the absolute value of the determinant of (a).

Theorem 9.3.2 (2×2 determinants). *Let $\mathbf{a} = a_1\mathbf{i} + a_2\mathbf{j}$ and $\mathbf{b} = b_1\mathbf{i} + b_2\mathbf{j}$ be a pair of plane vectors. Then the area of the parallelogram determined by these vectors is the absolute value of the determinant*

$$\begin{vmatrix} a_1 & b_1 \\ a_2 & b_2 \end{vmatrix}.$$

Proof. The area of the parallelogram is the absolute value of $\|\mathbf{a} \times \mathbf{b}\|$ from the definition of the vector product. Calculate

$$\mathbf{a} \times \mathbf{b} = \begin{vmatrix} a_1 & a_2 \\ b_1 & b_2 \end{vmatrix} \mathbf{k}.$$

The result now follows by the result that a matrix and its transpose have the same determinant. □

Theorem 9.3.3 (3 × 3 determinants). *Let*

$$\mathbf{a} = a_1\mathbf{i} + a_2\mathbf{j} + a_3\mathbf{k}, \ \mathbf{b} = b_1\mathbf{i} + b_2\mathbf{j} + b_3\mathbf{k} \ and \ \mathbf{c} = c_1\mathbf{i} + c_2\mathbf{j} + c_3\mathbf{k}$$

be three vectors. Then the volume of the parallelepiped[4] determined by these three vectors is the absolute value of the determinant

$$\begin{vmatrix} a_1 & b_1 & c_1 \\ a_2 & b_2 & c_2 \\ a_3 & b_3 & c_3 \end{vmatrix}$$

or its transpose.

Proof. We refer to the diagram below.

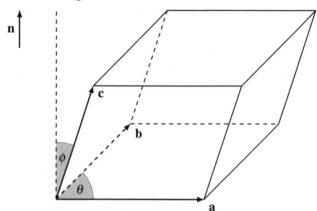

Denote by $\mathbf{n}$ the unit vector orthogonal to $\mathbf{a}$ and $\mathbf{b}$ pointing in the direction of $\mathbf{a} \times \mathbf{b}$. The volume of the box determined by the vectors $\mathbf{a}, \mathbf{b}, \mathbf{c}$ is equal to the base area times the vertical height. The base area is given by $\|\mathbf{a}\| \|\mathbf{b}\| \sin\theta$. The height is equal to the absolute value of $\|\mathbf{c}\| \cos\phi$. We have to use the absolute value of this expression because $\cos\phi$ can take negative values if $\mathbf{c}$ is below rather than above the plane of $\mathbf{a}$ and $\mathbf{b}$ as it is drawn here. Thus the volume is equal to the absolute value of

$$\|\mathbf{a}\| \|\mathbf{b}\| \sin\theta \|\mathbf{c}\| \cos\phi.$$

Now

- $\mathbf{a} \times \mathbf{b} = (\|\mathbf{a}\| \|\mathbf{b}\| \sin\theta)\mathbf{n}$. As expected, this has length equal to the area of the base parallelogram.

- $\mathbf{n} \cdot \mathbf{c} = \|\mathbf{c}\| \cos\phi$.

---

[4]This is a genuine word but to be honest not very common. Although not as accurate, I refer to this shape as a 'box' throughout the proof.

Thus
$$\|\mathbf{a}\| \|\mathbf{b}\| \sin \theta \|\mathbf{c}\| \cos \phi = (\mathbf{a} \times \mathbf{b}) \cdot \mathbf{c}.$$

By the properties of the inner product
$$(\mathbf{a} \times \mathbf{b}) \cdot \mathbf{c} = \mathbf{c} \cdot (\mathbf{a} \times \mathbf{b}) = [\mathbf{c}, \mathbf{a}, \mathbf{b}].$$

We now use properties of the determinant
$$[\mathbf{c}, \mathbf{a}, \mathbf{b}] = -[\mathbf{a}, \mathbf{c}, \mathbf{b}] = [\mathbf{a}, \mathbf{b}, \mathbf{c}].$$

It follows that the volume of the box is the absolute value of $[\mathbf{a}, \mathbf{b}, \mathbf{c}]$. $\qquad\square$

The sign of the determinant which seems an annoyance is actually a feature which plays a rôle in the further theory of matrices.

Example 9.3.4. We describe the geometrical intuition that lies behind Theorem 8.5.21. Let A be a 3×3 matrix regarded as a list of its column vectors $\mathbf{a}, \mathbf{b}, \mathbf{c}$. These three vectors determine a parallelepiped. Without loss of generality, suppose that $\mathbf{c}$ is a linear combination of $\mathbf{a}$ and $\mathbf{b}$. Then $\mathbf{c}$ is parallel to the plane determined by $\mathbf{a}$ and $\mathbf{b}$ and so the volume enclosed is zero. It follows that $\det(A)$ is zero. Suppose now that $\det(A)$ is zero and so the volume of the parallelepiped is zero. Assuming that the vectors are not zero, which is a harmless assumption, this means that the box is squashed flat. Thus one of the vectors must be parallel to the plane determined by the other two, and so is a linear combination of the other two.

9.4   GEOMETRY WITH VECTORS

We now have most of the technology needed to describe lines and planes. Although these do not sound very interesting, they are the essential building blocks of spatial geometry. For example, they are the basis of vector calculus. It only remains to describe some final concepts.

Freedom and bondage

The vectors we have been dealing with up to now are also known as *free vectors*. They can be moved around parallel to themselves as required, but for that very reason they cannot be used to describe the exact positions of points in space. To do this, we need to choose and fix one point O in space, called an *origin*. We now consider a new species of vectors, called *bound* or *position vectors*, whose initial points are rooted at O. Such vectors can obviously be used to determine the exact position of a point in space. A quiver of position vectors emanating from an origin O is visualized below.

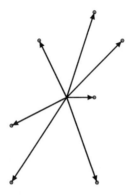

The notation $\mathbb{E}^3$ denotes the free vectors whereas the notation $\mathbb{R}^3$ denotes the position vectors. The distinction between free and position vectors is an important one and frequently a source of confusion.

Dependence and independence

In Section 8.5, linear combinations and the linear dependence and independence of column vectors were introduced. We can now shed some light on the geometric meaning of these notions.

- Let **a** be a non-zero vector. If **b** and **a** are linearly dependent then **b** is a scalar multiple of **a**. Such vectors **b** are precisely those which are parallel to **a**.

- Let **a** and **b** be vectors. They are linearly independent if neither is the zero vector and neither vector is a scalar multiple of the other. In this case, every vector **r** parallel to the plane determined by **a** and **b** can be written as a linear combination of **a** and **b**.

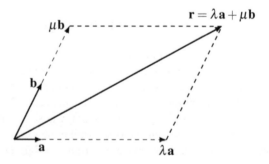

- Let **a**, **b** and **c** be vectors. By Theorem 8.5.22, they are linearly independent if and only if every vector in $\mathbb{R}^3$ can be written as a linear combination of them. They can therefore be used as a *coordinate system* or *basis*. Observe that we do not require that the vectors be pairwise orthogonal. If they are, they form an *orthogonal coordinate system* or *orthogonal basis*. If all the vectors, in addition, have unit length we get an *orthonormal coordinate system* or *orthonormal basis*. Thus $\mathbf{i}, \mathbf{j}, \mathbf{k}$ form an orthonormal coordinate system.

Parametric and non-parametric

Recall that there are two ways in which curves and surfaces ('geometric objects') in general can be described: parametrically and non-parametrically. Parametric equations generate points that belong to a geometric object by varying the value or values of a parameter or parameters. Non-parametric equations are used to check whether a point belongs or does not belong to a geometric object. They are essentially relations.

Example 9.4.1. Consider the circle of unit radius centred at the origin. This is described by the non-parametric equation $x^2 + y^2 = 1$. It is also described by the parametric equation $\theta \mapsto (\cos\theta, \sin\theta)$ where $0 \le \theta \le 2\pi$. The parameter in question is θ.

Lines

Intuitively, a line in space is determined by one of the following two pieces of information.

1. Two distinct points each described by a position vector.

2. One point and a direction, where the point is described by a position vector and the direction by a (free) vector.

Let $\mathbf{a}$ and $\mathbf{b}$ be the position vectors of two distinct points A and B, respectively. Then they determine the line indicated by the dashed line.

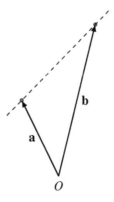

Let $\mathbf{r} = x\mathbf{i} + y\mathbf{j} + z\mathbf{k}$ be the position vector of an arbitrary point on this line.

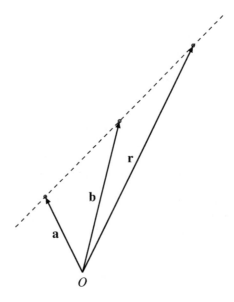

The line is parallel to the vector $\mathbf{b} - \mathbf{a}$. *This is the direction the line is parallel to.* The vectors $\mathbf{r} - \mathbf{a}$ and $\mathbf{b} - \mathbf{a}$ are parallel and by assumption $\mathbf{b} - \mathbf{a} \neq \mathbf{0}$. Thus there is a scalar λ such that $\mathbf{r} - \mathbf{a} = \lambda(\mathbf{b} - \mathbf{a})$. It follows that

$$\mathbf{r} = \mathbf{a} + \lambda(\mathbf{b} - \mathbf{a}).$$

This is called the *(vector form of) the parametric equation of the line*. The *parameter* in question is λ. We therefore have the following.

1. The equation of the line through the points described by position vectors $\mathbf{a}$ and $\mathbf{b}$ is $\mathbf{r} = \mathbf{a} + \lambda(\mathbf{b} - \mathbf{a})$.

2. The equation of the line through the point described by the position vector $\mathbf{a}$ and parallel to the free vector $\mathbf{c}$ is $\mathbf{r} = \mathbf{a} + \lambda\mathbf{c}$.

We can now easily derive the coordinate form of the parametric equation. Let

$$\mathbf{a} = a_1\mathbf{i} + a_2\mathbf{j} + a_3\mathbf{k} \text{ and } \mathbf{b} = b_1\mathbf{i} + b_2\mathbf{j} + b_3\mathbf{k}.$$

Substituting in our vector equation above and equating components we obtain

$$x = a_1 + \lambda(b_1 - a_1), \quad y = a_2 + \lambda(b_2 - a_2), \quad z = a_3 + \lambda(b_3 - a_3).$$

For convenience, put $c_i = b_i - a_i$. Thus the *coordinate form of the parametric equation of the line* is

$$x = a_1 + \lambda c_1, \quad y = a_2 + \lambda c_2, \quad z = a_3 + \lambda c_3.$$

There is a final equation of the line that we can now easily obtain. If $c_1, c_2, c_3 \neq 0$ then we can eliminate the parameters in the above equations to get the *non-parametric equations of the line*

$$\frac{x - a_1}{c_1} = \frac{y - a_2}{c_2}, \quad \frac{y - a_2}{c_2} = \frac{z - a_3}{c_3}.$$

Observe that these are two linear equations. They are often written in the form

$$\frac{x - a_1}{c_1} = \frac{y - a_2}{c_2} = \frac{z - a_3}{c_3}.$$

This is not a practice followed in this book.

Example 9.4.2. It was assumed above that $c_1, c_2, c_3 \neq 0$. We now deal with the cases where some of the $c_i = 0$. Suppose that $c_1 = 0$. Then we get the corresponding equation $x = a_1$ which is independent of λ. If $c_2, c_3 \neq 0$ then, in addition, we get the equation

$$\frac{y - a_2}{c_2} = \frac{z - a_3}{c_3}.$$

Thus

$$x = a_1 \text{ and } \frac{y - a_2}{c_2} = \frac{z - a_3}{c_3}.$$

Suppose that $c_1 = 0 = c_2$. Then $x = a_1$ and $y = a_2$. We must then assume that $c_3 \neq 0$ because otherwise our line contracts to a point. It follows that z can take any value. Thus the line is described by the equations $x = a_1$ and $y = a_2$.

The non-parametric equation of the line is more complicated than might have been expected. What it means will become apparent once we have derived the non-parametric equation of the plane.

Example 9.4.3. We derive the parametric and the non-parametric equations of the line through the point with position vector $\mathbf{i} + 2\mathbf{j} + 3\mathbf{k}$ and *parallel to* the vector $4\mathbf{i} + 5\mathbf{j} + 6\mathbf{k}$. Here we are given one point and the direction that the line is parallel to. Thus $\mathbf{r} - (\mathbf{i} + 2\mathbf{j} + 3\mathbf{k})$ is parallel to $4\mathbf{i} + 5\mathbf{j} + 6\mathbf{k}$. It follows that

$$\mathbf{r} = \mathbf{i} + 2\mathbf{j} + 3\mathbf{k} + \lambda(4\mathbf{i} + 5\mathbf{j} + 6\mathbf{k})$$

for some scalar λ is the vector form of the parametric equation of the line. We now find the cartesian form of the parametric equation. Put $\mathbf{r} = x\mathbf{i} + y\mathbf{j} + z\mathbf{k}$. Then

$$x\mathbf{i} + y\mathbf{j} + z\mathbf{k} = \mathbf{i} + 2\mathbf{j} + 3\mathbf{k} + \lambda(4\mathbf{i} + 5\mathbf{j} + 6\mathbf{k}).$$

These two vectors are equal if and only if their coordinates are equal. Thus

$$\begin{aligned} x &= 1 + 4\lambda \\ y &= 2 + 5\lambda \\ z &= 3 + 6\lambda. \end{aligned}$$

This is the cartesian form of the parametric equation of the line. Finally, we eliminate λ to get the non-parametric equation of the line

$$\frac{x-1}{4} = \frac{y-2}{5} \text{ and } \frac{y-2}{5} = \frac{z-3}{6}.$$

These two equations can be rewritten in the form

$$5x - 4y = -3 \text{ and } 6y - 5z = -3.$$

Planes

Intuitively, a plane in space is determined by one of the following three pieces of information.

1. Any three points that do not all lie in a straight line. That is, the points form the vertices of a triangle.

2. One point and two non-parallel directions.

3. One point and a direction which is perpendicular or *normal* to the plane.

We begin by finding the parametric equation of the plane P determined by the three points with position vectors $\mathbf{a}, \mathbf{b}$ and $\mathbf{c}$.

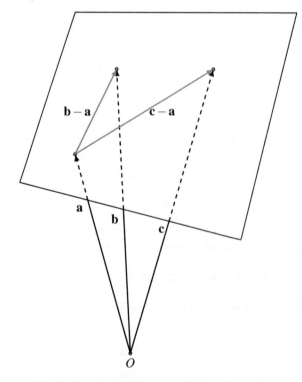

The free vectors $\mathbf{b} - \mathbf{a}$ and $\mathbf{c} - \mathbf{a}$ are both parallel to the plane P, but are not parallel to each other. Every vector parallel to the plane P therefore has the form

$$\lambda(\mathbf{b} - \mathbf{a}) + \mu(\mathbf{c} - \mathbf{a})$$

for some scalars λ and μ. If the position vector of an arbitrary point on the plane P is $\mathbf{r}$, then the free vector $\mathbf{r} - \mathbf{a}$ is parallel to the plane P. It follows that $\mathbf{r} - \mathbf{a} = \lambda(\mathbf{b} - \mathbf{a}) + \mu(\mathbf{c} - \mathbf{a})$. The *(vector form of) the parametric equation of the plane* is therefore

$$\mathbf{r} = \mathbf{a} + \lambda(\mathbf{b} - \mathbf{a}) + \mu(\mathbf{c} - \mathbf{a}).$$

This can easily be written in coordinate form by equating components.

Observe that the position vector $\mathbf{a}$ describes a point on the plane and the free vectors $\mathbf{c} = \mathbf{b} - \mathbf{a}$ and $\mathbf{d} = \mathbf{c} - \mathbf{a}$ are vectors parallel to the plane and not parallel to each other. In terms of these vectors, the equation of the plane P is

$$\mathbf{r} = \mathbf{a} + \lambda\mathbf{c} + \mu\mathbf{d}.$$

To find the non-parametric equation of a plane, we use the fact that the plane is determined once a point on the plane is known together with a vector normal to the plane. Let $\mathbf{n}$ be a vector normal to the plane and let $\mathbf{a}$ be the position vector of a point in the plane.

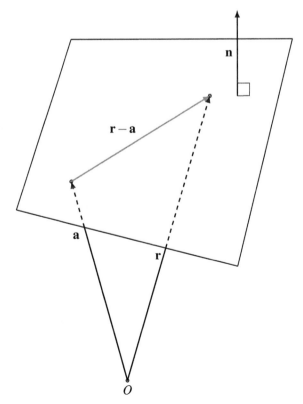

Then $\mathbf{r} - \mathbf{a}$ is orthogonal to $\mathbf{n}$. Thus $(\mathbf{r} - \mathbf{a}) \cdot \mathbf{n} = 0$. This is the *(vector form) of the non-parametric equation* of the plane. To find the coordinate form of the non-parametric equation, let

$$\mathbf{r} = x\mathbf{i} + y\mathbf{j} + z\mathbf{k}, \quad \mathbf{a} = a_1\mathbf{i} + a_2\mathbf{j} + a_3\mathbf{k}, \quad \mathbf{n} = n_1\mathbf{i} + n_2\mathbf{j} + n_3\mathbf{k}.$$

From $(\mathbf{r} - \mathbf{a}) \cdot \mathbf{n} = 0$ we get $(x - a_1)n_1 + (y - a_2)n_2 + (z - a_3)n_3 = 0$. Thus the *non-parametric equation of the plane* is

$$n_1 x + n_2 y + n_3 z = a_1 n_1 + a_2 n_2 + a_3 n_3.$$

There is one final question: given the parametric equation of the plane, how do we find the non-parametric equation? Let $\mathbf{a}, \mathbf{b}, \mathbf{c}$ be the position vectors of three points in the plane that do not lie in a straight line. Then the vectors $\mathbf{b} - \mathbf{a}$ and $\mathbf{c} - \mathbf{a}$ are parallel to the plane but not parallel to each other. The vector

$$\mathbf{n} = (\mathbf{b} - \mathbf{a}) \times (\mathbf{c} - \mathbf{a})$$

is therefore normal to the plane. It follows that with the help of the position vector $\mathbf{a}$ and the free vector $\mathbf{n}$, the non-parametric equation of the plane can now be written down.

Example 9.4.4. We derive the parametric and non-parametric equations of the plane containing the three points with position vectors

$$\mathbf{a} = \mathbf{j} - \mathbf{k}, \quad \mathbf{b} = \mathbf{i} + \mathbf{j}, \quad \mathbf{c} = \mathbf{i} + 2\mathbf{j}.$$

The free vectors

$$\mathbf{b} - \mathbf{a} = \mathbf{i} + \mathbf{k} \text{ and } \mathbf{c} - \mathbf{a} = \mathbf{i} + \mathbf{j} + \mathbf{k}$$

are parallel to the plane but not parallel to each other. Thus the parametric equation of the plane is

$$\mathbf{r} = \mathbf{j} - \mathbf{k} + \lambda(\mathbf{i} + \mathbf{k}) + \mu(\mathbf{i} + \mathbf{j} + \mathbf{k}).$$

To find the non-parametric equation, we need to find a vector normal to the plane. The vector

$$(\mathbf{b} - \mathbf{a}) \times (\mathbf{c} - \mathbf{a}) = \mathbf{k} - \mathbf{i}$$

is just the ticket. Thus

$$(\mathbf{r} - \mathbf{a}) \cdot (\mathbf{k} - \mathbf{i}) = 0.$$

That is

$$(x\mathbf{i} + (y - 1)\mathbf{j} + (z + 1)\mathbf{k}) \cdot (\mathbf{k} - \mathbf{i}) = 0.$$

This simplifies to

$$z - x = -1,$$

the non-parametric equation of the plane. We now check that our three original points satisfy this equation. The point $\mathbf{a}$ has coordinates $(0, 1, -1)$; the point $\mathbf{b}$ has coordinates $(1, 1, 0)$; the point $\mathbf{c}$ has coordinates $(1, 2, 0)$. It is easy to check that each set of coordinates satisfies the equation.

Example 9.4.5. The theory of planes we have developed enables us to interpret geometrically the solution of linear equations in three unknowns. From the non-parametric equation of the plane derived above, a linear equation in three unknowns

$$ax + by + cz = d$$

has a solution set in $\mathbb{R}^3$ which is one of the following possibilities.

1. The whole of $\mathbb{R}^3$ if $a = b = c = d = 0$.

2. The empty set if $a = b = c = 0$ and $d \neq 0$.

3. Otherwise, a plane in $\mathbb{R}^3$.

We say that cases (1) and (2) are *degenerate* and that case (3) is *non-degenerate*. We can now see that the non-parametric equation of the line in fact describes the line as the intersection of two planes.

Given three equations in three unknowns then, as long as the planes are angled correctly, they will intersect in a point, and so the equations will have a unique solution. However, there are cases where either the planes have no points in common, which means the equations have no solution, or have lines or indeed planes in common yielding infinitely many solutions. Thus the nature of the solutions of a system of linear equations in three unknowns is intimately bound up with the geometry of the planes they determine.

Exercises 9.4

1. (a) Find the parametric and the non-parametric equations of the line through the two points with position vectors $\mathbf{i} - \mathbf{j} + 2\mathbf{k}$ and $2\mathbf{i} + 3\mathbf{j} + 4\mathbf{k}$.

    (b) Find the parametric and the non-parametric equations of the plane containing the three points with position vectors $\mathbf{i} + 3\mathbf{k}$, $\mathbf{i} + 2\mathbf{j} - \mathbf{k}$ and $3\mathbf{i} - \mathbf{j} - 2\mathbf{k}$.

    (c) The plane $px + qy + rz = s$ contains the point with position vector $\mathbf{a} = 2\mathbf{i} + 2\mathbf{j} + \mathbf{k}$ and is orthogonal to the vector $\mathbf{b} = \mathbf{i} + \mathbf{j} - \mathbf{k}$. Find p, q, r and s.

2. *Intersection of two lines.*

    (a) Find the vector equation of the line L_1 through the point with position vector $4\mathbf{i} + 5\mathbf{j} + \mathbf{k}$ and parallel to the vector $\mathbf{i} + \mathbf{j} + \mathbf{k}$.

    (b) Find the vector equation of the line L_2 through the point with position vector $5\mathbf{i} - 4\mathbf{j}$ and parallel to the vector $2\mathbf{i} - 3\mathbf{j} + \mathbf{k}$.

    (c) Determine whether the two lines L_1 and L_2 intersect and if they do find the position vector of the point of intersection.

3. *Intersection of two planes.* Find the vector equation of the line of intersection of the planes

$$x - y - 2z = 3 \text{ and } 2x + 3y + 4z = -2.$$

4. *The distance of a point from a line* is defined to be the length of the perpendicular from the point to the line. Let the line in question have parametric equation $\mathbf{r} = \mathbf{p} + \lambda\mathbf{d}$ and let the position vector of the point be $\mathbf{q}$. Show that the distance of the point from the line is

$$\frac{\|\mathbf{d} \times (\mathbf{q} - \mathbf{p})\|}{\|\mathbf{d}\|}.$$

5. *The distance of a point from a plane* is defined to be the length of the perpendicular from the point to the plane. Let the position vector of the point be $\mathbf{q}$ and the equation of the plane be $(\mathbf{r} - \mathbf{p}) \cdot \mathbf{n} = 0$. Show that the distance of the point from the plane is

$$\frac{|(\mathbf{q} - \mathbf{p}) \cdot \mathbf{n}|}{\|\mathbf{n}\|}.$$

6. *Spheres.* Let $\mathbf{c}$ be the position vector of the centre of a sphere with radius R. Let an arbitrary point on the sphere have position vector $\mathbf{r}$. Why is $\|\mathbf{r} - \mathbf{c}\| = R$? Squaring both sides we get $(\mathbf{r} - \mathbf{c}) \cdot (\mathbf{r} - \mathbf{c}) = R^2$. If $\mathbf{r} = x\mathbf{i} + y\mathbf{j} + z\mathbf{k}$ and $\mathbf{c} = c_1\mathbf{i} + c_2\mathbf{j} + c_3\mathbf{k}$, deduce that the *equation of the sphere* with centre $c_1\mathbf{i} + c_2\mathbf{j} + c_3\mathbf{k}$ and radius R is $(x - c_1)^2 + (y - c_2)^2 + (z - c_3)^2 = R^2$.

   (a) Find the equation of the sphere with centre $\mathbf{i} + \mathbf{j} + \mathbf{k}$ and radius 2.

   (b) Find the centre and radius of the sphere with equation
   $$x^2 + y^2 + z^2 - 2x - 4y - 6z - 2 = 0.$$

7. Let $\mathbf{u}, \mathbf{v}, \mathbf{w}$ be three non-zero orthogonal vectors. Prove that they are linearly independent in the sense defined in Section 8.5.

8. *Calculate
   $$\mathbf{a} \times (\mathbf{b} \times \mathbf{c}) + \mathbf{b} \times (\mathbf{c} \times \mathbf{a}) + \mathbf{c} \times (\mathbf{a} \times \mathbf{b}).$$

9. *Prove that $\mathbf{u} \times (\mathbf{v} \times \mathbf{w}) = (\mathbf{u} \cdot \mathbf{w})\mathbf{v} - (\mathbf{u} \cdot \mathbf{v})\mathbf{w}$. Hint: Explain why $\mathbf{u} \times (\mathbf{v} \times \mathbf{w})$ is a linear combination of $\mathbf{v}$ and $\mathbf{w}$, and take an inner product with $\mathbf{u}$.

10. *Prove that
   $$(\mathbf{a} \times \mathbf{b}) \cdot (\mathbf{c} \times \mathbf{d}) = \begin{vmatrix} \mathbf{a} \cdot \mathbf{c} & \mathbf{a} \cdot \mathbf{d} \\ \mathbf{b} \cdot \mathbf{c} & \mathbf{b} \cdot \mathbf{d} \end{vmatrix}.$$

   Hint: Use the properties of the scalar triple product and the previous question.

9.5 LINEAR FUNCTIONS

In this section, our work on matrices and vectors will be combined with the theory of functions described in Section 3.4. This is the beginning of a subject known as linear algebra. Our aim is to describe some important motivating examples and essential ideas. Throughout this section, all vectors are position vectors. Such vectors can be written $a\mathbf{i} + b\mathbf{j} + c\mathbf{k}$ but they can equally well be written as column vectors

$$\begin{pmatrix} a \\ b \\ c \end{pmatrix}.$$

It will be more convenient here to regard them as column vectors. The set of such column vectors is denoted by $\mathbb{R}^3$. More generally, the set $\mathbb{R}^n$ is the set of real column vectors with n entries. We study the functions $\mathbf{x} \mapsto A\mathbf{x}$ mainly, but not solely, in the case where A is a 2×2 or 3×3 matrix because the primary intention is to develop geometric intuition. We use the column vectors $\mathbf{e}_i \in \mathbb{R}^n$. When $n = 3$, we have that $\mathbf{e}_1 = \mathbf{i}$, $\mathbf{e}_2 = \mathbf{j}$ and $\mathbf{e}_3 = \mathbf{k}$. Let $\mathbf{u} \in \mathbb{R}^n$, and let u_i be its ith component. Then

$$\mathbf{u} = \sum_{i=1}^{n} u_i \mathbf{e}_i.$$

We have therefore written $\mathbf{u}$ as a linear combination of the vectors $\mathbf{e}_i$.

The fundamental definition of this section is the following. A function $f \colon \mathbb{R}^m \to \mathbb{R}^n$ is said to be *linear* if

$$f(\lambda \mathbf{u} + \mu \mathbf{v}) = \lambda f(\mathbf{u}) + \mu f(\mathbf{v})$$

for all real numbers $\lambda, \mu \in \mathbb{R}$ and all column vectors $\mathbf{u}, \mathbf{v} \in \mathbb{R}^m$. Linear functions are usually called *linear transformations* for historical reasons. A simple induction argument shows that if f is linear then

$$f\left(\sum_{j=1}^{r} \lambda_j \mathbf{v}_j \right) = \sum_{j=1}^{r} \lambda_j f(\mathbf{v}_j).$$

Thus a linear function maps linear combinations of vectors to linear combinations of their images. We now connect linear functions with matrices.

Theorem 9.5.1 (Structure of linear transformations).

1. *A function $f \colon \mathbb{R}^m \to \mathbb{R}^n$ is linear if and only if there is an $n \times m$ real matrix A such that $f(\mathbf{u}) = A\mathbf{u}$.*

2. *A linear function $f \colon \mathbb{R}^m \to \mathbb{R}^m$ given by $\mathbf{x} \mapsto A\mathbf{x}$ is bijective if and only if A is invertible.*

Proof. (1) Let f be a linear function. Define the matrix A to have as ith column vector $f(\mathbf{e}_i)$. Let

$$\mathbf{u} = u_1 \mathbf{e}_1 + \ldots + u_m \mathbf{e}_m.$$

Then $f(\mathbf{u}) = A\mathbf{u}$. The proof of the converse is an immediate result of the properties of matrix multiplication described in Section 8.2.

(2) This is proved as part (6) of Theorem 8.5.22. The statement is repeated here because the context is different. □

Linear functions are of fundamental importance in mathematics. The following two examples give some inkling as to why.

Example 9.5.2. Let $f\colon \mathbb{R}^3 \to \mathbb{R}^3$ be a function which we assume can be partially differentiated with respect to x, y and z. We can write

$$f(\mathbf{x}) = (f_1(\mathbf{x}), f_2(\mathbf{x}), f_3(\mathbf{x}))^T.$$

Define

$$\begin{pmatrix} \frac{\partial f_1}{\partial x} & \frac{\partial f_1}{\partial y} & \frac{\partial f_1}{\partial z} \\ \frac{\partial f_2}{\partial x} & \frac{\partial f_2}{\partial y} & \frac{\partial f_2}{\partial z} \\ \frac{\partial f_3}{\partial x} & \frac{\partial f_3}{\partial y} & \frac{\partial f_3}{\partial z} \end{pmatrix}$$

called the *Jacobian matrix* of f. When this matrix is evaluated at a specific point it describes a linear function, which can be proved to be the best linear approximation at that point to the function f. Calculus, which started as constructing tangents to curves, is now revealed to be the study of functions which can be well-approximated at each point by linear functions. This shows that the notion of linear function, and so of matrix, is of wide applicability.

Example 9.5.3. Let P_2 denote all real polynomials of degree at most 2. Encode such a polynomial $a_0 + a_1 x + a_2 x^2$ by means of a column vector

$$\begin{pmatrix} a_2 \\ a_1 \\ a_0 \end{pmatrix}.$$

Define the function $D\colon P_2 \to P_2$ by $D(a_0 + a_1 x + a_2 x^2) = a_1 + 2a_2 x$. Thus D is nothing other than differentiation and is clearly linear. The associated matrix is

$$\begin{pmatrix} 0 & 0 & 0 \\ 2 & 0 & 0 \\ 0 & 1 & 0 \end{pmatrix}.$$

In the Preface we mentioned that mathematics does not consist of watertight compartments. Nowhere is this more true than in the theory of linear functions.

The notion of linearity can be extended to functions of more than one variable.

Example 9.5.4. The inner product is a function from $\mathbb{R}^3 \times \mathbb{R}^3$ to $\mathbb{R}$. Rather than write $(\mathbf{u}, \mathbf{v}) \mapsto \mathbf{u} \cdot \mathbf{v}$ we write $(\mathbf{u}, \mathbf{v}) \mapsto \langle \mathbf{u}, \mathbf{v} \rangle$. By Theorem 9.1.5, we see that

$$\langle \mathbf{u}, \lambda \mathbf{v} + \mu \mathbf{w} \rangle = \lambda \langle \mathbf{u}, \mathbf{v} \rangle + \mu \langle \mathbf{u}, \mathbf{w} \rangle$$

and

$$\langle \lambda \mathbf{u} + \mu \mathbf{v}, \mathbf{w} \rangle = \lambda \langle \mathbf{u}, \mathbf{w} \rangle + \mu \langle \mathbf{v}, \mathbf{w} \rangle.$$

Thus the inner product is linear in each variable and, accordingly, is an example of a *bilinear function*. In the next section, we shall investigate some *multilinear functions*.

Box 9.1: Dimension

You might wonder why we did not discuss bijective linear functions between $\mathbb{R}^m$ and $\mathbb{R}^n$ when $m \neq n$ in Theorem 9.5.1. The reason is: there are none. To prove this, however, requires a notion that is lurking under the surface of our work: that of dimension. This is best handled in the more general context of vector spaces, but we shall say something about it here. If we consider the solutions of a system of homogeneous equations then there is either exactly one solution or infinitely many solutions. Although saying there are infinitely many solutions certainly tells you the size of the solution set, it is not a very helpful measure of size. For example, the solution set might be a line through the origin or it might be a plane. They both describe infinitely many solutions, but we would also want to say that the first was *one-dimensional* and the second was *two-dimensional*. These are much more informative. Intuitively, we know what these terms mean, but how do we define them exactly? We use the notion of linear independence. Given m linearly independent vectors the set of all their linear combinations is said to describe an *m-dimensional space*.

Linear functions are rigid. Once you know what a linear function does to the vectors $e_1, \ldots, e_n$, you know what it does everywhere. Thus in the plane, the effect of a linear function is determined by what it does to the vectors $\mathbf{i}$ and $\mathbf{j}$ and so, by extension, to the square with vertices $\mathbf{0}, \mathbf{i}, \mathbf{j}$ and $\mathbf{i} + \mathbf{j}$. In space, the effect of a linear function is determined by what it does to the vectors $\mathbf{i}, \mathbf{j}$ and $\mathbf{k}$ and so, by extension, to the cube determined by these three vectors. Non-linear functions, on the other hand, will do different things in different places.

Examples 9.5.5.

1. The matrix
$$A = \begin{pmatrix} \cos\theta & -\sin\theta \\ \sin\theta & \cos\theta \end{pmatrix}$$

   represents a rotation about the origin by θ in an anticlockwise direction. Observe that its determinant is 1. In the plane, a rotation fixes the origin, preserves all distances and has positive determinant. In space, a rotation fixes a line through the origin, preserves all distances and has positive determinant.

2. The matrix
$$A = \begin{pmatrix} \cos\theta & \sin\theta \\ \sin\theta & -\cos\theta \end{pmatrix}$$

   represents a reflection in the line through the origin at an angle of $\frac{\theta}{2}$. Observe that its determinant is -1. In the plane, a reflection fixes a line through the origin, and then points on either side of this line, which are joined by a perpendicular bisector of this line, are interchanged. So mathematical reflections are a two-way street.

3. The matrix
$$A = \begin{pmatrix} \lambda & 0 \\ 0 & \mu \end{pmatrix}$$

   scales by λ in the x-direction and by μ in the y-direction.

Box 9.2: Vector Spaces

There is an analogy between matrices and differentiation. If A is a matrix then $A(\lambda\mathbf{u} + \mu\mathbf{v}) = \lambda A\mathbf{u} + \mu A\mathbf{v}$, whereas if we represent differentiation by x by the symbol D, we have that $D(af(x) + bg(x)) = aDf(x) + bDg(x)$. We can make this analogy precise by defining systems in which it is possible to construct linear combinations and talk about linear dependence and linear independence. Such systems are called *vector spaces*. The elements of a vector space need not be vectors in the sense used in this book: they simply have to behave like vectors in that they can be added and multiplied by scalars. One of the great advances made in the nineteenth century was the understanding that mathematics does not deal in essences but only in form. This insight became the foundation of all subsequent developments. Once we have vector spaces, we define linear transformations as those functions between them that preserve linear combinations. Everything in Chapters 8, 9 and 10 can then be carried out in this setting, so we end up being able to prove theorems of much greater generality. For example, Fourier discovered that periodic phenomena could be analysed into sines and cosines of different frequencies and amplitudes. This has important applications in science and is the basis of electronic sound production. It is also analogous to writing a vector as a linear combination of basis vectors, with the sines and cosines playing the rôle of basis vectors. Within the theory of vector spaces this analogy can be made precise.

Exercises 9.5

1. Determine the eigenvalues and eigenvectors of

$$A = \begin{pmatrix} \cos\theta & -\sin\theta \\ \sin\theta & \cos\theta \end{pmatrix}.$$

2. Let

$$A = \begin{pmatrix} \cos\theta & \sin\theta \\ \sin\theta & -\cos\theta \end{pmatrix}.$$

Show that

$$\mathbf{u} = \begin{pmatrix} \cos\frac{\theta}{2} \\ \sin\frac{\theta}{2} \end{pmatrix} \text{ and } \mathbf{v} = \begin{pmatrix} -\sin\frac{\theta}{2} \\ \cos\frac{\theta}{2} \end{pmatrix}$$

are both eigenvectors of A. What are the corresponding eigenvalues? How do these results relate to the fact that A represents a reflection?

3. *This question is about matrices that arise naturally from elementary row operations.

    (a) An $n \times n$ matrix E is called an *elementary matrix* if it is obtained from the $n \times n$ identity matrix by means of a single elementary row operation. Show that the 2×2 elementary matrices are of the form

    $$\begin{pmatrix} 0 & 1 \\ 1 & 0 \end{pmatrix}, \quad \begin{pmatrix} \lambda & 0 \\ 0 & 1 \end{pmatrix}, \quad \begin{pmatrix} 1 & 0 \\ 0 & \lambda \end{pmatrix}, \quad \begin{pmatrix} 1 & \lambda \\ 0 & 1 \end{pmatrix}, \quad \begin{pmatrix} 1 & 0 \\ \lambda & 1 \end{pmatrix}$$

    where $\lambda \neq 0$.

    (b) Prove that each elementary row matrix is invertible and describe them geometrically.

(c) Let B be obtained from A by means of a single elementary row operation ρ. Thus $B = \rho(A)$. Let $E = \rho(I)$. Prove that $B = EA$.

(d) Prove that each invertible 2×2 matrix can be written as a product of elementary matrices.

4. *By considering cases, describe geometrically the linear function determined by the matrix

$$\begin{pmatrix} a & b \\ c & d \end{pmatrix}$$

assuming that it is singular.

9.6 ALGEBRAIC MEANING OF DETERMINANTS

Determinants were defined algebraically in Section 8.4 and then interpreted geometrically in Section 9.3. In this section, they are described in a way that unifies the algebraic and geometric points of view. We only do this for 3×3 determinants but the arguments easily generalize. The starting point is the list of properties given in Theorem 8.4.3. We prove that some of these properties completely characterize determinants. To do this, we need some notation. In this section, we regard $\mathbb{R}^3$ as being the set of 3×1 real column vectors. As usual, such vectors will be denoted by bold lower case letters. Amongst these are the column vectors $\mathbf{i}, \mathbf{j}$ and $\mathbf{k}$. We are interested in a function from $\mathbb{R}^3 \times \mathbb{R}^3 \times \mathbb{R}^3$ to $\mathbb{R}$ whose value on the input $(\mathbf{a}, \mathbf{b}, \mathbf{c})$ is denoted by $[\mathbf{a}, \mathbf{b}, \mathbf{c}]$. Observe that this is just convenient notation at this point though we shall soon see that it really is the scalar triple product. The vectors $\mathbf{a}, \mathbf{b}$ and $\mathbf{c}$ are called the *arguments* of this function. We assume that this function has the following three properties.

(D1) $[\mathbf{i}, \mathbf{j}, \mathbf{k}] = 1$. This is just a *normalization* condition.

(D2) If any two arguments are swapped then the value of the function is multiplied by -1. Thus, for example, $[\mathbf{a}, \mathbf{b}, \mathbf{c}] = -[\mathbf{b}, \mathbf{a}, \mathbf{c}]$. This is the *alternating* condition.

(D3) If any two arguments are fixed then the function is linear in the remaining argument. Thus, for example,

$$[\mathbf{a}, \lambda \mathbf{b}_1 + \mu \mathbf{b}_2, \mathbf{c}] = \lambda [\mathbf{a}, \mathbf{b}_1, \mathbf{c}] + \mu [\mathbf{a}, \mathbf{b}_2, \mathbf{c}].$$

This is the *multilinear condition*.

Lemma 9.6.1. *Suppose that only (D3) holds. Then (D2) is equivalent to the property that the function is zero when any two arguments are equal.*

Proof. Suppose that (D2) and (D3) hold. Consider $[\mathbf{a}, \mathbf{a}, \mathbf{c}]$ where the first two arguments are equal. If we interchange them then the sign should change but we still have $[\mathbf{a}, \mathbf{a}, \mathbf{c}]$. It follows that $[\mathbf{a}, \mathbf{a}, \mathbf{c}] = 0$. The other cases follow by similar arguments. We now prove the converse. Assume that our function vanishes when any two arguments are equal and that (D3) holds. Consider $[\mathbf{a}+\mathbf{b},\mathbf{a}+\mathbf{b},\mathbf{c}]$. By assumption this is zero and by (D3) it simplifies to

$$0 = [\mathbf{a},\mathbf{b},\mathbf{c}] + [\mathbf{b},\mathbf{a},\mathbf{c}].$$

Thus $[\mathbf{a},\mathbf{b},\mathbf{c}] = -[\mathbf{b},\mathbf{a},\mathbf{c}]$. The other cases follow by similar arguments.[5] □

We already have an example of a function that satisfies (D1), (D2) and (D3): the determinant. Here we regard the determinant as a function of the three column vectors of the matrix it is defined by, so we can write $\det(\mathbf{a},\mathbf{b},\mathbf{c})$.

Theorem 9.6.2 (Characterization of determinants). *The only function from $\mathbb{R}^3 \times \mathbb{R}^3 \times \mathbb{R}^3$ to $\mathbb{R}$ which is normalized, alternating and multilinear is the determinant.*

Proof. We show that we can evaluate $[\mathbf{a},\mathbf{b},\mathbf{c}]$ exactly using only the three properties (D1), (D2) and (D3). We write $\mathbf{a} = a_1\mathbf{i} + a_2\mathbf{j} + a_3\mathbf{k}$. Thus

$$[\mathbf{a},\mathbf{b},\mathbf{c}] = [a_1\mathbf{i} + a_2\mathbf{j} + a_3\mathbf{k},\mathbf{b},\mathbf{c}].$$

We repeatedly apply multilinearity to get

$$[\mathbf{a},\mathbf{b},\mathbf{c}] = a_1[\mathbf{i},\mathbf{b},\mathbf{c}] + a_2[\mathbf{j},\mathbf{b},\mathbf{c}] + a_3[\mathbf{k},\mathbf{b},\mathbf{c}].$$

We now evaluate $[\mathbf{i},\mathbf{b},\mathbf{c}]$, $[\mathbf{j},\mathbf{b},\mathbf{c}]$ and $[\mathbf{k},\mathbf{b},\mathbf{c}]$. We simply report the values leaving the details as exercises. We have that

$$[\mathbf{i},\mathbf{b},\mathbf{c}] = b_1[\mathbf{i},\mathbf{i},\mathbf{c}] + b_2[\mathbf{i},\mathbf{j},\mathbf{c}] + b_3[\mathbf{i},\mathbf{k},\mathbf{c}]$$

and

$$[\mathbf{j},\mathbf{b},\mathbf{c}] = b_1[\mathbf{j},\mathbf{i},\mathbf{c}] + b_2[\mathbf{j},\mathbf{j},\mathbf{c}] + b_3[\mathbf{j},\mathbf{k},\mathbf{c}]$$

and

$$[\mathbf{k},\mathbf{b},\mathbf{c}] = b_1[\mathbf{k},\mathbf{i},\mathbf{c}] + b_2[\mathbf{k},\mathbf{j},\mathbf{c}] + b_3[\mathbf{k},\mathbf{k},\mathbf{c}].$$

By Lemma 9.6.1

$$0 = [\mathbf{i},\mathbf{i},\mathbf{c}] = [\mathbf{j},\mathbf{j},\mathbf{c}] = [\mathbf{k},\mathbf{k},\mathbf{c}].$$

Thus it remains to evaluate the following

$$[\mathbf{i},\mathbf{j},\mathbf{c}] \text{ and } [\mathbf{j},\mathbf{i},\mathbf{c}],$$

$$[\mathbf{i},\mathbf{k},\mathbf{c}] \text{ and } [\mathbf{k},\mathbf{i},\mathbf{c}],$$

$$[\mathbf{j},\mathbf{k},\mathbf{c}] \text{ and } [\mathbf{k},\mathbf{i},\mathbf{c}].$$

Because of the alternating property, it is enough to evaluate just once in each pair and the value of the other one will be -1 times that. In doing so, we use Lemma 9.6.1 a number of times. We get that

$$[\mathbf{i},\mathbf{j},\mathbf{c}] = c_3[\mathbf{i},\mathbf{j},\mathbf{k}]$$

---

[5]In advanced work, it is more convenient to define alternating to be the property of this lemma rather than (D2).

and

$$[\mathbf{i},\mathbf{k},\mathbf{c}] = c_2[\mathbf{i},\mathbf{k},\mathbf{j}]$$

and

$$[\mathbf{j},\mathbf{k},\mathbf{c}] = c_1[\mathbf{j},\mathbf{k},\mathbf{i}].$$

By using normalization and the alternating property, we can compute explicit values for the following $6 = 3!$ expressions

$$[\mathbf{i},\mathbf{j},\mathbf{k}], \ [\mathbf{i},\mathbf{k},\mathbf{j}], \ [\mathbf{j},\mathbf{i},\mathbf{k}], \ [\mathbf{j},\mathbf{k},\mathbf{i}], \ [\mathbf{k},\mathbf{i},\mathbf{j}], \ [\mathbf{k},\mathbf{j},\mathbf{i}].$$

We deduce that

$$[\mathbf{a},\mathbf{b},\mathbf{c}] = a_1 b_2 c_3 - a_1 b_3 c_2 - a_2 b_1 c_3 + a_2 b_3 c_1 + a_3 b_1 c_2 - a_3 b_2 c_1.$$

A routine calculation shows that this is exactly

$$\begin{vmatrix} a_1 & b_1 & c_1 \\ a_2 & b_2 & c_2 \\ a_3 & b_3 & c_3 \end{vmatrix}.$$

□

The argument above generalizes and can be used to characterize all determinants.

We can now explain the origin of the sign changes in the definition of the determinant. Take for example the expression $[\mathbf{i},\mathbf{k},\mathbf{j}]$. This evaluates to -1 because one transposition is needed to convert it to $[\mathbf{i},\mathbf{j},\mathbf{k}]$. On the other hand $[\mathbf{k},\mathbf{i},\mathbf{j}]$ evaluates to 1 because we first interchange $\mathbf{i}$ and $\mathbf{k}$, and then we interchange $\mathbf{k}$ and $\mathbf{j}$. Thus we need two transpositions. Thus the signs that occur are precisely the signs of the corresponding permutation as defined in Section 7.5. This motivates the following. Let $A = (a_{ij})$ be an $n \times n$ matrix. We define the *determinant* of A, denoted $\det(A)$, to be the number

$$\boxed{\det(A) = \sum_{f \in S_n} \text{sgn}(f) a_{1f(1)} \cdots a_{nf(n)}.}$$

With this definition, we then have the general version of Theorem 9.6.2. It can be proved that the definition of determinant given in Section 8.4 is equivalent to this one. We also have the correct definition to prove Theorem 8.4.3 in full generality.

Example 9.6.3. As a check on our reasoning, we calculate

$$\begin{vmatrix} a_{11} & a_{12} & a_{13} \\ a_{21} & a_{22} & a_{23} \\ a_{31} & a_{32} & a_{33} \end{vmatrix}$$

using the definition of determinant above. The table below lists the permutations of

S_3, their corresponding signs and the corresponding (signed) terms in the definition of the determinant.

f	sgn(f)	Term
ι	1	$a_{11}a_{22}a_{33}$
(12)	-1	$-a_{12}a_{21}a_{33}$
(23)	-1	$-a_{11}a_{23}a_{32}$
(13)	-1	$-a_{13}a_{22}a_{31}$
(123)	1	$a_{12}a_{23}a_{31}$
(132)	1	$a_{13}a_{21}a_{32}$

Add up the terms to get

$$a_{11}a_{22}a_{33} - a_{11}a_{23}a_{32} - a_{12}a_{21}a_{33} + a_{12}a_{23}a_{31} + a_{13}a_{22}a_{32} - a_{13}a_{22}a_{31}.$$

This is precisely the determinant when computed along the first row.

Example 9.6.4. Let A be a 3×3 matrix. As we have seen this determines a function f from $\mathbb{R}^3$ to itself given by $\mathbf{x} \mapsto A\mathbf{x}$. This function is bijective if and only if A is invertible by Theorem 8.5.21, and A is invertible if and only if $\det(A) \neq 0$ by Theorem 8.5.13. We can use our work on determinants and linear functions to gain some insight into why $\det(A) \neq 0$ is necessary and sufficient for A to be invertible. Linear functions behave the same way everywhere, so we need only look at the way f treats the unit cube determined by $\mathbf{i}, \mathbf{j}$ and $\mathbf{k}$. The signed volume of the cube determined by these vectors is $[\mathbf{i}, \mathbf{j}, \mathbf{k}] = 1$ and the signed volume of the image of this cube is $[A\mathbf{i}, A\mathbf{j}, A\mathbf{k}]$. This is equal to $\det(A)$ by Theorem 9.3.3 since $A\mathbf{i}$ is the first column of A, $A\mathbf{j}$ is the second column of A and $A\mathbf{k}$ is the final column of A.

Suppose that $\det(A) = 0$. Then the image of the unit cube has zero volume which means that collapsing takes place under f and so different vectors in the domain are mapped to the same vector in the image. Thus f is not injective.

Conversely, suppose that f is not injective. Then vectors are identified by f. But if $f(\mathbf{a}) = f(\mathbf{b})$ then $f(\mathbf{a} - \mathbf{b}) = \mathbf{0}$. Thus if f is not injective, non-zero vectors are mapped to zero by f. Let $f(\mathbf{a}) = \mathbf{0}$ where $\mathbf{a} \neq \mathbf{0}$. Now choose vectors $\mathbf{b}$ and $\mathbf{c}$ such that $\mathbf{a}, \mathbf{b}, \mathbf{c}$ are linearly independent. Each of the vectors $\mathbf{i}, \mathbf{j}$ and $\mathbf{k}$ can be written as linear combinations of $\mathbf{a}, \mathbf{b}, \mathbf{c}$. By linearity, this implies that the images of $\mathbf{i}, \mathbf{j}$ and $\mathbf{k}$ under f lie in the plane determined by $\mathbf{b}$ and $\mathbf{c}$. It follows that the image of the unit cube has zero volume and so $\det(A) = 0$.

Thus $\det(A) \neq 0$ is necessary and sufficient for f to be injective. Surjectivity comes along for free because we are only dealing with finite-dimensional spaces. Dimension is discussed in one of the boxes but an analogy will help. A function from a *finite* set to itself is injective if and only if it is surjective. The result we need is similar except that finite cardinality is replaced by finite dimensionality.

9.7 QUATERNIONS

We met Sir William Rowan Hamilton for the first time in Section 6.4 when we sketched his approach to showing that complex numbers exist. On the 16th October 1843, he had a mathematical epiphany and as a result committed a small act of vandalism on Brougham Bridge in Dublin. For the previous decade and a half, Hamilton had been trying to generalize the complex numbers, which are inherently 2-dimensional, to something 3-dimensional. His aim was to make all the useful properties of complex numbers available for studying the geometry of space. The idea he had been trying to develop was to replace $a + bi$ by triples like $a + bi + cj$. The problem was that he could not make any definition of multiplication work in a way that gave him the properties he was looking for. His epiphany was that triples would not work but, amazingly, quadruples $a + bi + cj + dk$ would. The system he obtained was called the *quaternions*, denoted by $\mathbb{H}$. In his excitement, he scratched onto the bridge the multiplication law

$$i^2 = j^2 = k^2 = ijk = -1.$$

The quaternions had all the properties of the complex numbers he wanted except that they were non-commutative. In fact, this was the first algebraic system that was non-commutative. Just a couple of months after his discovery, John T. Graves (1806–1870) constructed the *octonions*, denoted by $\mathbb{O}$, which are 8-tuples. Not only are these not commutative they are not even associative.[6] It was from the theory of quaternions that the modern theory of vectors with inner and vector products developed and is the source of the notation **i**,**j** and **k** for the orthonormal unit vectors. To describe what the quaternions are, history will be reversed and quaternions will be derived from vectors.

Define the matrices $I, X, Y, Z, -I, -X, -Y, -Z$ as follows

$$X = \begin{pmatrix} 0 & 1 \\ -1 & 0 \end{pmatrix}, Y = \begin{pmatrix} i & 0 \\ 0 & -i \end{pmatrix} \text{ and } Z = \begin{pmatrix} 0 & -i \\ -i & 0 \end{pmatrix}$$

where i is the complex number i. We are interested in how any two of the above matrices multiply. The following portion of their Cayley table contains everything we need.

	X	Y	Z
X	$-I$	Z	$-Y$
Y	$-Z$	$-I$	X
Y	Y	$-X$	$-I$

Consider matrices of the form $\lambda I + \alpha X + \beta Y + \gamma Z$ where $\lambda, \alpha, \beta, \gamma \in \mathbb{R}$. We calculate the product of two such matrices using matrix algebra and the above multiplication table. The product

$$(\lambda I + \alpha X + \beta Y + \gamma Z)(\mu I + \alpha' X + \beta' Y + \gamma' Z)$$

can be written in the form $aI + bX + cY + dZ$ where $a, b, c, d \in \mathbb{R}$ although we write it in a slightly different form

$$(\lambda \mu - \alpha \alpha' - \beta \beta' - \gamma \gamma')I +$$
$$\lambda(\alpha' X + \beta' Y + \gamma' Z) + \mu(\alpha X + \beta Y + \gamma Z) +$$
$$(\beta \gamma' - \gamma \beta')X + (\gamma \alpha' - \alpha \gamma')Y + (\alpha \beta' - \beta \alpha')Z.$$

---

[6]There is a notation clash between octonions and odd numbers. This is not an issue.

This looks complicated but contains familiar things: the first term contains what looks like an inner product and the last term contains what looks like a vector product.

The above calculation motivates the following construction. Recall that $\mathbb{R}^3$ denotes the set of all 3-dimensional vectors. A typical element of $\mathbb{R}^3$ is $\alpha\mathbf{i} + \beta\mathbf{j} + \gamma\mathbf{k}$. Put $\mathbb{H} = \mathbb{R} \times \mathbb{R}^3$. The elements of $\mathbb{H}$ are therefore ordered pairs $(\lambda, \mathbf{a})$ consisting of a real number λ and a vector $\mathbf{a}$. We define the sum of two elements of $\mathbb{H}$ in a very simple way

$$(\lambda, \mathbf{a}) + (\mu, \mathbf{a}') = (\lambda + \mu, \mathbf{a} + \mathbf{a}').$$

The product is defined in a way that mimics what we did above

$$(\lambda, \mathbf{a})(\mu, \mathbf{a}') = (\lambda\mu - \mathbf{a} \cdot \mathbf{a}', \lambda\mathbf{a}' + \mu\mathbf{a} + (\mathbf{a} \times \mathbf{a}')).$$

This product is associative because matrix multiplication is associative. We now investigate what we can do with $\mathbb{H}$. We deal with multiplication because addition poses no problems.

- Consider the subset $\mathscr{R}$ of $\mathbb{H}$ which consists of elements of the form $(\lambda, \mathbf{0})$. You can check that $(\lambda, \mathbf{0})(\mu, \mathbf{0}) = (\lambda\mu, \mathbf{0})$. Thus $\mathscr{R}$ mimics the real numbers.

- Consider the subset $\mathscr{C}$ of $\mathbb{H}$ which consists of the elements of the form $(\lambda, a\mathbf{i})$. You can check that

$$(\lambda, a\mathbf{i})(\mu, a'\mathbf{i}) = (\lambda\mu - aa', (\lambda a' + \mu a)\mathbf{i}).$$

In particular, $(0, \mathbf{i})(0, \mathbf{i}) = (-1, \mathbf{0})$. Thus $\mathscr{C}$ mimics the set of complex numbers.

- Consider the subset $\mathscr{E}$ of $\mathbb{H}$ which consists of elements of the form $(0, \mathbf{a})$. You can check that
$$(0, \mathbf{a})(0, \mathbf{a}') = (-\mathbf{a} \cdot \mathbf{a}', \mathbf{a} \times \mathbf{a}').$$

Thus $\mathscr{E}$ mimics vectors, the inner product and the vector product.

The set $\mathbb{H}$ with the above operations of addition and multiplication is called the *quaternions*. The proof of the following will be completed in the exercises.

Theorem 9.7.1 (Properties of quaternions). *The quaternions satisfy all the axioms for a field except commutativity of multiplication.*

In working with quaternions, it is more convenient to write $(\lambda, \mathbf{a})$ as $\lambda + \mathbf{a}$, which we shall do in the exercises. Thus a quaternion can be regarded as the sum of a scalar and a vector, which is analogous to the way that a complex number can be written as the sum of a real number and a purely imaginary number.

Exercises 9.7

1. *We construct a counterexample to show that the product of two natural numbers, each of which can be written as the sum of three squares, need not itself be written as the sum of three squares.

   (a) Prove that if a natural number n can be written as a sum of three squares then n cannot be of the form $8m + 7$.

   (b) Prove that both 3 and 21 can be written as a sum of three squares but that their product 63 cannot.

2. *We complete the proof of Theorem 9.7.1. Let $u = a + bi + cj + dk$ be a quaternion. Define $u^* = a - bi - cj - dk$.

   (a) Calculate uu^*.

   (b) Prove that if $u \neq 0$ then u is invertible.

3. *For the quaternion $\lambda + \mathbf{a}$ define $|\lambda + \mathbf{a}| = \sqrt{\lambda^2 + \mathbf{a} \cdot \mathbf{a}}$.

   (a) Let $\mathbf{u}, \mathbf{v} \in \mathbb{H}$ where $\mathbf{u} = \lambda + u_1 \mathbf{i} + u_2 \mathbf{j} + u_3 \mathbf{k}$ and $\mathbf{v} = \mu + v_1 \mathbf{i} + v_2 \mathbf{j} + v_3 \mathbf{k}$. Prove that $|\mathbf{u}\mathbf{v}| = |\mathbf{u}|\,|\mathbf{v}|$.

   (b) Prove that the product of two natural numbers, each of which can be written as the sum of four squares, can itself be written as the sum of four squares.

4. *We showed in Section 6.4 that complex numbers could be constructed from ordered pairs of real numbers. The quaternions can be constructed from ordered pairs of complex numbers. Let $(a,b), (c,d) \in \mathbb{C}^2$. Define

$$(a,b)(c,d) = (ac - \bar{d}b, da + b\bar{c}).$$

Show that this can be interpreted as the multiplication of two quaternions.

Box 9.3: Geometric Algebra

We described the theory of vectors as portrayed in this chapter as being a practical tool for studying spatial geometry. And so it is. But that does not mean this is the end of the story. Mathematics is a living subject and not a mere fossil. There are a couple of problems with vector algebra. To begin with, vector products cannot be defined in higher dimensions. This does not make doing higher-dimensional geometry impossible but it is less convenient. But there is also what might be called a philosophical problem. To understand what this is, we compare scalars and vectors. A scalar is just a number and has magnitude but no direction. A vector has magnitude and direction. These are represented by directed line segments. What might come next? Following the internal logic, we would want a directed area. Now we get something like that with the vector product of two vectors. This is because $\mathbf{a} \times \mathbf{b}$ has magnitude the area of the parallelogram enclosed and the orientation of this parallelogram is specified by choosing one of the two unit vectors orthogonal to both $\mathbf{a}$ and $\mathbf{b}$. Thus $\mathbf{a} \times \mathbf{b}$ is a way of describing an oriented area. But an oriented area should be something new, not just a vector. Viewed in this light, $\mathbf{a} \times \mathbf{b}$ is a vectorial way of representing something that should perhaps be represented directly by means of a new concept. These ideas go back to the work of Hermann Grassmann [a] (1809–1877), but were most fully realized in the geometric algebra of William Kingdon Clifford (1845–1879). These provide the perspective for viewing everything in this chapter. I cannot resist ending with a quote from an article by John Baez [6].

> "There are exactly four normed division algebras [...] The real numbers are the dependable breadwinner of the family, the complete ordered field you can rely on. The complex numbers are a slightly flashier but still respectable younger brother: not ordered, but algebraically complete. The quaternions, being noncommutative, are the eccentric cousin who is shunned at important family gatherings. But the octonions are the crazy old uncle nobody lets out of the attic: they are *nonassociative*".

---
[a]Grassmann was also a noted scholar of Indo-European languages [127].

CHAPTER 10

The principal axes theorem

"Only connect." – Edward Morgan Forster

To translate geometry into algebra requires a coordinate system to be chosen. Unless chosen carefully, there is no reason why it should be well adapted to the geometry that interests us. This means that nice geometry can lead to nasty algebra. This sounds like a problem but it is one we can turn to our advantage: it means that nasty algebra can actually arise from nice geometry. This is the idea behind this chapter. It will be used to help us study equations of degree 2 in two or three variables called, respectively, conics and quadrics. Equations such as these can be handled using the methods of this book, but those of higher degree cannot, so they are a fitting place to end. Not surprisingly, the theme that permeates this book — the relationship between algebra and geometry — will be uppermost: we perform algebraic operations motivated by geometric ideas which we ultimately interpret geometrically. This chapter makes heavy use of Section 8.6 and Section 9.5. All the matrices in this chapter will either be 2×2 or 3×3. Thus when we write that a matrix is an $n \times n$ matrix we mean that $n = 2$ or $n = 3$.

10.1 ORTHOGONAL MATRICES

In Section 8.5, we proved that a square matrix A is invertible if and only if its columns are linearly independent. In the 2×2 and 3×3 cases, this means that the columns form a basis for $\mathbb{R}^2$ and $\mathbb{R}^3$, respectively. We now focus on those matrices A where the columns form an *orthonormal* basis. This is equivalent to saying that $A^T A = I$. We accordingly make the following definition. A real square matrix A is *orthogonal* if $A^T A = I$. It readily follows that a matrix A is orthogonal if and only if A is invertible with inverse A^T. The importance of this class of matrices is explained by Proposition 10.1.2 below. To prove it, we need the following lemma left as an exercise.

Lemma 10.1.1 (Polarization identity). *The identity*

$$\mathbf{u} \cdot \mathbf{v} = \tfrac{1}{2} \left(\mathbf{u}^2 + \mathbf{v}^2 - (\mathbf{u} - \mathbf{v})^2 \right)$$

holds for all vectors $\mathbf{u}$ *and* $\mathbf{v}$.

Recall that if $\mathbf{a}$ is a vector then $\mathbf{a}^2 = \mathbf{a} \cdot \mathbf{a}$, the square of the length of $\mathbf{a}$. Thus Lemma 10.1.1 shows that the inner product is determined by lengths.

Proposition 10.1.2. *The following are equivalent for an $n \times n$ matrix A.*

1. *A is orthogonal.*

2. $\|A\mathbf{x}\| = \|\mathbf{x}\|$ *for all* $\mathbf{x} \in \mathbb{R}^n$.

3. $(A\mathbf{x}) \cdot (A\mathbf{y}) = \mathbf{x} \cdot \mathbf{y}$.

Proof. Throughout the proof, we use the fact that $(\mathbf{u} \cdot \mathbf{v}) = \mathbf{u}^T \mathbf{v}$ where $\mathbf{u}$ and $\mathbf{v}$ are column vectors.

$(1) \Rightarrow (2)$. This follows from the matrix calculations

$$(A\mathbf{x})^T A\mathbf{x} = \mathbf{x}^T (A^T A)\mathbf{x} = \mathbf{x}^T \mathbf{x}.$$

$(2) \Rightarrow (3)$. By Lemma 10.1.1

$$(A\mathbf{x}) \cdot (A\mathbf{y}) \quad = \quad \tfrac{1}{2} \left((A\mathbf{x})^2 + (A\mathbf{y})^2 - (A\mathbf{x} - A\mathbf{y})^2 \right). \tag{10.1}$$

By distributivity $A\mathbf{x} - A\mathbf{y} = A(\mathbf{x} - \mathbf{y})$. By assumption the righthand side of equation (10.1) is

$$\tfrac{1}{2} \left(\mathbf{x}^2 + \mathbf{y}^2 - (\mathbf{x} - \mathbf{y})^2 \right)$$

which in turn is equal to $\mathbf{x} \cdot \mathbf{y}$ by Lemma 10.1.1 again.

$(3) \Rightarrow (1)$. By assumption,

$$\mathbf{x}^T \mathbf{y} = \mathbf{x}^T A^T A\mathbf{y}$$

for all $\mathbf{x}, \mathbf{y} \in \mathbb{R}^n$. This can be rewritten as

$$\mathbf{x}^T \left(A^T A - I \right) \mathbf{y} = \mathbf{0}.$$

Define $B = A^T A - I$ and choose $\mathbf{x} = B\mathbf{y}$. Then $\mathbf{y}^T B^T B\mathbf{y} = \mathbf{0}$. Hence $(B\mathbf{y}) \cdot (B\mathbf{y}) = 0$. It follows that $B\mathbf{y} = \mathbf{0}$ for all $\mathbf{y}$. By choosing for $\mathbf{y}$ the vectors $\mathbf{i}, \mathbf{j}$ and $\mathbf{k}$, respectively, we deduce that $B = O$. It follows that $A^T A = I$, and so A is orthogonal. □

By Proposition 10.1.2 the function from $\mathbb{R}^n$ to itself defined by $\mathbf{x} \mapsto A\mathbf{x}$, where A is orthogonal, does not change angles or lengths and so, more informally, does not change shapes. This is crucial to our project. The proofs of the following are all straightforward and left as exercises.

Lemma 10.1.3.

1. *The determinant of an orthogonal matrix is ± 1.*

2. *If A is an orthogonal matrix then A^{-1} exists and $A^{-1} = A^{T}$.*

3. *A is orthogonal if and only if A^{T} is orthogonal.*

4. *If A and B are orthogonal then AB is orthogonal.*

5. *If A is symmetric and B is a matrix the same size as A then $B^{T}AB$ is symmetric.*

Part (1) of Lemma 10.1.3 tells us that there are two classes of orthogonal matrices: those with determinant 1 and those with determinant -1. We shall see that those with determinant 1 have a simple geometric description. In the case of 2×2 orthogonal matrices, we can also easily describe those with determinant -1. The geometric interpretations of the functions below were described in Section 9.5.

Theorem 10.1.4 (Planar rotations and reflections). *Let A be a 2×2 orthogonal matrix.*

1. *If $\det(A) = 1$ then*
$$A = \begin{pmatrix} \cos\theta & -\sin\theta \\ \sin\theta & \cos\theta \end{pmatrix}$$

and represents a rotation about the origin by θ in an anticlockwise direction.

2. *If $\det(A) = -1$ then*
$$A = \begin{pmatrix} \cos\theta & \sin\theta \\ \sin\theta & -\cos\theta \end{pmatrix}$$

and represents a reflection in the line through the origin at an angle of $\frac{\theta}{2}$.

Proof. We prove (1) and (2) together. Let
$$A = \begin{pmatrix} a & b \\ c & d \end{pmatrix}$$

be an orthogonal matrix. Then
$$a^2 + c^2 = 1 = b^2 + d^2 \text{ and } ab + cd = 0.$$

By trigonometry there is an angle α such that
$$a = \cos\alpha \text{ and } c = \sin\alpha$$

and an angle β such that
$$b = \sin\beta \text{ and } d = \cos\beta.$$

With these choices, the condition $ab + cd = 0$ translates into
$$\sin(\alpha + \beta) = 0.$$

Thus $\alpha + \beta = n\pi$ where $n \in \mathbb{Z}$. There are now two cases to consider depending on whether n is even or odd.

(1) n is even. Then $\beta = 2m\pi - \alpha$ for some m. Thus

$$A = \begin{pmatrix} \cos\alpha & \sin(2m\pi - \alpha) \\ \sin\alpha & \cos(2m\pi - \alpha) \end{pmatrix} = \begin{pmatrix} \cos\alpha & -\sin\alpha \\ \sin\alpha & \cos\alpha \end{pmatrix}$$

which has determinant 1.

(2) n is odd. Then $\beta = (2m+1)\pi - \alpha$ for some m. Thus

$$A = \begin{pmatrix} \cos\alpha & \sin((2m+1)\pi - \alpha) \\ \sin\alpha & \cos((2m+1)\pi - \alpha) \end{pmatrix} = \begin{pmatrix} \cos\alpha & \sin\alpha \\ \sin\alpha & -\cos\alpha \end{pmatrix}$$

which has determinant -1. □

The case of 3×3 orthogonal matrices is more complex. We only consider here those with determinant 1. Before we get down to business, we need to say more about a matrix function that made a brief appearance in Section 8.6. If A is a square matrix, the *trace of A*, denoted by $\mathrm{tr}(A)$, is the sum of the entries in the leading diagonal. Thus

$$\mathrm{tr}\begin{pmatrix} a & b \\ c & d \end{pmatrix} = a + d.$$

Unlike the determinant, the trace does not have a straightforward geometrical interpretation. The properties we need are listed below and the proofs are left as exercises.

Lemma 10.1.5 (Properties of the trace).

1. $\mathrm{tr}(A + B) = \mathrm{tr}(A) + \mathrm{tr}(B)$.

2. $\mathrm{tr}(\lambda A) = \lambda \mathrm{tr}(A)$.

3. $\mathrm{tr}(AB) = \mathrm{tr}(BA)$.

4. *If* $A = C^{-1}BC$ *then* $\mathrm{tr}(A) = \mathrm{tr}(B)$.

The following uses the theory described in Section 8.6.

Theorem 10.1.6 (Spatial rotations). *A 3×3 orthogonal matrix A with determinant 1 describes a rotation through an angle θ where* $\mathrm{tr}(A) = 1 + 2\cos\theta$.

Proof. We prove first that there is a unit vector $\mathbf{n}$ such that $A\mathbf{n} = \mathbf{n}$. We do this by showing that $A - I$ is singular. We can compute $\det(A - I)$ as follows

$$\det(A - I) = \det(A - AA^T) = \det(A(I - A^T)) = \det(A)\det(I - A^T)$$

by Theorem 8.4.3. Thus

$$\det(A - I) = \det(I - A^T)$$

because we are assuming that $\det(A) = 1$. Also

$$\det(I - A^T) = \det(I - A)^T = \det(I - A)$$

again using Theorem 8.4.3. However $\det(I - A) = -\det(A - I)$ and so we have shown that $\det(A - I) = -\det(A - I)$. It follows that $\det(A - I) = 0$, and so the matrix $A - I$ is singular. By Theorem 8.5.22 there is therefore a non-zero vector $\mathbf{v}$ such that $(A - I)\mathbf{v} = \mathbf{0}$. This is equivalent to $A\mathbf{v} = \mathbf{v}$. Define $\mathbf{n}$ to be the normalization of $\mathbf{v}$. Then $\mathbf{n}$ is a unit vector such that $A\mathbf{n} = \mathbf{n}$.

We are interested in the behaviour of the function $f(\mathbf{x}) = A\mathbf{x}$. There is a unique plane P through the origin orthogonal to $\mathbf{n}$. The line L determined by $\mathbf{n}$, that is all scalar multiples of $\mathbf{n}$, is fixed under f. The plane P is mapped to itself under f because A is an orthogonal matrix and so, by Proposition 10.1.2, preserves angles between vectors, which means that vectors orthogonal to L are mapped to vectors orthogonal to L. The function f induces on the vectors in the plane P by restriction of the domain (see the end of Section 3.4) is linear. This means that it can be described by a 2×2 matrix A' on P by Theorem 9.5.1. In addition it preserves lengths. This means that the matrix A' is orthogonal by Proposition 10.1.2.

Choose two unit orthogonal vectors $\mathbf{a}$ and $\mathbf{b}$ in P such that $\mathbf{a} \times \mathbf{b} = \mathbf{n}$. In the plane P, you can regard $\mathbf{a}$ as $\mathbf{i}$ and $\mathbf{b}$ as $\mathbf{j}$. Now $A\mathbf{a} = a_{11}\mathbf{a} + a_{21}\mathbf{b}$ and $A\mathbf{b} = a_{12}\mathbf{a} + a_{22}\mathbf{b}$. Thus restricted to the plane P the function f is described by

$$A' = \begin{pmatrix} a_{11} & a_{12} \\ a_{21} & a_{22} \end{pmatrix}.$$

Define U to be the matrix with column vectors $\mathbf{n}$, $\mathbf{a}$ and $\mathbf{b}$. It is orthogonal with determinant 1. Put

$$B = \begin{pmatrix} 1 & 0 & 0 \\ 0 & a_{11} & a_{12} \\ 0 & a_{21} & a_{22} \end{pmatrix}.$$

Observe that $AU = UB$. Thus $A = UBU^T$. Hence $\det(A) = \det(B)$ by Lemma 8.6.5. It follows that A' has determinant 1. By our arguments above and Theorem 10.1.4, this means that the function determined by A' rotates the vectors in the plane by some angle θ. Summing up: the function f determined by A rotates space about the axis determined by the vector $\mathbf{n}$ by an angle θ. Observe that $\mathrm{tr}(A) = \mathrm{tr}(B) = 1 + \mathrm{tr}(A')$ by Lemma 10.1.5. But $\mathrm{tr}(A') = 2\cos\theta$ by Theorem 10.1.4. □

Arbitrary spatial rotations are complicated to describe but the following theorem, important in solid mechanics, shows that they may be accomplished by combining simpler ones.

Theorem 10.1.7 (Euler angles). *Every 3×3 orthogonal matrix A with determinant 1 can be written in the form*

$$A = \begin{pmatrix} \cos\gamma & -\sin\gamma & 0 \\ \sin\gamma & \cos\gamma & 0 \\ 0 & 0 & 1 \end{pmatrix} \begin{pmatrix} 1 & 0 & 0 \\ 0 & \cos\beta & -\sin\beta \\ 0 & \sin\beta & \cos\beta \end{pmatrix} \begin{pmatrix} \cos\alpha & -\sin\alpha & 0 \\ \sin\alpha & \cos\alpha & 0 \\ 0 & 0 & 1 \end{pmatrix}$$

for some α, β and γ called the Euler angles.

Proof. The proof is a procedure for finding the Euler angles.[1] Let $A = (a_{ij})$ be an arbitrary 3×3 orthogonal matrix. To prove the theorem, it is enough to find angles ϕ, ψ and β such that

$$\begin{pmatrix} \cos \phi & -\sin \phi & 0 \\ \sin \phi & \cos \phi & 0 \\ 0 & 0 & 1 \end{pmatrix} A \begin{pmatrix} \cos \psi & -\sin \psi & 0 \\ \sin \psi & \cos \psi & 0 \\ 0 & 0 & 1 \end{pmatrix} = \begin{pmatrix} 1 & 0 & 0 \\ 0 & \cos \beta & -\sin \beta \\ 0 & \sin \beta & \cos \beta \end{pmatrix}$$

since then $\gamma = -\phi$ and $\alpha = -\psi$. With reference to the above equation, we proceed as follows.

1. The entry in the lefthand side of the above equation in the third row and third column is a_{33}. Because A is an orthogonal matrix, we know that $|a_{33}| \leq 1$. We can therefore find β such that $a_{33} = \cos \beta$ where $0 \leq \beta < 2\pi$.

2. The entry in the lefthand side of the above equation in row one and column three is $a_{13} \cos \phi - a_{23} \sin \phi = 0$. If $a_{23} \neq 0$ then $\frac{a_{13}}{a_{23}} = \tan \phi$. The function $\tan \colon (-\frac{\pi}{2}, \frac{\pi}{2}) \to \mathbb{R}$ is surjective and so ϕ exists. If $a_{23} = 0$ then we can choose $\phi = \frac{\pi}{2}$. In either event, we can solve for ϕ.

3. We now focus on the first two rows and columns of the lefthand side of the above equation. We have that

$$\begin{pmatrix} \cos \phi & -\sin \phi \\ \sin \phi & \cos \phi \end{pmatrix} \begin{pmatrix} a_{11} & a_{12} \\ a_{21} & a_{22} \end{pmatrix} \begin{pmatrix} \cos \psi & -\sin \psi \\ \sin \psi & \cos \psi \end{pmatrix} = \begin{pmatrix} 1 & 0 \\ 0 & \cos \beta \end{pmatrix}$$

which can be rewritten as

$$\begin{pmatrix} \cos \phi & -\sin \phi \\ \sin \phi & \cos \phi \end{pmatrix} \begin{pmatrix} a_{11} & a_{12} \\ a_{21} & a_{22} \end{pmatrix} = \begin{pmatrix} 1 & 0 \\ 0 & \cos \beta \end{pmatrix} \begin{pmatrix} \cos \psi & \sin \psi \\ -\sin \psi & \cos \psi \end{pmatrix}.$$

In particular

$$\begin{pmatrix} \cos \phi & -\sin \phi \end{pmatrix} \begin{pmatrix} a_{11} & a_{12} \\ a_{21} & a_{22} \end{pmatrix} = \begin{pmatrix} \cos \psi & \sin \psi \end{pmatrix}$$

which we have to solve for ψ. Observe that the product

$$\begin{pmatrix} \cos \phi & -\sin \phi & 0 \\ \sin \phi & \cos \phi & 0 \\ 0 & 0 & 1 \end{pmatrix} \begin{pmatrix} a_{11} & a_{12} & a_{13} \\ a_{21} & a_{22} & a_{23} \\ a_{31} & a_{32} & a_{33} \end{pmatrix}$$

is an orthogonal matrix. Thus the top row of this product is a unit vector and, from the choice of ϕ above, its third component is zero. Its remaining two components are computed from

$$\begin{pmatrix} \cos \phi & -\sin \phi \end{pmatrix} \begin{pmatrix} a_{11} & a_{12} \\ a_{21} & a_{22} \end{pmatrix}$$

which therefore has unit length. It follows that ψ exists.

---

[1] I am following the proof described in [43].

It can be finally checked that the values we have found satisfy all conditions. □

We now spell out what Theorem 10.1.7 is saying. Let

$$A_1 = \begin{pmatrix} \cos\alpha & -\sin\alpha & 0 \\ \sin\alpha & \cos\alpha & 0 \\ 0 & 0 & 1 \end{pmatrix}.$$

Then $A_1\mathbf{k} = \mathbf{k}$, where we regard $\mathbf{k}$ as a column vector. Thus A_1 represents a rotation about the z-axis by an angle α. Let

$$A_2 = \begin{pmatrix} 1 & 0 & 0 \\ 0 & \cos\beta & -\sin\beta \\ 0 & \sin\beta & \cos\beta \end{pmatrix}.$$

Then $A_2\mathbf{i} = \mathbf{i}$. It follows that A_2 represents a rotation about the x-axis by an angle β. To visualize Theorem 10.1.7 imagine a sphere that is free to rotate about its centre, and choose x-, y- and z-axes fixed in the sphere. Any rotation of the sphere can be accomplished by carrying out the following sequence of rotations.

1. Rotate by some angle α about the z-axis.

2. Rotate by some angle β about the x-axis.

3. Rotate by some angle γ about the z-axis again.

We conclude this section with a theorem that is the starting point for investigations of symmetry in space, and is also useful in the mechanics of solid bodies. A bijective function $f : \mathbb{R}^3 \to \mathbb{R}^3$ is called an *isometry* if $\|f(\mathbf{u}) - f(\mathbf{v})\| = \|\mathbf{u} - \mathbf{v}\|$ for all $\mathbf{u}, \mathbf{v}$. Thus isometries are precisely the functions which preserve the distance between points. We can describe exactly what these functions look like.

Theorem 10.1.8 (Description of isometries). *The isometries of $\mathbb{R}^3$ are precisely the functions of the form $\mathbf{x} \mapsto A\mathbf{x} + \mathbf{a}$ where A is an orthogonal matrix and $\mathbf{a}$ is any vector.*

Proof. The fact that the functions $\mathbf{x} \mapsto A\mathbf{x} + \mathbf{a}$ are all isometries is left as an exercise. We prove the converse. Let f be an isometry. Put $f(\mathbf{0}) = -\mathbf{a}$. Then the function $g(\mathbf{x}) = f(\mathbf{x}) + \mathbf{a}$ is also an isometry but $g(\mathbf{0}) = \mathbf{0}$. Thus it is enough to restrict to the case where $f(\mathbf{0}) = \mathbf{0}$ and prove that $f(\mathbf{x}) = A\mathbf{x}$ for some orthogonal matrix A. By Lemma 10.1.1 and the fact that $f(\mathbf{0}) = \mathbf{0}$

$$f(\mathbf{u}) \cdot f(\mathbf{v}) = \mathbf{u} \cdot \mathbf{v} \tag{10.2}$$

for all $\mathbf{u}, \mathbf{v} \in \mathbb{R}^3$. Put

$$\mathbf{w} = f(\lambda\mathbf{u} + \mu\mathbf{v}) - \lambda f(\mathbf{u}) - \mu f(\mathbf{v}).$$

Then

$$\mathbf{w} \cdot f(\mathbf{x}) = (\lambda\mathbf{u} + \mu\mathbf{v}) - \lambda\mathbf{u} \cdot \mathbf{x} - \mu\mathbf{v} \cdot \mathbf{x} = 0$$

by equation (10.2). It follows that $\mathbf{w} \cdot f(\mathbf{x}) = 0$ for all $\mathbf{x}$. By assumption, f is bijective and so it is surjective. It follows that $\mathbf{w} = f(\mathbf{x})$ for some $\mathbf{x}$. But then $\mathbf{w}^2 = 0$ and so $\mathbf{w} = 0$. We have therefore proved that f is linear. By Theorem 9.5.1, $f(\mathbf{x}) = A\mathbf{x}$ for some matrix A. By equation (10.2) and Proposition 10.1.2, we deduce that A is orthogonal. □

Examples 10.1.9.

1. A special class of isometries is obtained by taking the orthogonal matrix in Theorem 10.1.8 to be the identity matrix. We then get functions of the form $\mathbf{x} \mapsto \mathbf{x} + \mathbf{a}$ called *translations*, for obvious reasons.

2. Another special class of isometries is obtained by considering those functions $\mathbf{x} \mapsto A\mathbf{x}$ where the matrix A is obtained from the identity matrix I by permuting its columns. The determinant of such a matrix is 1 or -1 depending on whether the permutation is even or odd. The effect of such a matrix is simply to interchange variables.

Exercises 10.1

1. Prove Lemma 10.1.1.

2. Prove Lemma 10.1.3.

3. (a) Prove Lemma 10.1.5.

   (b) Let A be a fixed $n \times n$ matrix. Find all solutions to $AX - XA = I$ where I is the $n \times n$ identity matrix.

4. Denote by $SO(3)$ the set of 3×3 real orthogonal matrices with determinant 1. Prove that this set is closed under matrix multiplication and under inverses.

5. Prove that the function from $\mathbb{R}^3$ to itself given by $\mathbf{x} \mapsto A\mathbf{x} + \mathbf{a}$, where A is orthogonal, is an isometry.

6. *The crystallographic restriction. Suppose that the trace of the matrix below is an integer

$$\begin{pmatrix} \cos\phi & -\sin\phi & 0 \\ \sin\phi & \cos\phi & 0 \\ 0 & 0 & 1 \end{pmatrix}.$$

Prove that $\phi = \frac{2\pi}{k}$ where $k = 1, 2, 3, 4, 6$ and observe that 5 is not in this list.

7. *Let A be a square real matrix. Let $\varepsilon > 0$ be small so that ε^2 is really small. Prove that $\det(I + \varepsilon A) \approx 1 + \varepsilon \operatorname{tr}(A)$ where $\approx$ means *approximately equal to*. (A better result is obtained if the dual numbers of Exercises 7.11 are used instead.)

10.2 ORTHOGONAL DIAGONALIZATION

We prove one theorem in this section. It has important applications in mechanics to the study of solid bodies and to oscillations as well as to the classification of conics and quadrics described in the next section.

Theorem 10.2.1 (Principal axes theorem).

1. *Let A be a* 2×2 *symmetric matrix. Then there is an orthogonal matrix B, in fact a rotation, such that* $B^T A B$ *is a diagonal matrix. In addition, the characteristic polynomial of A has only real roots.*

2. *Let A be a* 3×3 *symmetric matrix. Then there is an orthogonal matrix B, in fact a rotation, such that* $B^T A B$ *is a diagonal matrix. In addition, the characteristic polynomial of A has only real roots.*

Proof. (1) The proof is a bare-handed calculation. Let

$$A = \begin{pmatrix} a & b \\ b & c \end{pmatrix}$$

be the real symmetric matrix in question and let

$$B = \begin{pmatrix} \cos\theta & -\sin\theta \\ \sin\theta & \cos\theta \end{pmatrix}$$

be an orthogonal matrix representing a rotation. Our goal is to prove that there is a value of θ such that $B^T A B$ is a diagonal matrix. We therefore calculate $B^T A B$ explicitly using standard trigonometric formulae to get

$$\begin{pmatrix} a\cos^2\theta + c\sin^2\theta + b\sin 2\theta & \frac{c-a}{2}\sin 2\theta + b\cos 2\theta \\ \frac{c-a}{2}\sin 2\theta + b\cos 2\theta & a\sin^2\theta + c\cos^2\theta - b\sin 2\theta \end{pmatrix}.$$

As expected, the resulting matrix is also symmetric. There is therefore only one entry we need concentrate on and that is

$$\frac{c-a}{2}\sin 2\theta + b\cos 2\theta.$$

The question is whether we can find a value of θ that makes this zero. We therefore need to solve the equation

$$\frac{c-a}{2}\sin 2\theta + b\cos 2\theta = 0. \qquad (10.3)$$

We deal with a special case first. If $c = a$ then $\theta = \frac{\pi}{4}$ works so in what follows we can assume that $c \neq a$. With a little rearrangement, equation (10.3) becomes

$$\tan 2\theta = \frac{2b}{a-c}.$$

At this point we are home and dry because the function $\tan: (-\frac{\pi}{2}, \frac{\pi}{2}) \to \mathbb{R}$ is surjective. Thus equation (10.3) has a solution $2\theta \in (-\frac{\pi}{2}, \frac{\pi}{2})$. Finally, we know from the theory developed in Section 8.6 that when $B^T AB$ is a diagonal matrix its diagonal entries are the eigenvalues of A, which are equal to the roots of the characteristic polynomial of A. Clearly these are real numbers.

(2) Let A be a 3×3 real symmetric matrix. Its characteristic polynomial is a real polynomial of degree three and therefore has at least one real root λ by Section 7.3. By Section 8.6, there is a non-zero vector $\mathbf{x}_1$ such that $A\mathbf{x}_1 = \lambda\mathbf{x}_1$. We can normalize $\mathbf{x}_1$ and so we can assume that $\mathbf{x}_1$ has unit length. Unit vectors $\mathbf{x}_2$ and $\mathbf{x}_3$ can be found so that $\mathbf{x}_1, \mathbf{x}_2, \mathbf{x}_3$ form an orthonormal basis. To see why, let P be the plane through the origin orthogonal to $\mathbf{x}_1$. Let $\mathbf{x}_2$ be any unit vector in that plane. Then $\mathbf{x}_1$, $\mathbf{x}_2$, and $\mathbf{x}_3 = \mathbf{x}_1 \times \mathbf{x}_2$ are three orthonormal vectors. Let B be the matrix with columns $\mathbf{x}_1, \mathbf{x}_2, \mathbf{x}_3$. Then B is an orthogonal matrix. The matrix $B^T AB$ is symmetric and has the following form

$$\begin{pmatrix} \lambda & 0 & 0 \\ 0 & a & b \\ 0 & b & c \end{pmatrix}.$$

The entries in the first row and first column arise from calculations such as

$$\mathbf{x}_1^T A\mathbf{x}_2 = (A^T\mathbf{x}_1)^T\mathbf{x}_2 = \lambda\mathbf{x}_1^T\mathbf{x}_2 = (0)$$

where we have used the fact that $A^T = A$. We denote the 2×2 symmetric submatrix in the bottom righthand corner by A_1. By part (1), there is a 2×2 orthogonal matrix

$$B_1 = \begin{pmatrix} p & q \\ r & s \end{pmatrix}$$

such that $B_1^T A_1 B_1$ is a diagonal matrix. Define

$$C = \begin{pmatrix} 1 & 0 & 0 \\ 0 & p & q \\ 0 & r & s \end{pmatrix}.$$

Then C is an orthogonal matrix and $C^T(B^T AB)C$ is a diagonal matrix. By Lemma 10.1.3, the matrix BC is orthogonal and $(BC)^T A(BC)$ is diagonal. Thus A can be diagonalized and its eigenvalues are real. □

Example 10.2.2. We should say something about the name 'principal axes theorem'. A special case will suffice. Consider an ellipse in the plane. There are two axes of symmetry denoted l_1 and l_2 which intersect at a point P. Call these the principal axes of the ellipse. If the coordinate system is chosen with origin P and with the x-axis lined up with l_1 and the y-axis lined up with l_2 then the equation of the ellipse in this system has the form $ax^2 + by^2 = c$. If on the other hand coordinate axes are chosen which are rotated with respect to l_1 and l_2 then in general cross-terms will be introduced in the equation of the ellipse. More generally, the principal axes of a quadratic form are the mutually orthogonal eigenvectors of its associated symmetric matrix.

Box 10.1: Symmetric Matrices

Theorem 10.2.1 is in fact true for all real symmetric matrices and can be proved by induction using the ideas contained in the proof of part (2). We prove here two general results about symmetric matrices because they are easy and striking. Let A be any symmetric matrix.

1. All the roots of its characteristic polynomial are real.

2. Eigenvectors belonging to distinct eigenvalues are orthogonal.

Proof of (1). We work with column vectors $\mathbf{u}$ whose entries might be complex which is a new departure. Write $\bar{\mathbf{u}}$ to mean the vector obtained by taking the complex conjugate of each entry. Suppose that $\lambda \in \mathbb{C}$ is such that

$$A\mathbf{u} = \lambda \mathbf{u}$$

for some non-zero $\mathbf{u}$. In this equation, take the complex conjugate of every element. Then $A\bar{\mathbf{u}} = \bar{\lambda}\bar{\mathbf{u}}$ since A is real. Take the transpose of both sides to get

$$\bar{\mathbf{u}}^T A = \bar{\lambda}\bar{\mathbf{u}}^T$$

using the fact that A is symmetric. From the two displayed equations we deduce that

$$\bar{\mathbf{u}}^T A\mathbf{u} = \bar{\lambda}\bar{\mathbf{u}}^T \mathbf{u} = \lambda \bar{\mathbf{u}}^T \mathbf{u}.$$

Now $\bar{\mathbf{u}}^T \mathbf{u}$ is a sum of real squares so that if $\mathbf{u} \neq \mathbf{0}$ then $\bar{\mathbf{u}}^T \mathbf{u} \neq 0$. By cancelling, we get that $\bar{\lambda} = \lambda$.

Proof of (2). We need to explain what we mean by saying the eigenvectors are orthogonal because we are not assuming that our vectors are spatial. Let $\mathbf{u}, \mathbf{v} \in \mathbb{R}^n$. Define $\mathbf{u} \cdot \mathbf{v}$ to be the inner product as defined in Section 8.1. Thus $\mathbf{u}^T \mathbf{v} = (\mathbf{u} \cdot \mathbf{v})$. We *define* $\mathbf{u}$ and $\mathbf{v}$ to be *orthogonal* if $\mathbf{u} \cdot \mathbf{v} = 0$.

Let $A\mathbf{u} = \lambda \mathbf{u}$ and $A\mathbf{v} = \mu \mathbf{v}$ where $\lambda \neq \mu$ and $\mathbf{u} \neq \mathbf{0} \neq \mathbf{v}$. Then

$$\lambda(\mathbf{u}^T \mathbf{v}) = (A\mathbf{u})^T \mathbf{v} = \mathbf{u}^T A^T \mathbf{v} = \mathbf{u}^T A\mathbf{v} = \mu(\mathbf{u}^T \mathbf{v}).$$

It follows that $\mathbf{u} \cdot \mathbf{v} = 0$.

Let A be a symmetric matrix. The process of finding an orthogonal matrix B such that $B^T AB$ is a diagonal matrix is called *orthogonal diagonalization*. In the following two examples, we describe the practical method for orthogonally diagonalizing a symmetric matrix.

Example 10.2.3. The symmetric matrix

$$A = \begin{pmatrix} 2 & 2 & -10 \\ 2 & 11 & 8 \\ -10 & 8 & 5 \end{pmatrix}$$

has the characteristic polynomial

$$-x^3 + 18x^2 + 81x - 1458$$

and three distinct eigenvalues $9, -9, 18$. The case of distinct eigenvalues is the easiest one to calculate with. Three corresponding eigenvectors are

$$\begin{pmatrix} 2 \\ 2 \\ -1 \end{pmatrix}, \quad \begin{pmatrix} 2 \\ -1 \\ 2 \end{pmatrix} \text{ and } \begin{pmatrix} 1 \\ -2 \\ -2 \end{pmatrix}.$$

These three vectors are pairwise orthogonal as the theory predicted. The next step is to normalize these vectors to get

$$\frac{1}{3}\begin{pmatrix} 2 \\ 2 \\ -1 \end{pmatrix}, \quad \frac{1}{3}\begin{pmatrix} 2 \\ -1 \\ 2 \end{pmatrix} \quad \text{and} \quad \frac{1}{3}\begin{pmatrix} 1 \\ -2 \\ -2 \end{pmatrix}.$$

The following is therefore an orthogonal matrix

$$U = \frac{1}{3}\begin{pmatrix} 2 & 2 & 1 \\ 2 & -1 & -2 \\ -1 & 2 & -2 \end{pmatrix}$$

and

$$U^T A U = \begin{pmatrix} 9 & 0 & 0 \\ 0 & -9 & 0 \\ 0 & 0 & 18 \end{pmatrix}.$$

Repeated eigenvalues cause us more work and we need the following lemma whose proof is left as an exercise. It is a special case of a general procedure for converting linearly independent vectors into orthogonal vectors.

Lemma 10.2.4 (Gram-Schmidt process). *Let* $\mathbf{u}, \mathbf{v} \in \mathbb{R}^3$ *be non-zero linearly independent vectors. Define*

$$\mathbf{u}' = \mathbf{u} - \left(\frac{\mathbf{u} \cdot \mathbf{v}}{\mathbf{v} \cdot \mathbf{v}}\right)\mathbf{v}.$$

Then $\mathbf{u}'$ *and* $\mathbf{v}$ *are orthogonal and the set of linear combinations of* $\{\mathbf{u}, \mathbf{v}\}$ *is the same as the set of linear combinations of* $\{\mathbf{u}', \mathbf{v}\}$.

The geometry that lies behind the construction used in Lemma 10.2.4 is illustrated in the following diagram where

$$\mathbf{v}' = \left(\frac{\mathbf{u} \cdot \mathbf{v}}{\mathbf{v} \cdot \mathbf{v}}\right)\mathbf{v}.$$

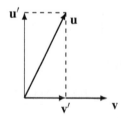

Example 10.2.5. The symmetric matrix

$$A = \begin{pmatrix} 2 & 2 & 1 \\ 2 & 5 & 2 \\ 1 & 2 & 2 \end{pmatrix}$$

has the characteristic polynomial

$$-x^3 + 9x^2 - 15x + 7$$

and three eigenvalues $7, 1, 1$ with two repeats. In this case, we have to do a bit more work. An eigenvector corresponding to 7 is

$$\begin{pmatrix} 1 \\ 2 \\ 1 \end{pmatrix}.$$

We have to find two orthogonal eigenvectors belonging to the repeated eigenvalue 1. This means solving the homogeneous equations

$$\begin{pmatrix} 1 & 2 & 1 \\ 2 & 4 & 2 \\ 1 & 2 & 1 \end{pmatrix} \begin{pmatrix} x \\ y \\ z \end{pmatrix} = \begin{pmatrix} 0 \\ 0 \\ 0 \end{pmatrix}.$$

The solution set is

$$\begin{pmatrix} -2\mu - \lambda \\ \mu \\ \lambda \end{pmatrix}$$

where $\lambda, \mu \in \mathbb{R}$. The vectors

$$\mathbf{u} = \begin{pmatrix} 1 \\ 0 \\ -1 \end{pmatrix} \text{ and } \mathbf{v} = \begin{pmatrix} 0 \\ 1 \\ -2 \end{pmatrix}$$

are linearly independent solutions but they are not orthogonal. We can find an orthogonal pair of solutions using Lemma 10.2.4. This works because linear combinations of eigenvectors belonging to a given eigenvalue also belong to that eigenvalue by Lemma 8.6.11. This procedure replaces the vector $\mathbf{u}$ by

$$\begin{pmatrix} 1 \\ -\frac{2}{5} \\ -\frac{1}{5} \end{pmatrix} \text{ or the more convenient } \begin{pmatrix} 5 \\ -2 \\ -1 \end{pmatrix}.$$

Thus two orthogonal eigenvectors belonging to the eigenvalue 1 are

$$\mathbf{u} = \begin{pmatrix} 5 \\ -2 \\ -1 \end{pmatrix} \text{ and } \mathbf{v} = \begin{pmatrix} 0 \\ 1 \\ -2 \end{pmatrix}.$$

Once we have reached this point, we continue as in the case of distinct eigenvalues. The required orthogonal matrix is therefore

$$U = \begin{pmatrix} \frac{1}{\sqrt{6}} & \frac{5}{\sqrt{30}} & 0 \\ \frac{2}{\sqrt{6}} & -\frac{2}{\sqrt{30}} & \frac{1}{\sqrt{5}} \\ \frac{1}{\sqrt{6}} & -\frac{1}{\sqrt{30}} & -\frac{2}{\sqrt{5}} \end{pmatrix}$$

and

$$U^T A U = \begin{pmatrix} 7 & 0 & 0 \\ 0 & 1 & 0 \\ 0 & 0 & 1 \end{pmatrix}.$$

<hr>

Exercises 10.2

1. Prove Lemma 10.2.4.

2. For each symmetric matrix A below find an orthogonal matrix U such that $U^T A U$ is a diagonal matrix.

   (a)
$$\begin{pmatrix} 10 & -14 & -10 \\ -14 & 7 & -4 \\ -10 & -4 & 19 \end{pmatrix}.$$

   (b)
$$\begin{pmatrix} 1 & 4 & 2 \\ 4 & 1 & 2 \\ 2 & 2 & -2 \end{pmatrix}.$$

10.3 CONICS AND QUADRICS

A *conic* is an equation of the form

$$ax^2 + 2bxy + cy^2 + dx + ey + f = 0$$

where a, b, c, d, e, f are real numbers. We exclude the case where $a = b = c = 0$ because we then have just a linear equation. Observe that the term $2bxy$ is used so that the associated matrix does not contain explicit fractions. There is no loss in generality. Conics are the most general polynomials of degree 2 in two variables. The part

$$ax^2 + 2bxy + cy^2$$

is called a *quadratic form* because every monomial has degree 2. The solutions of the conic describe a curve in the plane. Our goal is to use the theory we have developed to describe those curves. We can rewrite the above equation as a matrix equation

$$\mathbf{x}^T A \mathbf{x} + J^T \mathbf{x} + (f) = (0)$$

where

$$A = \begin{pmatrix} a & b \\ b & c \end{pmatrix} \quad \mathbf{x} = \begin{pmatrix} x \\ y \end{pmatrix} \quad J = \begin{pmatrix} d \\ e \end{pmatrix}.$$

The matrix A is symmetric. By Theorem 10.2.1, there is a rotation matrix U such that $U^T A U = D$ where

$$D = \begin{pmatrix} \lambda_1 & 0 \\ 0 & \lambda_2 \end{pmatrix}.$$

Define $\mathbf{y} = U^T\mathbf{x}$ to be new variables. If we replace $\mathbf{x}$ by $U\mathbf{y}$ and relabel the variables afterwards we obtain an equation of the form

$$\lambda_1 x^2 + \lambda_2 y^2 + gx + hy + k = 0.$$

The shape of the set of solutions of this equation will be identical to the shape of the solutions of the original equation because all we have done is rotate. Thus the quadratic form part of the equation has had its *cross-term* in xy removed. It is in this way that geometrical considerations lead to a simplification of algebra. This is the heavy lifting part. What remains is routine algebra.

We now proceed to analyse conics without cross-terms. Reorder the terms of the equation to get

$$\left(\lambda_1 x^2 + gx\right) + \left(\lambda_2 y^2 + hy\right) + k = 0.$$

There are a number of cases to consider.

1. $\lambda_1 = 0 = \lambda_2$. This case can only occur if the matrix A is the zero matrix which means that our original equation was really a linear equation. This case is therefore excluded.

2. $\lambda_1 \neq 0$ and $\lambda_2 = 0$. We divide through by λ_1. By relabelling coefficients, we therefore have an equation of the form

$$x^2 + gx + hy + k = 0.$$

We now complete the square by Section 4.3 to obtain

$$\left(x + \frac{g}{2}\right)^2 + hy + \left(k - \frac{g^2}{4}\right) = 0.$$

By changing the variable x to $x + \frac{g}{2}$, which is simply a translation, and relabelling, we obtain an equation of the form

$$x^2 + hy + k = 0.$$

There are now three subcases.

(a) $h = 0 = k$. The equation assumes the form $\boxed{x^2 = 0.}$ Then $x = 0$ and y can take any value and so we get a line in the plane.

(b) $h = 0$ and $k \neq 0$. Then $x^2 = -k$. If k is positive then this equation has no real solutions and we get the empty set. We can express this mathematically by putting $k = a^2$. We therefore have the equation $\boxed{x^2 = -a^2.}$ If k is negative then x has two real values and y can take any values. We therefore get two parallel lines. We can express this mathematically by putting $k = -a^2$. We therefore have the equation $\boxed{x^2 = a^2.}$

(c) $h \neq 0$. Then $y = -\frac{1}{h}x^2 - \frac{k}{h}$. This is a parabola. We can equivalently write this equation as $x^2 = -h(y - \frac{k}{h})$. Define a new variable to be $y - \frac{k}{h}$, by translation, and then relabel to get $x^2 = -hy$. It is conventional to put $-h = 4a$. We can then interchange the rôles of x and y to get $\boxed{y^2 = 4ax.}$

3. $\lambda_1 = 0$ and $\lambda_2 \neq 0$. This is the same analysis as that carried out in (2) but with the rôles of x and y interchanged.

4. $\lambda_1 \neq 0$ and $\lambda_2 \neq 0$. We rewrite the equation in the form

$$\lambda_1 \left(x^2 + \frac{g}{\lambda_1} x \right) + \lambda_2 \left(y^2 + \frac{h}{\lambda_2} y \right) + k = 0.$$

Completing the square where necessary and relabelling, our equation assumes the form

$$\lambda_1 x^2 + \lambda_2 y^2 = -k.$$

There are now two subcases.

(a) $\lambda_1 \lambda_2 > 0$. By multiplying though by -1 if necessary, we can assume that both λ_1 and λ_2 are positive. By relabelling, the equation can be written in the form $\lambda_1 x^2 + \lambda_2 y^2 = k$.

   i. If $k > 0$ the equation describes an ellipse. Divide both sides of the equation by k and define $\frac{\lambda_1}{k} = \frac{1}{a^2}$ and $\frac{\lambda_2}{k} = \frac{1}{b^2}$. Then we get the equation

$$\boxed{\frac{x^2}{a^2} + \frac{y^2}{b^2} = 1.}$$

   ii. If $k = 0$ then the equation describes a point. In a similar way to the above, we get the equation

$$\boxed{\frac{x^2}{a^2} + \frac{y^2}{b^2} = 0.}$$

   iii. If $k < 0$ then the equation describes the empty set. In a similar way to the above, we get the equation

$$\boxed{\frac{x^2}{a^2} + \frac{y^2}{b^2} = -1.}$$

(b) $\lambda_1 \lambda_2 < 0$. This means that λ_1 and λ_2 have opposite signs. By relabelling so that λ_1 and λ_2 are positive, the equation can be written in the form $\lambda_1 x^2 - \lambda_2 y^2 = k$.

   i. If $k \neq 0$. This is a hyperbola. In a similar way to the above, and possibly by interchanging x and y, we get the equation

$$\boxed{\frac{x^2}{a^2} - \frac{y^2}{b^2} = 1.}$$

   ii. If $k = 0$. This is two intersecting lines. In a similar way to the above, and possibly by interchanging x and y, we get the equation

$$\boxed{\frac{x^2}{a^2} - \frac{y^2}{b^2} = 0.}$$

We have therefore proved the following theorem due to Euler. The proof does not require the derivations that led to the boxed equations above, we shall explain those anon. Ellipses, hyperbolas and parabolas all have interesting properties in their own right and their study goes back to the work of Apollonius.

Theorem 10.3.1 (Classification of conics I). *The solution set of the equation*

$$ax^2 + 2bxy + cy^2 + dx + ey + f = 0,$$

where a,b,c,d,e,f are real numbers where we exclude $a = b = c = 0$, and

$$ac - b^2 = \begin{vmatrix} a & b \\ b & c \end{vmatrix},$$

is one the following.

	Nondegenerate	Degenerate
$ac - b^2 = 0$	parabola	line, two parallel lines, empty set
$ac - b^2 > 0$	ellipse	empty set, point
$ac - b^2 < 0$	hyperbola	empty set, pair of crossed lines

If we also consider the derivations that led to the boxed equations above, we arrive at the following. Observe that this may also require a reflection in the $y = x$ axis in order to interchange x and y.

Theorem 10.3.2 (Classification of conics II). *Every conic can be reduced to one of the following nine canonical forms by means of rotations, translations and reflections, the reflections enabling us to interchange the rôles of x and y. We write a^2, etc, as a way of indicating that a number is positive.*

	Equation	Description
1.	$\frac{x^2}{a^2} + \frac{y^2}{b^2} = 1$	Ellipse
2.	$\frac{x^2}{a^2} + \frac{y^2}{b^2} = -1$	Empty set
3.	$\frac{x^2}{a^2} + \frac{y^2}{b^2} = 0$	Point
4.	$\frac{x^2}{a^2} - \frac{y^2}{b^2} = 1$	Hyperbola
5.	$\frac{x^2}{a^2} - \frac{y^2}{b^2} = 0$	Pair of intersecting lines
6.	$y^2 = 4ax$	Parabola
7.	$x^2 = a^2$	Pair of parallel lines
8.	$x^2 = -a^2$	Empty set
9.	$x^2 = 0$	A line

Example 10.3.3. Let $f: \mathbb{R}^2 \to \mathbb{R}$ be a function. This can be viewed as a surface and we are interested in the humps and bumps on this surface. Assuming the function f can be differentiated twice with respect to all variables, we obtain the matrix

$$A = \begin{pmatrix} \frac{\partial^2 f}{\partial x^2} & \frac{\partial^2 f}{\partial x \partial y} \\ \frac{\partial^2 f}{\partial y \partial x} & \frac{\partial^2 f}{\partial y^2} \end{pmatrix}.$$

Under mild conditions on f, this matrix is symmetric. The nature of the environment of the surface determined by f around a particular point (x_1, y_1) is determined by evaluating the matrix A at this point. If $\det(A) \neq 0$ then the nature of this environment is determined by the sign of the determinant and the sign of $\mathrm{tr}(A)$. This shows how far-ranging the ideas of this chapter are.

Exercises 10.3

1. Prove that if A is a diagonalizable matrix then $\mathrm{tr}(A)$ is the sum of the roots of the characteristic polynomial of A.

2. This question looks harder than it is. Repeat the analysis carried out in this section for quadrics instead of conics to derive the table below.[2] The term *one sheet* below means that the surface is all in one piece. The term *two sheets* means that the surface is in two disjoint pieces. *Ellipsoids* look like rugby balls, a *hyperboloid of one sheet* is like a cooling tower, a *hyperboloid of two sheets* consists of two bowls facing each other, an *elliptic paraboloid* is a bowl with all cross sections ellipses but with a profile that is a parabola, a *hyperbolic paraboloid* looks like a horse saddle and *cylinders* have horizontal slices all the same.

---

[2]This is adapted from [1].

	Equation	Description
1.	$\frac{x^2}{a^2} + \frac{y^2}{b^2} + \frac{z^2}{c^2} = 1$	Ellipsoid
2.	$\frac{x^2}{a^2} + \frac{y^2}{b^2} + \frac{z^2}{c^2} = -1$	Empty set
3.	$\frac{x^2}{a^2} + \frac{y^2}{b^2} - \frac{z^2}{c^2} = 1$	Hyperboloid of one sheet
4.	$\frac{x^2}{a^2} + \frac{y^2}{b^2} - \frac{z^2}{c^2} = -1$	Hyperboloid of two sheets
5.	$\frac{x^2}{a^2} + \frac{y^2}{b^2} - \frac{z^2}{c^2} = 0$	Pair of cones point to point
6.	$\frac{x^2}{a^2} + \frac{y^2}{b^2} + \frac{z^2}{c^2} = 0$	Empty set
7.	$\frac{x^2}{a^2} + \frac{y^2}{b^2} - 2cz = 0$	Elliptic paraboloid
8.	$\frac{x^2}{a^2} - \frac{y^2}{b^2} - 2cz = 0$	Hyperbolic paraboloid
9.	$\frac{x^2}{a^2} + \frac{y^2}{b^2} = 1$	Elliptic cylinder
10.	$\frac{x^2}{a^2} + \frac{y^2}{b^2} = -1$	Empty set
11.	$\frac{x^2}{a^2} + \frac{y^2}{b^2} = 0$	Empty set
12.	$\frac{x^2}{a^2} - \frac{y^2}{b^2} = 1$	Hyperbolic cylinder
13.	$\frac{x^2}{a^2} - \frac{y^2}{b^2} = 0$	Pair of intersecting planes
14.	$y^2 = 2ax$	Parabolic cylinder
15.	$x^2 = a^2$	Pair of parallel planes
16.	$x^2 = -a^2$	Empty set
17.	$x^2 = 0$	A plane

3. Find the normal forms corresponding to the following equations.

(a) $5x^2 - 4xy + 8y^2 - 36 = 0$.

(b) $3x^2 - 10xy + 3y^2 + 14x - 2y + 3 = 0$.

(c) $2x^2 + 2xy + 2y^2 - 6xz - 6yz + 6z^2 = 0$.

Epilegomena

> **Box the last: Emmy Noether (1882–1935)**
>
> There are not many mathematicians who have made important contributions to both pure and applied mathematics, but Emmy Noether is one of them. Her approach to algebra was incorporated into a two-volume textbook on abstract algebra by B. L. van der Waerden [126], one of her students. This became the prototype for all subsequent algebra books, and it is in this way that Noether has influenced generations of mathematicians. But she also profoundly influenced modern physics [91] by proving a theorem that related the symmetries of a system to the quantities conserved by that system. Thus the conservation of linear momentum, angular momentum and energy are manifestations, respectively, of invariance of physical systems under translations, rotations and time shifts. As a Jewish woman living and working in Germany in the early decades of the twentieth century, she was the subject of extreme prejudice, and like many of her generation, she found a new home in the United States, working at Bryn Mawr College until her untimely death.

Here are some highly partial thoughts on what to study next.

In Chapters 1 and 2, I tried to give some flavour of what kind of subject mathematics is. For proper accounts of the history of mathematics, see [17, 44, 75, 92, 117, 119]. Neugebauer [92] is the classic text on Mesapotamian science but remember that history does not stand still: there are always new interpretations and new texts. My interest in the subject was transformed by Fowler's book [44] which was a revelation about the nature of Greek mathematics. In particular, it made me realize the signal importance of Euclid's algorithm in all its incarnations.

The philosophy of mathematics is not as well served by books as the history of mathematics. Whereas many mathematics degrees include modules on history, I think it rare for them to include ones on philosophy. A starting point at least is [33]. One book on mathematical philosophy that has been influential is Lakatos [82], but otherwise I wonder if mathematicians believe the subject virtually begins and ends with Bertrand Russell.

In Chapter 3, I touched on both logic and set theory. My favourite introduction to first-order logic is Smullyan's wonderfully user-unfriendly book [113]. A more mainstream and rather more extensive introduction is [125]. Once you have a smattering of logic you have the background needed to understand the full axiomatic development of set theory [41, 49]. The apotheosis of logic and set theory is the work of Kurt Gödel, the greatest logician in history [13].

This book is not about analysis, but analysis, both real and complex, is the companion of algebra and geometry. For example, to prove deep results about numbers, such as the prime number theorem, analysis is essential. Classic accounts of analysis can be found in [21, 53, 56, 107, 114], whereas Needham's book [90] emphasizes the geometry in complex analysis.

Chapters 4, 5, 6 and 7 formed an introduction to algebra. The next step is abstract algebra. The prototype for books in this subject was Van der Waerden's [126] which led to English language analogues such as [15, 63]. All of these are still worth reading but I would begin with Artin [5] and Clark [28], Artin's book [5] being the more comprehensive and Clark's [28] the more concise. At its core abstract algebra has not changed that much, where it has changed is in the development of applications. To many people's surprise, algebra turned out to be useful, particularly in manipulating information in various ways such as in cryptography and error-correcting codes. A book on algebra which emphasizes applications is Childs' [25] and a good overview of how to use algebra to build codes of various kinds is [14]. Cryptography, in particular, has become big business. Koblitz [77] takes a mathematician's perspective whereas [76] treats the subject more on its own terms. The apotheosis of algebra is Galois theory. This is usually taught after abstract algebra because modern presentations need the full panoply of groups, rings and fields [106, 115]. This is the right approach for more advanced developments but misses the concrete connections with equations. For this reason, I like Tignol's book [120] because it is focused on equations and tries to recapture the spirit of what Galois was actually doing. If you want to see where algebra goes after this then a book that surveys the field is [111].

Number theory is in fact a separate subject that interacts with just about every other part of mathematics. Perhaps because it is so well established, there is no shortage of classic books at every level. The following is a list roughly in order of difficulty [32, 58, 87, 93] with the last book making connections between the primes and particle physics.

Geometry has fared badly in schools with the demise of Euclid. This is a loss to human culture. If we can read the *Illiad* with enjoyment today, and we can, why not the *Elements*? Book I alone is worth reading for the intellectual pleasure of seeing how a handful of axioms that appear to say almost nothing lead step by step to the statement and proof of Pythagoras' theorem. A clear edition of the *Elements* without excessive footnotes is [35].[3] By not teaching Euclid, we have lost the context for much mathematics. Some of this context is described in [61, 94]. Important though Euclid is, it should be remembered that he stands at the beginning of the subject and not the end. A modern approach to Euclid via vectors is [103] whereas [18, 60, 86] take you to the world of non-Euclidean geometries.

Mathematicians invented computer science though are sometimes slow to admit maternity. Despite this, theoretical computer science is as much a part of applied mathematics as electromagnetic theory. The theory of algorithms is not just about programming [19, 78, 83, 95] and has ramifications in pure mathematics and in the security of your bank account.

Combinatorics is what grown-ups call counting and Peter Cameron [22] is the best person at counting I know.

Probability theory is applied counting that offers a unique perspective on the world. It is not just about calculating the odds in gambling. It contains challenges to our intuition such as the Birthday Paradox and the Monty Hall Problem and provides

---

[3]But if you do want to know more then look up the work of Sir Thomas Little Heath (1861–1940).

tools for the fast solution of otherwise intractable problems. I know that probability theorists swear by [42].

Chapters 8, 9 and 10 are different from the preceding ones in that they are an introduction to one subject: linear algebra. Good starting points are the more elementary [39, 45, 108] and the more advanced [54, 69]. But this is a subject that is ineluctably connected to its applications and it is necessary to see linear algebra in action to fully appreciate the power of its ideas [66, 74, 80]. I would also recommend [7] for taking a more advanced perspective on vector calculus by using alternating forms. Their theory supercedes the usual approach via div, curl and grad and is ultimately superior. This is one of the reasons why I decided to treat determinants in this book in more depth than is usual.

"Und da dies wirklich das Ende der Geschichte ist und ich nach dem letzten Satz sehr müde bin, werde ich, glaube ich, hier aufhören." P. Baer, *Gesamtausgabe*.

Bibliography

[1] A. D. Aleksandrov et al., *Mathematics: its content, methods and meaning*, Dover, 1999.

[2] A. Alexander, *Infinitesimal*, Oneworld, 2014.

[3] R. C. Alperin, A mathematical theory of origami constructions and numbers, *New York Journal of Mathematics* **6** (2000), 119–133.

[4] J. W. Archbold, *Algebra*, fourth edition, Pitman Paperbacks, 1970.

[5] M. Artin, *Algebra*, Prentice-Hall, 1991.

[6] J. C. Baez, The octonians, *Bulletin of the American Mathematical Society* **39** (2002), 145–205.

[7] P. Bamberg and S. Sternberg, *A course in mathematics for students of physics 1*, CUP, 1998.

[8] D. Bayer and P. Diaconis, Trailing the dovetail shuffle to its lair, *The Annals of Applied Probability* **2** (1992), 294–313.

[9] R. A Beaumont and H. S. Zuckerman, A characterization of the subgroups of the additive rationals, *Pacific Journal of Mathematics* **1** (1951), 169–177.

[10] J. L. Bell, *A primer of infinitesimal analysis*, CUP, 1998.

[11] O. Benson and J. Stangroom, *Why truth matters*, Continuum, 2006.

[12] A. J. Berrick and M. E. Keating, Rectangular invertible matrices, *The American Mathematical Monthly* **104** (1997), 297–302.

[13] F. Berto, *There's something about Gödel*, Wiley-Blackwell, 2009.

[14] N. L. Biggs, *Codes*, Springer, 2008.

[15] G. Birkhoff and S. Mac Lane, *A survey of modern algebra*, third edition, Macmillan, 1965.

[16] W. A. Blankinship, A new version of the Euclidean algorithm, *The American Mathematical Monthly* **70** (1963), 742–745.

[17] C. B. Boyer and U. Merzbach, *History of mathematics*, third edition, Jossey Bass, 2011.

[18] D. A. Brannan, M. F. Esplen and J. J. Gray, *Geometry*, second edition, CUP, 2012.

[19] G. S. Boolos, J. P. Burgess and R. C. Jeffrey, *Computability and logic*, fifth edition, CUP, 2010.

[20] J. L. Borges, *Labyrinths*, Penguin Books, 1976.

[21] J. C. Burkill, *A first course in mathematical analysis*, CUP, 1970.

[22] P. J. Cameron, *Combinatorics*, CUP, 1994.

[23] B. Casselman, If Euclid had been Japanese, *Notices of the American Mathematical Society* **54** (2007), 626–628.

[24] A. Cayley, A memoir on matrices, *Philosophical Transactions of the Royal Society of London* **148** (1858), 17–37.

[25] L. N. Childs, *A concrete introduction to higher algebra*, second edition, Springer, 1995.

[26] G. Chrystal, *Introduction to algebra*, Adam and Charles Black, 1902.

[27] G. Chrystal, *Algebra* parts I and II, Adam and Charles Black, 1900, 1904.

[28] A. Clark, *Elements of abstract algebra*, Dover, 1984.

[29] D. Corfield, *Towards a philosophy of real mathematics*, CUP, 2003.

[30] R. Courant, *Differential and integral calculus*, Volume 1, Blackie and Son Limited, 1945.

[31] R. Courant and H. Robbins, *What is mathematics?*, OUP, 1978.

[32] H. Davenport, *The higher arithmetic*, eighth edition, CUP, 2012.

[33] P. Davis, R. Hersh and E. A. Marchisotto, *The mathematical experience*, Birkhäuser, 2012.

[34] H. F. Davis and A. D. Snider, *Introduction to vector analysis*, fifth edition, William C. Brown, 1988.

[35] D. Densmore (editor), *Euclid's Elements*, Green Lion Press, Santa Fe, New Mexico, 2013.

[36] K. Devlin, *The Millennium problems*, Granta Books, 2002.

[37] W. Diffie and M. Hellman, New directions in cryptography, *IEEE Transactions on Information Theory* **22** (1976), 644–654.

[38] M. du Sautoy, *The music of the primes*, Harper Perennial, 2004.

[39] D. Easdown, *A first course in linear algebra*, second edition, Pearson Education Australia, 2008.

[40] H.-D. Ebbinghaus et al., *Numbers*, Springer, 1996.

[41] H. B. Enderton, *Elements of set theory*, Academic Press, 1977.

[42] W. Feller, *An introduction to probability theory and its applications, volume 1*, third edition, TBS, 1968.

[43] G. Fischer, *Analytische Geometrie*, Friedr. Vieweg & Sohn, Braunschweig/Wiesbaden, 1985.

[44] D. H. Fowler, *The mathematics of Plato's academy*, Clarendon Press, Oxford, 2011.

[45] J. B. Fraleigh and R. A. Beauregard, *Linear algebra*, third edition, Addison-Wesley, 1995.

[46] M. Gardner, *Hexaflexagons, probability paradoxes, and the Tower of Hanoi: Martin Gardner's first book of mathematical puzzles and games*, CUP, 2002.

[47] T. A. Garrity, *All the mathematics you missed (but need to know for graduate school)*, CUP, 2002.

[48] S. Givant and P. Halmos, *Introduction to Boolean algebras*, Springer, 2009.

[49] D. Goldrei, *Classic set theory*, Chapman & Hall/CRC, 1998.

[50] T. Gowers, *Mathematics: a very short introduction*, Oxford University Press, 2002.

[51] A. Granville, It is easy to determine whether a given integer is prime, *Bulletin of the American Mathematical Society* **42** (2004), 3–38.

[52] A. C. Grayling, *Russell: a very short introduction*, OUP, 2002.

[53] E. Hairer and G. Wanner, *Analysis by its history*, Springer, 2008.

[54] P. Halmos, *Finite-dimensional vector spaces*, Springer-Verlag, 1974.

[55] R. W. Hamming, Error detecting and error correcting codes, *Bell Systems Technical Journal* **29** (1950), 147–160.

[56] G. H. Hardy, *A course of pure mathematics*, tenth edition, CUP, 1967.

[57] G. H. Hardy, *A mathematician's apology*, CUP, 2012.

[58] G. H. Hardy and E. M. Wright, *An introduction to the theory of numbers*, sixth edition, OUP, 2008.

[59] D. Harel, *Computers Ltd. What they really can't do*, OUP, 2012.

[60] R. Hartshorne, *Geometry: Euclid and beyond*, Springer, 2000.

[61] J. L. Heilbron, *Geometry civilized*, Clarendon Press, Oxford, 2000.

[62] R. Hersh, *What is mathematics really?*, Vintage, 1998.

[63] I. N. Herstein, *Topics in algebra*, second edition, John Wiley & Sons, 1975.

[64] D. Hilbert, *Grundlagen der Geometrie*, Teubner Studienbücher, 13 Auflage, 1987.

[65] L. S. Hill, Cryptography in an algebraic alphabet, *The American Mathematical Monthly* **36** (1929), 306–312.

[66] M. W. Hirsch and S. Smale, *Differential equations, dynamical systems, and linear algebra*, Academic Press, 1974.

[67] A. Hodges, *Alan Turing: the enigma*, Vintage Books, 1992.

[68] J. Hoffman, Q & A: the mathemagician, *Nature* **478** (2011), 457.

[69] K. Hoffman and R. Kunze, *Linear algebra*, second edition, Prentice-Hall, 1971.

[70] C. Hollings, *Mathematics across the Iron Curtain*, American Mathematical Society, 2014.

[71] N. Jacobson, *Lectures in abstract algebra, Volume 1*, Van Nostrand Reinhold Company, 1951.

[72] J. P. Jones, D. Sato, H. Wada and D. Wiens, Diophantine representation of the set of prime numbers, *The American Mathematical Monthly* **83** (1976), 449–464.

[73] D. Kalman, Fractions with cycling digit patterns, *The College Mathematics Journal* **27** (1996), 109–115.

[74] W. Kaplan, *Advanced calculus*, third edition, Addison-Wesley, 1984.

[75] V. J. Katz, *A history of mathematics*, third edition, Pearson, 2013.

[76] J. Katz and Y. Lindell, *Introduction to modern cryptography*, Chapman & Hall/CRC, 2008.

[77] N. Koblitz, *A course in number theory and cryptography*, second edition, Springer, 1994.

[78] D. C. Kozen, *Automata and computability*, Springer, 2007.

[79] S. G. Krantz and H. R. Parks, *A mathematical odyssey*, Springer, 2014.

[80] D. L. Kreider, R. G. Kuller, D. R. Ostberg and F. W. Perkins, *An introduction to linear analysis*, Addison-Wesley, 1966.

[81] J. C. Lagarias, The $3x + 1$ problem and its generalizations, *The American Mathematical Monthly* **92** (1985), 3–23.

[82] I. Lakatos, *Proofs and refutations: the logic of mathematical discovery*, CUP, 1976.

[83] M. V. Lawson, *Finite automata*, Chapman and Hall/CRC, 2004.

[84] D. E. Littlewood, *A university algebra*, second edition, Heinemann, 1958.

[85] J. E. Littlewood, *A mathematician's miscellany*, Nabu Press, 2011.

[86] J. McCleary, *Geometry from a differentiable viewpoint*, second edition, CUP, 2012.

[87] S. J. Miller and R. Takloo-Bighash, *An invitation to modern number theory*, Princeton University Press, 2006.

[88] A. A. Milne, *Winnie-the-Pooh*, Methuen, 2000.

[89] F. Mosteller, R. E. K. Rourke and G. B. Thomas, *Probability with statistical applications*, Addison-Wesley, second edition, 1970.

[90] T. Needham, *Visual complex analysis*, Clarendon Press, Oxford, 2002.

[91] D. E. Neuenschwander, *Emmy Noether's wonderful theorem*, The Johns Hopkins University Press, 2011.

[92] O. Neugebauer, *The exact sciences in antiquity*, second edition, Dover, 1969.

[93] O. Ore, *Number theory and its history*, Dover, 1948.

[94] A. Ostermann and G. Wanner, *Geometry by its history*, Springer, 2012.

[95] C. H. Papadimitriou, *Computational complexity*, Addison-Wesley, 1993.

[96] A. J. Pettofrezzo, *Vectors and their applications*, Prentice-Hall, Inc., 1966.

[97] A. J. Pettofrezzo, *Matrices and transformations*, Dover, 1966.

[98] C. Petzold, *Codes: the hidden language of computer hardware and software*, Microsoft Press, 2000.

[99] C. Petzold, *The annotated Turing*, Wiley, 2008.

[100] R. I. Rivest, A. Shamir and L. Adleman, A method for obtaining digital signatures and public-key cryptosystems, *Communications of the ACM* **21** (1978), 120–126.

[101] E. Robson, Neither Sherlock Holmes nor Babylon: a reassessment of Plimpton 322, *Historia Mathematica* **28** (2001), 167–206.

[102] E. Robson, Words and pictures: new light on Plimpton 322, *The American Mathematical Monthly* **109** (2002), 105–119.

[103] J. Roe, *Elementary geometry*, OUP, 1993.

[104] M. Ronan, *Symmetry and the Monster*, OUP, 2006.

[105] J. C. Rosales and P. A. García-Sánchez, *Numerical semigroups*, Springer, 2012.

[106] J. Rotman, *Galois theory*, second edition, Springer, 1998.

[107] W. Rudin, *Principles of mathematical analysis*, McGraw-Hill, second edition, 1964.

[108] W. W. Sawyer, *An engineering approach to linear algebra*, CUP, 1972.

[109] W. W. Sawyer, *Prelude to mathematics*, Dover, 1983.

[110] R. Séroul, *Programming for mathematicians*, Springer, 2000.

[111] I. R. Shafarevich, *Basic notions of algebra*, Springer, 1997.

[112] G. Simmons, *Calculus gems*, McGraw-Hill, Inc., New York, 1992.

[113] R. M. Smullyan, *First-order logic*, Dover Publications Inc, 1995.

[114] M. Spivak, *Calculus*, third edition, CUP, 2010.

[115] I. Stewart, *Galois theory*, second edition, Chapman & Hall, 1998.

[116] J. Stillwell, *Elements of algebra*, Springer, 1994.

[117] J. Stillwell, *Mathematics and its history*, third edition, Springer, 2010.

[118] R. R. Stoll, *Set theory and logic*, W. H. Freeman and Company, 1963.

[119] D. J. Struik, *A concise history of mathematics*, fourth revised edition, Dover, 1987.

[120] J.-P. Tignol, *Galois' theory of algebraic equations*, World Scientific, 2011.

[121] C. J. Tranter, *Advanced level pure mathematics*, fourth edition, Hodder and Stoughton, 1978.

[122] J. V. Uspensky, *Theory of equations*, McGraw-Hill, 1948.

[123] L. M. Wapner, *The pea and the sun*, A. K. Peters, Ltd, 2005.

[124] C. Villani, *Birth of a theorem: the story of a mathematical adventure*, Bodley Head, 2015.

[125] D. van Dalen, *Logic and structure*, fifth edition, Springer, 2013.

[126] B. L. Van der Waerden, *Algebra* (two volumes), Springer-Verlag, 1966.

[127] C. Watkins, *How to kill a dragon*, OUP, 2001.

[128] D. Zagier, A one-sentence proof that every prime $p \equiv 1 \pmod 4$ is a sum of two squares, *The American Mathematical Monthly* **97** (1990), 144.

Index